PRINCIPLES OF ANIMAL VIROLOGY

PRINCIPLES OF ANIMAL BIOLOGY

Principles of Animal Virology

edited by

Wolfgang K. Joklik, D. Phil.

*James B. Duke Distinguished Professor of Microbiology and Immunology
Chairman, Department of Microbiology and Immunology
Duke University Medical Center*

APPLETON-CENTURY-CROFTS / New York

Copyright © 1980 by APPLETON-CENTURY-CROFTS
A Publishing Division of Prentice-Hall, Inc.

80 81 82 83 84 / 10 9 8 7 6 5 4 3 2 1

Prentice-Hall International, Inc., London
Prentice-Hall of Australia, Pty. Ltd., Sydney
Prentice-Hall of India Private Limited, New Delhi
Prentice-Hall of Japan, Inc., Tokyo
Prentice-Hall of Southeast Asia (Pte.) Ltd., Singapore
Whitehall Books Ltd., Wellington, New Zealand

Library of Congress Cataloging in Publication Data
Main entry under title:

Principles of animal virology.

Includes index.
1. Virology. I. Joklik, Wolfgang K. [DNLM:
1. Vertebrate viruses.   2. Virus diseases. QW164
P957]
QR360.P695 1980   616'.0194   80-17809
ISBN 0-8385-7920-5

Cover design: Jacquelyn Munn
Cover micrograph: Courtesy of Dr. Erskine Palmer,
Center for Disease Control, Atlanta, Georgia

PRINTED IN THE UNITED STATES OF AMERICA

CONTRIBUTORS

Wolfgang K. Joklik, D. Phil.

*James B. Duke Distinguished Professor of
Microbiology and Immunology; Chairman,
Department of Microbiology and Immunology,
Duke University Medical Center*

David W. Barry, M.D.

*Head, Department of Clinical Investigation,
Burroughs-Wellcome Company and Associate Professor
of Medicine, Duke University Medical Center*

Thomas E. Frothingham, M.D.

*Professor of Pediatrics and Professor of
Community Health Sciences, Duke University
Medical Center*

Samuel L. Katz, M.D.

*Wilburt C. Davison Professor of Pediatrics;
Chairman, Department of Pediatrics,
Duke University Medical Center*

David J. Lang, M.D.

*Professor of Pediatrics and Associate Professor
of Microbiology, Duke University Medical Center;
present affiliation: Chairman, Department of
Pediatrics, University of Maryland School of
Medicine, Baltimore, Maryland*

Catherine M. Wilfert, M.D.

*Professor of Pediatrics and Professor of
Virology, Duke University Medical Center*

CONTENTS

PREFACE

Virology occupies a unique position in contemporary biology. It is the discipline *par excellence* for studies at the molecular level: the opportunity to observe the functioning of small genomes encouraged the development of sophisticated genetic and biochemical techniques the exploitation of which has yielded a rich harvest of fascinating fundamental discoveries ranging from the very existence of messenger RNA in 1959 to recently acquired totally unexpected and surprising insight into the arrangement of genetic material. Viruses also represent the last major challenge in infectious disease: although many human viral pathogens were isolated and characterized during the first half of this century culminating in a golden period during the early 1950's, viruses that cause important human diseases are still being discovered nowadays—such as rotavirus and the virus that causes African hemorrhagic fever. Further, our gradually developing knowledge concerning the nature of latent and persistent viral infections may soon provide clues regarding the causes of chronic debilitating conditions such as diabetes, lupus, multiple sclerosis and Creutzfeld-Jakob disease, and the involvement of viruses in cancer.

In view of its central role in contemporary biology and medicine, virology provides an important conceptual framework that must not only be mastered by graduate and medical students, but is also becoming increasingly important in undergraduate curricula. This textbook is designed to fill several clearly defined needs. It is intended to be a text for advanced undergraduates who intend to proceed to graduate and medical school, as well as for graduate students in cell and molecular biology, genetics, biochemistry, microbiology and immunology, and related disciplines. It is also written for medical students experiencing their first exposure to medical microbiology, as well as for advanced medical students, house officers and practicing physicians. Finally, the book is designed as a reference source for instructors. To achieve these aims, the book comprises two sections, one that discusses the biochemistry, molecular biology and genetics of animal viruses (and also includes a chapter on the molecular virology of bacteriophages), and another that describes the viruses that are pathogenic for man, the nature and symptoms of the diseases that they cause, and how to treat and prevent such diseases.

In reference to the bibliography, we have elected not to reference specific statements in the text, but have appended to each chapter a list of recent reviews and important original papers. The former will quickly guide the reader to any specific aspect of virology that he wishes to pursue; the latter makes available the detailed considerations and circumstances that generated key discoveries. Many of the papers that are cited already are, or no doubt will soon become, "classics."

The list of individuals who have helped produce this volume is long and we are profoundly indebted to them. We would especially like to thank our many colleagues who permitted us to use illustrative material and who almost invariably supplied us with original photographs, and the many publishers who allowed us to reproduce previously published material. We would also like to thank Lynda Frejlach, who did a superb job in drawing the numerous charts and diagrams, and all the secretaries who cheerfully typed and retyped the manuscript. Finally, we wish to express our appreciation to the staff of Appleton-Century-Crofts for their efficient cooperation in producing this book.

Wolfgang K. Joklik
David W. Barry
Thomas E. Frothingham
Samuel L. Katz
David J. Lang
Catherine M. Wilfert

BASIC VIROLOGY

CHAPTER 1

The Nature, Isolation, and Measurement of Animal Viruses

Many important infectious diseases that afflict mankind are caused by viruses. Some are important because they are frequently fatal; among such are rabies, smallpox, poliomyelitis, hepatitis, yellow fever, and various encephalitic diseases. Others are important because they are very contagious and create acute discomfort; among such are influenza, the common cold, measles, mumps, and chickenpox, as well as respiratory-gastrointestinal disorders. Still other viruses, such as rubella and cytomegalovirus, can cause congenital abnormalities; and finally there are viruses that can cause tumors and cancer in animals and perhaps also in humans.

There is little that can be done to interfere with the growth of viruses, since they multiply within cells, using the cells' synthetic capabilities. Only a limited number of highly specialized reactions are under their own control. It is hoped that their selective inhibition will form the basis of a rational system of antiviral chemotherapy, thereby permitting virus diseases to be brought under effective control, just as antibiotics have brought most bacterial diseases under control.

In addition to their medical importance viruses provide the simplest model systems for many basic problems in biology. The reason is that viruses are essentially small segments of genetic material encased in protective shells. Since the information encoded in viral genomes differs from that in host cell genomes, viruses afford unrivaled opportunities for the study of the mechanisms that control the replication, transcription, and translation of genetic information. Knowledge of these mechanisms is fundamental to an understanding of the development and operation of differentiated functions in higher organisms and is therefore directly applicable to the practice of medicine and the improvement of human welfare.

Historical background

There are three major classes of viruses: animal viruses, plant viruses, and bacterial viruses. Since knowledge concerning each of these

classes has accumulated along distinctive lines, extensive specialization has developed. Bacterial viruses are, therefore, dealt with only briefly in this book, and plant viruses are not considered at all. Yet discoveries made concerning each of these classes of viruses have influenced profoundly our understanding of the nature of each of the others.

The existence of viruses became evident during the closing years of the nineteenth century when, as the result of newly acquired expertise in the handling of bacteria, the infectious agents of numerous diseases were being isolated. For some infectious diseases this proved to be an elusive task until it was realized that the agents causing them were smaller than bacteria. Iwanowski in 1892 was probably the first to record the transmission of an infection (tobacco mosaic disease) by a suspension filtered through a bacteria-proof filter. This was followed in 1898 by a similar report by Loeffler and Frosch concerning foot-and-mouth disease of cattle. Beijerinck (1898) considered the infectious agents in bacteria-free filtrates to be living but fluid—that is, nonparticulate—and introduced the term "virus" (Latin, poison) to describe them. It quickly became clear, however, that viruses were particulate, and the term "virus" became the operational definition of infectious agents smaller than bacteria and unable to multiply outside living cells. In 1911 Rous discovered a virus that produced malignant tumors in chickens, and during World War I Twort and d'Herelle independently discovered the viruses growing in bacteria, the bacteriophages.

During the next 25 years the experimental approaches in the three areas of virology diverged. Plant viruses proved easy to obtain in large amounts, thus permitting extensive chemical and physical studies. This work first led to the demonstration that plant viruses consisted only of nucleic acid and protein, and culminated in the crystallization of tobacco mosaic virus by Stanley in 1935. This feat evoked great astonishment, since it cut across preconceived ideas concerning the attributes of living organisms and demonstrated that agents able to reproduce in living cells behaved under certain conditions as typical macromolecules.

Work with bacteriophages concentrated on their clinical application. It was hoped that bacteria could be destroyed inside the body by injecting appropriate bacteriophages. Their activity in vivo, however, never matched their activity in vitro, most probably because they are eliminated efficiently from the bloodstream.

Work with animal viruses concentrated on the pathogenesis of viral infections and on epidemiology. Throughout this period, fundamental studies on animal cell-virus interactions were severely hampered by the absence of rapid and efficient techniques for quantitating viruses. The only method then available was the expensive and time-consuming serial end point dilution method, using animals (p. 11).

Around the year 1940 came several breakthroughs. First, the advent of electron microscopy permitted visualization of viruses for the first time. As will become evident, not only is morphology an important criterion of virus classification, but the study of the morphology of viruses has also had a profound impact on our understanding of their behavior and function. Second, techniques for purifying certain animal viruses were being perfected, and a group of workers at the Rockefeller Institute headed by Rivers carried out some excellent chemical studies on vaccinia virus. Third, Hirst discovered that influenza virus agglutinates chicken red cells. This phenomenon, hemagglutination, was rapidly developed into an accurate method for quantitating myxoviruses, as a result of which this group of viruses became in the 1940's the most intensively investigated group of animal viruses. Finally, this period marked the beginning of the modern era of bacterial virology. Until then the interaction of bacteriophages with bacteria had been analyzed principally in terms of populations, rather than at the level of a single virus particle interacting with a single cell. This conceptual block was removed by Ellis and Delbrück's study of the one-step growth cycle, as a result of which the bacteriophage-bacterium system became extraordinarily amenable to experimentation. Indeed, during the last three decades, many of the major advances in molecular biology have resulted from work in the bacteriophage field. Among these are the demonstration that initiation of virus infection involves the separation of virus nucleic acid and protein, the demonstration that the virus genome can become integrated into the genome of the host cell, the discovery of messenger RNA, and elucidation of the factors that control initiation and termination of both transcription and translation of genetic information.

In animal virology, rapid advances followed the development, in the late 1940's, of techniques for growing animal cells in vitro. Strains of many types of mammalian cells can now be grown in media of defined composition. As a result, animal cell-virus interactions can now be analyzed with the same techniques that have proved so powerful in the case of bacteriophages.

The nature of viruses

Viruses are a heterogeneous class of agents. They vary in size and morphology; they vary in chemical composition; they vary in host range and in the effect that they have on their hosts. There are certain characteristics, however, that are shared by all viruses:

1. Viruses consist of a genome, either RNA or DNA, that is surrounded by a protective protein shell. Frequently this shell is itself enclosed within an envelope that contains both protein and lipid.
2. Viruses multiply only inside cells. They are absolutely dependent on the host cells' synthetic and energy-yielding apparatus. They are parasites at the genetic level.
3. The multiplication of viruses involves as an initial step the separation of either their genomes or their nucleocapsids from their protective shells (see below).

In essence, therefore, viruses are nucleic acid molecules that can enter cells, replicate in them, and code for proteins capable of forming protective shells around them.

Given this definition of viruses, are they to be regarded as living organisms or as lifeless arrangements of molecules? The answer to this question depends on whether one is concerned with viruses as extracellular suspensions of particles or as infectious agents. Isolated virus particles are arrangements of nucleic and protein molecules with no metabolism of their own; they are no more active than isolated chromosomes. Within cells, however, virus particles are capable of reproducing their own kind manyfold by virtue of precisely regulated sequences of reactions. Considered in this light, viruses may indeed be said to possess at least some of the attributes of life. Such terms as "organism" and "living," however, are not really applicable to viruses; it is preferable to refer to viruses as being functionally active or inactive, rather than living or dead.

The origin of viruses

The question of the origin of viruses poses a fascinating problem. The two likeliest hypotheses are (1) viruses are the products of regressive evolution of free-living cells. An evolutionary pathway of this type has been suggested for mitochondria, which still retain vestiges of cellular organization, as well as a mechanism for replicating, transcribing, and translating genetic information. The largest animal viruses, the poxviruses, are so complex that one could imagine them also to be derived from a cellular ancestor. (2) Viruses are derived from cellular genetic material that has acquired the capacity to exist and function independently. Nowadays the latter hypothesis is considered much more likely for all viruses (with the possible exception of poxviruses).

The characteristics of cultured animal cells

The medical practitioner should understand not only how viruses affect the patient as a whole but also how viruses interact with cells. This understanding can be acquired far more readily by studying isolated infected cells than by examining infected cells in the intact organism. Animal virology provided the main impetus for the development of tissue culture—the technique of growing cells in vitro. Tissue culture is now used extensively for fundamental studies in areas ranging from growth, differentiation, and aging to molecular biology and genetics. Since knowledge concerning the normal cell is crucial to an understanding of virus-cell interaction, we will first examine briefly the characteristics of animal cells cultured in vitro.

The Establishment of Animal Cell Strains

Cells of many organs can be grown in vitro. As a rule, small pieces of the tissue in question are dissociated into single cells by treatment with a dilute solution of trypsin, and a suspension of the cells is then placed into a flask, bottle, or petri dish. There the cells attach to the flat surface, and provided that they are supplied with a growth medium, they multiply. The essential constituents of a growth medium are physiologic amounts of 13 essential amino acids and 9 vitamins, salts, glucose, and a buffering system that generally consists of bicarbonate in equilibrium with an atmosphere containing about 5 percent carbon dioxide. This medium is supplemented to the extent of about 5 percent with serum, the source of which is not predicated by the species from which the cells were derived; calf and fetal calf serum are the two most commonly employed. Antibiotics, such as penicillin and streptomycin, are also usually added in order to minimize the growth of bacterial contaminants, and a dye, such as phenol red, is generally included as a pH indicator. This medium, or more complex versions of it, will permit most cell types to multiply with a division time of 24 to 48 hours.

When cells are brought into contact with a surface, they generally attach firmly and flatten so as to occupy the maximum surface area. The only time when they are not maximally extended is during mitosis, when they become round and are therefore easily dislodged from the substratum. Cells multiply until they occupy all available surface area—that is, until they are confluent, but no further. The reason for this is the cells cease dividing when they make contact with neighboring cells, a phenomenon known as "contact inhibition" (see Chap. 9).

Animal cells can be cloned just like bacterial

cells, although the efficiency of cloning is frequently less than 100 percent. Numerous genetically pure cell strains are now available. They fall into two morphologic categories, epithelial cells with a polygonal outline, and fibroblasts with a narrow spindlelike shape (Fig. 1-1).

The first cultures after tissue dispersion are known as "primary cultures." When such cultures are confluent they are passaged by dislodgment from the surface by means of trypsin or the chelating agent ethylene diamine tetraacetate (EDTA) and reseeded into several new containers, in which they form secondary cultures. Passaging can then be continued in this manner, provided that an adequate supply of growth medium is supplied at regular intervals.

The overall properties of cell strains are generally stable on continuous culturing. However, mutations occur constantly, so that one particular mutant, or variant, usually emerges as the dominant population component under any given set of conditions. As a result, the same cell strain cultured in two different laboratories may exhibit detectable phenotypic differences.

The Multiplication Cycle

The multiplication of each individual cell conforms to a regular pattern which can be thought of as a cycle (Fig. 1-2). According to this scheme, the interval between successive mitoses is divided into three periods: the G1 period that precedes DNA replication, the S period during which DNA replicates, and the G2 period during which the cell prepares for the next mitosis. RNA and protein are not synthesized while mitosis proceeds—that is, during metaphase—but are otherwise synthesized throughout the multiplication cycle. Nongrowing cells are usually arrested in the G1 period; the resting state is often referred to as G0 (G zero). The relative durations of the periods are quite variable, but metaphase rarely occupies more than one hour.

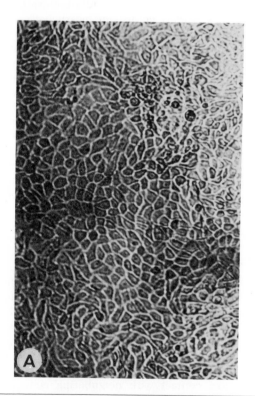
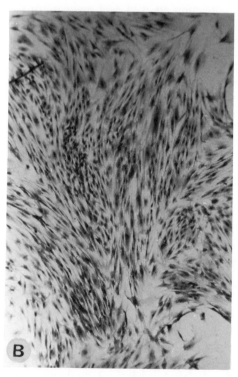

Fig. 1-1 Cultured mammalian cells. **A.** Unstained monkey kidney cells, which exhibit a typical epithelioid morphology. **B.** Chick embryo fibroblasts (Giemsa stain). Note characteristic spindle shape and orderly alignment. (A, from Eagle and Foley, Cancer Res 18:1017, 1958. B, courtesy of Dr. R. E. Smith.)

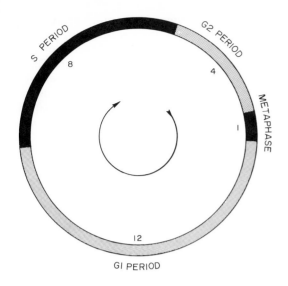

FIG. 1-2. The multiplication cycle of mammalian cells. The duration of the cycle illustrated here is 25 hours; the average lengths (in hours) of the individual periods is indicated by the numbers inside the cycle.

Under conditions of normal growth, the individual cells of a growing culture pass through this multiplication cycle in an unsynchronized fashion, so that cells at all stages of the cycle are always present. It is, however, possible to synchronize cells so that they multiply in step for several generations. Synchronized cell cultures are useful in studies of the reaction that are essential for progression through the multiplication cycle.

The Aging of Cell Strains

Cells derived from normal tissues cannot be passaged indefinitely. Instead, after about 50 passages, which generally occupy about one year, their growth rate inevitably begins to slow. The amount of time that they spend in G0 following each mitosis gradually increases, fewer and fewer cells enter the S period, and the cells' karyotype, that is, their chromosomal complement, changes from the euploid (diploid) pattern characteristic of normal cells to an aneuploid one, characterized by the appearance of supernumerary chromosomes, chromosome fragments and chromosomal aberrations, that is, changes in the structure of individual chromosomes. Finally, the cell strain dies out.

Loss of cell strains in this manner is generally guarded against by growing large numbers of cells during the early passages and storing them at −196C, the boiling point of liquid nitrogen.

Continuous Cell Lines

While cells derived from normal tissues have the properties described thus far, malignant tissues give rise to aneuploid cell lines that have an infinite life span and are referred to as "established cell lines. " Infrequently such cell lines seem to arise from euploid cell strains, but the possibility that malignant or premalignant cells were not present originally is difficult to rule out. In addition to being aneuploid and immortal, such cell lines usually have two other significant properties: they form tumors when transplanted into animals, and they can grow in suspension culture like bacteria. Cells growing in suspension are used extensively for studies of virus multiplication, since they are easier to handle experimentally than cells growing as monolayers.

Patterns of Macromolecular Biosynthesis

Since virus multiplication consists essentially of nucleic acid and protein synthesis, a brief description of the patterns of macromolecular synthesis in animal cells is relevant. The essential feature of the animal cell is its compartmentalization. The DNA of the animal cell is restricted to the nucleus at all stages of the cell cycle except during metaphase, when no nucleus exists. All RNA is synthesized in the nucleus. Most of it remains there, but messenger RNA and transfer RNA migrate to the cytoplasm. Ribosomal RNA is synthesized in the nucleolus; the two ribosomal subunits are assembled partly in the nucleolus and partly in the nucleus, and then they also migrate to the cytoplasm. All protein synthesis proceeds in the cytoplasm. The only exception to this brief summary concerns the mitochondria, which contain DNA-, RNA-, and protein-synthesizing systems of their own and which are located only in the cytoplasm.

The detection of animal viruses

The presence of viruses is recognized by the manifestation of some abnormality in host organisms or host cells. In the organism, symptoms of virus infection vary widely, from unapparent infections (detectable only by the formation of antibody), the development of local lesions, or mild disease characterized by light febrile response, to progressively more severe disease culminating in death. In cells, the symptoms of viral infection vary from changes in morphology and growth patterns to cytopathic effects, such as rounding, breakdown of cell organelles, the development of inclusion bodies, and general necrotic reactions, finally resulting in complete disintegration.

The isolation of animal viruses

Many techniques have been developed for isolating viruses. The source of virus may be excreted or secreted material, the bloodstream, or some tissue. Samples are collected and, unless processed immediately, sheltered from heat, preferably by storage at −70C, the temperature of dry ice. If necessary, a suspension is then prepared by grinding or sonicating in the presence of cold buffer solution, and this is then centrifuged in order to remove large debris and contaminating microorganisms.

This suspension is then tested for the presence of virus in several ways. First, it is injected back into the original host species in order to determine whether the first noted abnormality is produced. Second, the suspension is injected into other animals in order to establish whether there exist more susceptible hosts in which the disease develops more rapidly, more severely, or in a more easily recognizable manner. Newborn or suckling animals (often mice or ham-

sters) or developing chick embryos are hosts which permit many viruses to multiply more extensively than do adult animals and are accordingly widely used for virus isolation. Third, a search is conducted for an optimal cultured animal cell strain or line in which the virus will multiply and in which it may actually be isolated and also assayed. The cells which will eventually be chosen will usually be ones in which the virus rapidly elicits readily observable cytopathic effects.

The final stage of the isolation procedure is passage at limiting dilution in order to ensure that only a single unique virus is being isolated. This may be accomplished either by limiting serial dilution, when the virus suspension is diluted to such an extent that only one out of several aliquots inoculated gives a positive response, or by plaque isolation (p. 9). The latter is preferable wherever possible, since plaques originate from single virus particles, just as bacterial colonies originate from single bacterial cells.

While virus isolation from severely diseased hosts may present no difficulty, it may be a formidable task if the original source is merely suspected of containing a small amount of virus. As a result, no symptoms may result when the initial virus suspension is inoculated into the various test systems. In such cases one generally resorts to so-called blind passaging, in the hope that gradual enrichment of virus will occur. In this procedure, cells are disrupted several days after inoculation even if they appear healthy and unaltered, and an extract of them is inoculated into fresh cells. This is repeated several times until symptoms appear. It is important that this procedure be adequately controlled by passaging extracts of uninfected cells under the same conditions, since animal cells are known to harbor latent viruses which may be induced to multiply and which may then be mistaken for the etiologic agent of whatever condition is under study.

Adaptation and Virulence

During the isolation of viruses, there may emerge variants capable of multiplying more efficiently in the host cells used for this purpose than the original wild-type virus. This phenomenon, which is known as "adaptation," has as its basis the selection of spontaneous mutants, which constantly arise during virus multiplication. These mutants multiply more efficiently in the cells used for isolating the virus than in the cells of the original infected tissue. Such variants damage the original host less severely than the wild-type virus and are therefore said to be less "virulent." Viruses are often purposely adapted in order to alter growth and virulence characterisitics. An example is provided by the attenuated vaccine virus strains, which are obtained by repeated passaging of virus virulent for one host in some different host, until virus strains with decreased virulence for the original host are selected.

The measurement of animal viruses

Viruses are measured by several methods that can be divided into two categories. First, viruses may be measured as infectious units—that is, in terms of their ability to infect, multiply, and produce progeny. Second, viruses may be measured in terms of the total number of virus particles, irrespective of their function as infectious agents.

MEASUREMENT OF VIRUSES AS INFECTIOUS UNITS

Measurement of the amount of virus in terms of the number of infectious units per unit volume is known as titration. There are several ways of determining the titer of a virus suspension, all of them involving infection of host or target cells in such a way that each particle that causes productive infection elicits a recognizable reaction.

Plaque Formation

In this method a series of monolayers of susceptible cells are inoculated with small aliquots of serial dilutions of the virus suspension to be titrated. Wherever virus particles infect cells, progeny virus particles are produced and released and then immediately infect adjoining

cells. This process is repeated until, after a period ranging from 2 to 12 days or more, there develop areas of infected cells which can be seen with the naked eye. These are called plaques. In order to ensure that progeny virus particles liberated into the medium do not diffuse away and initiate separate (or secondary) plaques, agar is frequently incorporated into the medium.

The fundamental prerequisite for this method of enumerating infectious units is that the infected cells must differ in some way from noninfected cells: for example, they must either be completely destroyed, become detached from the surface on which they grow, or possess staining properties different from those of normal cells. In practice, the most common method of visualizing plaques is to apply the vital stain neutral red to infected cell monolayers after a certain number of days and to count the number of areas that do not stain (Fig. 1-3). Titers are expressed in terms of numbers of plaque-forming units (PFU) per ml.

There is a linear relationship between the amount of virus and the number of plaques produced; that is, the dose-response curve is linear. This indicates that each plaque is formed by a single virus particle. The virus progeny in each plaque therefore are clones, and virus stocks derived from single plaques are said to be plaque purified. Plaque purification is an important technique for the isolation of genetically pure virus strains.

Plaque formation is often the most desirable method of titrating viruses. It is economical of cells and virus, as well as technically simple. However, not all viruses can be measured in this way, because there may be no cells that develop the desired cytopathic effects. For these viruses, alternate titration methods must be used.

Pock Formation

Many viruses cause macroscopically recognizable foci of infection or lesions on the chorioallantoic membrane of the developing chick embryo; these lesions may be used in a manner similar to the cell monolayers employed for plaque assay. The main advantage is ready availability, wide virus susceptibility, and ease of handling. The main disadvantage is variation in virus susceptibility among different eggs of

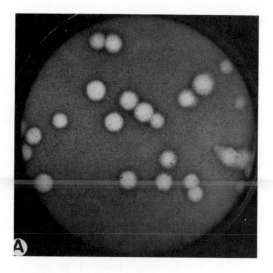
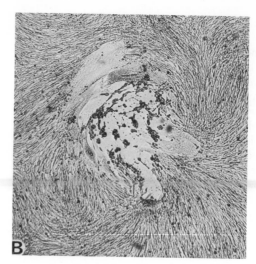

FIG. 1-3. Virus plaques. **A.** Plaques of influenza virus on monolayers of chick embryo cells, 4 days after inoculation. The monolayers were stained with neutral red on day 3. **B.** Photograph showing the microanatomy of a herpesvirus plaque on BHK 21 cells. (A, courtesy of Dr. G. Appleyard. B, courtesy of Dr. S. Moira Brown.)

even the same hatch, so that larger numbers of eggs than cell tissue culture monolayers are necessary in order to attain the same level of statistical significance. The lesions caused by viruses are known as pocks and are generally recognizable as opaque white or red areas caused by cell disintegration, migration, and proliferation, as well as edema and hemorrhage (in the case of red pocks) (Fig. 1-4). The actual

titration is carried out as described for plaques, with enumeration of pocks taking the place of plaque counting.

Focus Formation

Certain tumor viruses do not destroy the cells in which they multiply and therefore produce

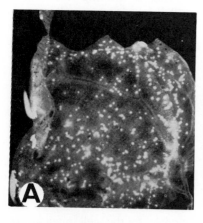
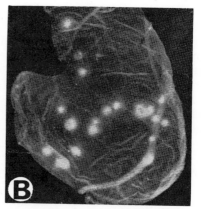

FIG. 1-4. Pocks on the chorioallantoic membrane of the developing chick embryo. The membrane is cut out two or three days after inoculation, washed and spread on a flat surface. **A.** Variola. **B.** Vaccinia. (From Kempe: Fed Proc 14:468, 1955.)

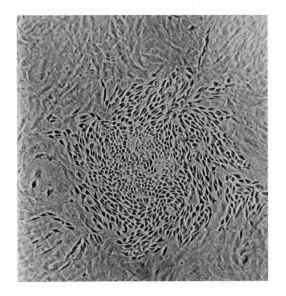

FIG. 1-5. Focus of NRK (normal rat kidney) cells transformed by Kirsten murine sarcoma virus. × 200. (Courtesy of Dr. S. A. Aaronson.)

no plaques. However, they cause cells to change morphology and to multiply at a faster rate than uninfected cells. As a result, transformed cells develop foci which gradually become large enough to be visible to the naked eye (Fig. 1-5). Assay by focus formation (counting the number focus-forming units, or FFU) is analogous to assay by plaque and pock formation.

Plaque and focus formation assays are generally performed on monolayers of cells growing in vitro but may be carried out in intact animals under special circumstances. For example, fowlpox virus may be assayed by inoculating the scalp of chickens and enumerating the number of local lesions produced, and certain mouse leukemia viruses may be assayed by injection into mice and counting the number of foci of transformed cells produced on the spleen.

The Serial Dilution End Point Method

Although many viruses destroy cells, they do not produce the type of cytopathic effects necessary for visible plaque formation. Such viruses may be titered by means of the serial dilution end point method. In this method serial dilutions of virus suspensions are inoculated into cell monolayers, which are then incubated until the cell sheet shows clear signs of cell destruction (Fig. 1-6). The end point is that dilution which gives a positive (cell-destroying) reaction, and the titer is calculated assuming that the last positive dilution originally contained at least one infectious unit. Considerable accuracy can be attained by the use of statistical methods of treating results.

The dilution end point method is also employed when virus is titrated in laboratory animals. Examples are the titration of togaviruses (arboviruses) and Group A Coxsackie viruses in the brains of suckling mice, with death as the end point.

ENUMERATION OF THE TOTAL NUMBER OF VIRUS PARTICLES

It is universally true for animal viruses that even though one virus particle is capable of causing infection, not all particles in a population actually do so. The total number of virus particles in a given preparation can be determined by either direct or indirect methods.

Counting by Means of Electron Microscopy

Direct counting of virus particles by means of electron microscopic examination is carried out according to either of two methods. The first involves mixing virus preparations with suspensions of latex spheres of similar size and known concentration, and spraying the mixture onto coated electron microscope grids. The number of virus particles and spheres in individual spray droplets is then counted; knowing the concentration of the spheres, the number of virus particles can be calculated (Fig. 1-7). The second method involves centrifuging virus preparations onto electron microscope grids and counting the virus particles; knowing what volume of the virus suspension was centrifuged, the virus concentration can again be calculated.

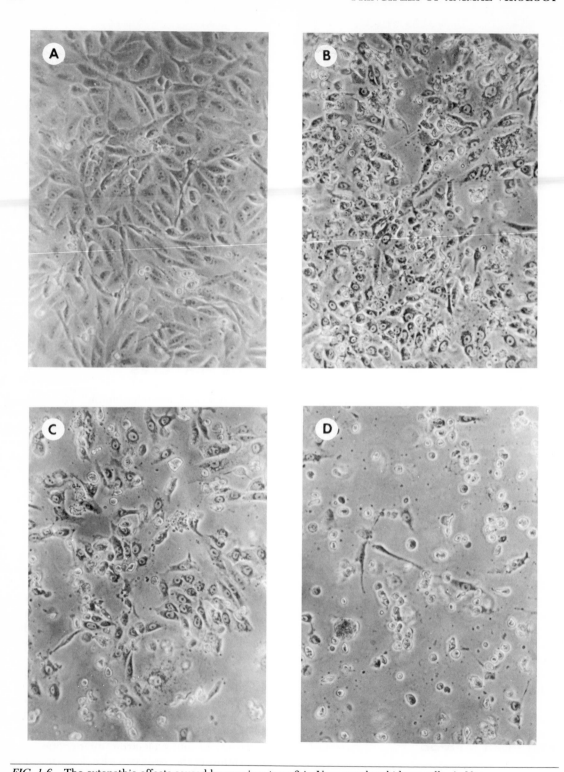

FIG. 1-6. The cytopathic effects caused by reovirus type 3 in Vero monkey kidney cells. **A.** Normal cell sheet. **B.** Partial cell destruction at 20 hours after infection. **C.** 36 hours after infection. **D.** 48 hours after infection. × 125. (Courtesy of Dr. E. C. Hayes.)

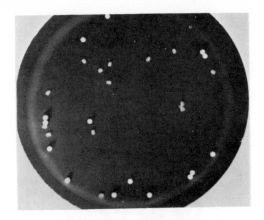

FIG. 1-7. A spray droplet containing 15 latex particles (spheres) and 14 vaccinia virus particles (slightly smaller brick-shaped particles). $\times$ 6,500. (From Dumbell, Downie, and Valentine: Virol 4:467, 1957.)

Measurement of Optical Density

The concentration of highly purified virus preparations can be routinely determined by very simple methods once they have been standardized by electron microscopy. One of these methods is measurement of the optical density. For example, 1 ml of a suspension of reovirus particles which absorbs 90 percent of incident light at a wavelength of 260 nm (that is, 1 Optical Density unit or 1 OD_{260nm}) contains 2.1×10^{12} virus particles, and 1 OD_{260nm} of vaccinia virus corresponds to 1.2×10^{10} virus particles.

The Hemagglutination Assay

The most common indirect method of measuring the number of virus particles is the hemagglutination assay. Many animal viruses adsorb to the red blood cells of various animal species. Each virus particle is multivalent in this regard; that is, it can adsorb to more than one cell at a time. In practice, the maximum number of cells with which any particular virus particle can combine is two, because red cells are far bigger than viruses. In a virus-cell mixture in which the number of cells exceeds the number of virus particles, the small number of cell dimers that may be formed is generally not detectable, but if the number of virus particles exceeds the number of cells, a lattice of cells is formed which settles out in a highly characteristic manner readily distinguishable from the settling pattern exhibited by unagglutinated cells.

The hemagglutination assay is performed by determining the virus dilution that will just hemagglutinate a given number of red cells (Fig. 1-8). Since the number of virus particles necessary for this is readily calculated, hemagglutination serves as a highly accurate and rapid method of quantitating virus particles. It was and still is particularly useful in studies with myxoviruses, particularly influenza virus, and many others.

The Significance of the Infectious Unit: Virus Particle Ratio

For all animal viruses the number of virus particles in any given preparation exceeds the number of demonstrably infectious units; usually the ratio of infectious units to particles is in the range of 1:10 to 1:1,000 or even less. There are two possible explanations for this. The first is that virus preparations contain a majority of noninfectious particles. Although this may be so sometimes, it is unlikely to be the general rule. It is more likely that although all virus particles in a given preparation are capable of causing productive infection, only a small proportion of them are actually successful in doing so. Two lines of evidence support this view. The first is that the titer of a given virus preparation varies markedly, depending on the nature of the assay system. For example, the titer often differs with the route of inoculation if the virus is assayed in whole animals, and with the type of cell if it is assayed in cultured cells. Second, before a virus particle can manifest itself as a plaque, pock, focus, and so on, it must initiate a productive infection cycle that requires numerous reactions, many of which have a low probability of occurring (see Chap. 5). Therefore, the number of infectious units cannot equal the total number of virus particles, and the ratio of the two may generally be regarded as a measure of the probability with which virus particles accomplish productive infection.

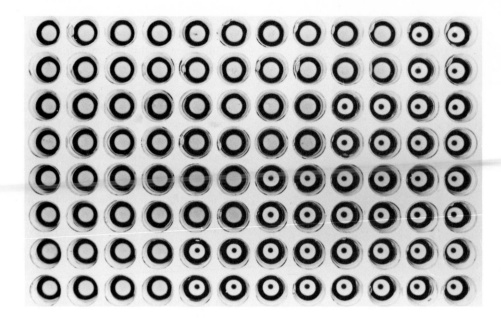

FIG. 1-8. Hemagglutination titration of influenza virus. In the top two rows a sample of influenza virus was diluted in serial twofold steps from left to right; in the next two rows the amount of virus in the first well was the same as in the third well in the top row, and so on down. The same number of red blood cells was then added to all wells, and after mixing, the tray was placed at 4C for two hours. Unagglutinated cells form a dark button; where the virus has agglutinated cells, the resulting lattice has prevented button formation. The pattern developed in this tray attests to the reproducibility of the technique. (Courtesy of Dr. E. C. Hayes.)

FURTHER READING

Books and Reviews

VIRUSES

Andrewes C, Pereira HG, Wildy P: Viruses of Vertebrates. 4th ed. London, Bailliere Tindall, 1978

Fenner FJ, McAuslan BR, Mims CA, Sambrook J, White DO: Animal Viruses, 2nd ed. New York, Academic Press, 1974

Fenner FJ, White DO: Medical Virology, 2nd ed. New York, San Francisco, London, Academic Press, 1976

Hughes SS: The Virus: A History of the Concept. New York, Neale Watson Academic Publication, 1977

Joklik WK: Evolution in Viruses. Symp Soc Gen Microbiol 24:293, 1974

Kalter SS, Heberling RL: Comparative virology of primates. Bacteriol Rev 35:310, 1971

Knight CA: Chemistry of Viruses, 2nd ed. New York Heidelberg, Berlin, Springer-Verlag, 1975

Luria SE, Darnell JE Jr, Baltimore D, Campbell A: General Virology, 3rd ed. New York, Santa Barbara, Chichester, Brisbane, Toronto, Wiley, 1978

Wishnow RN, Steinfeld JL: The conquest of the major infectious diseases in the United States: A bicentennial retrospect. Ann Rev Microbiol 30:427, 1976

CELLS

Altman PL, Katz DD (eds): Cell Biology. Bethesda, Fed Amer Soc Exp Biol, 1978

Barigozzi C (ed): Origin and Natural History of Cell Lines. New York, Liss, 1978

Bautz EKF, Karlson P, Kerstein H: Regulation of Transcription and Translation in Eukaryotes. New York, Heidelberg and Berlin, Springer-Verlag, 1974

Brinkley BR, Porter KR (eds): International Cell Biology, 1976-77. New York, Rockefeller Univ Press, 1977

Clarkson B, Baserga R (eds): Control of Proliferation in Animal Cells. New York, Cold Spring Harbor, 1974

Clarkson B, Marks PA, Till JE (eds): Differentiation of Normal and Neoplastic Hematopoietic Cells. New York, Cold Spring Harbor, 1979

Giese AC: Cell Physiology, 4th ed. Philadelphia, Saunders, 1973

Goldstein L, Prescott DM (eds): Genetic Mechanisms of Cells. New York, Academic Press, 1978

Ham RG: Unique requirements for clonal growth. J Natl Cancer Inst 53:1459, 1974

Harris H: Nucleus and Cytoplasm. Oxford, Clarendon Press, 1974

Kaighn ME: "Birth of a culture"—Source of postpartum anomalies. J Natl Cancer Inst 53:1437, 1974

Lerner RA, Bergsma D (eds): The Molecular Basis of Cell-Cell Interaction. New York, Liss, 1978

Pollack R (ed): Readings in Mammalian Cell Culture. New York, Cold Spring Harbor Lab, 1975

Taylor WG: "Feeding the baby"—Serum and other supplements to chemically defined medium. J Natl Cancer Inst 53-1449, 1974

Selected Papers

Amsterdam A, Jamieson JT: Techniques for dissociating pancreatic exocrine cells. J Cell Biol 63:1037, 1974

Bretscher MS, Raff MC: Mammalian plasma membranes. Nature 258:43, 1975

Chen, TR: In situ detection of mycoplasma contamination in cell cultures by fluorescent Hoechst 33258 stain. Exp Cell Res 104:255, 1977

Conrad GW, Hart GW, Chen Y: Differences in vitro between fibroblast-like cells from cornea, heart and skin and embryonic chicks. J Cell Sci 26:119, 1977

Crissman HA, Tobey RA: Cell-cycle analysis in 20 minutes. Science 184:1297, 1974

Goldman RD, Lazarides E, Pollack R, Weber K: The distribution of actin in non-muscle cells. Exp Cell Res 90:333, 1975

Gospodarowicz D, Moran J: Effect of a fibroblast growth factor, insulin, dexamethasone, and serum on the morphology of BALB/c 3T3 cells. Proc Natl Acad Sci USA 71-4648, 1974

Igarashi A, Mantani M: Rapid titration of dengue virus type 4 infectivity by counting fluorescent foci. Biken J 17:87, 1974

Kuroki T: Colony formation of mammalian cells on agar plates and its application to Lederberg's replica plating. Exp Cell Res 80:55, 1973

Osborn M, Franke WW, Weber K: Visualization of a system of filaments 7-10nm thick in cultured cells of an epithelioid line (PtK2) by immunofluorescence microscopy. Proc Natl Acad Sci USA 74:2490, 1977

Rheinwald JG, Green H: Growth of cultured mammalian cells on secondary glucose sources. Cell 2:287, 1974

Rubin H: Central role for magnesium in coordinate control of metabolism and growth in animal cells. Proc Natl Acad Sci USA 72:3551, 1975

Smith JR, Hayflick L: Variation in the life span of clones derived from human diploid strains. J Cell Biol 62:48, 1974

Stack SN, Brown DB, Dewey WC: Visualization of interphase chromosomes. J Cell Sci 26:281, 1977

CHAPTER 2

The Structure, Components, and Classification of Viruses

The morphology of animal viruses

Although animal viruses differ widely in shape and size, they are nevertheless constructed according to certain common principles. Basically, viruses consist of nucleic acid and protein. The nucleic acid is the genome which contains the information necessary for virus multiplication; the protein is arranged around the genome in the form of a layer or shell that is termed the capsid. The structure consisting of shell plus nucleic acid is the nucleocapsid. Many animal virus particles consist of naked nucleocapsids, while others possess an additional envelope which is usually acquired as the nucleocapsids bud from the host cell. The complete virus particle is known as the virion, a term that denotes both intactness of structure and the property of infectiousness.

Capsids

The essential feature of capsids is that they are composed of numerous repeating subunits—identical or belonging to only a few different species—arranged in precisely defined patterns. The simplest subunits are single protein molecules; more complex forms are morphologic subunits termed capsomers that can be seen with the electron microscope and that consist of several either identical or different protein molecules. The use of only a few types of subunits for capsid construction has two noteworthy consequences: (1) it minimizes the amount of genetic information necessary to specify capsids, and (2) it assures that they will be assembled efficiently. Capsid proteins exhibit a strong tendency to bind to one another, and much of the information necessary for the morphogenesis of nucleocapsids is inherent in their amino acid sequence.

Capsids (and envelopes) have a dual function. The first is to protect viral genomes from potentially destructive agents in the extracellular environment, such as enzymes, and the second is to introduce viral genomes into host cells. The need for this latter function stems from the fact that viral nucleic acids are often longer than cell diameters and cannot penetrate into cells by themselves. Capsids (and envelopes), on the other hand, adsorb readily to cell surfaces and can enter cells by several mechanisms (see Chap. 5).

Envelopes

Only six families of animal viruses exist as naked nucleocapsids. In all others the nucleocapsids are enclosed by membrane-containing envelopes that are acquired as the nucleocapsids bud through special patches of cell membrane—either in outer plasma cell membranes or in vacuolar membranes—on their way to the exterior of the cell. The membrane patches through which nucleocapsids bud are virus-modified: usually the cell-specified proteins in them are completely replaced by virus-specified proteins, and virus-specified glycoprotein spikes are attached to their outer surface. However, there are exceptions: the envelopes of herpesviruses and RNA tumor viruses probably still contain some host-coded proteins, and although herpesvirus envelopes contain glycoproteins, they do not possess obvious spikes.

Viral envelopes generally lack rigidity. As a consequence they usually appear heterogeneous in shape and size when fixed for electron microscopy, and enveloped viruses are therefore often said to be "pleomorphic." There is little doubt, however, that in their native state most enveloped viruses are spherical, enclosing either icosahedral nucleocapsids or spherically coiled helical nucleocapsids. However, two types of enveloped viruses are not spherical. These are the rhabdoviruses, which possess a highly characteristic bulletlike shape, rounded at one end and flat at the other (Fig. 2-5), and certain strains of influenza virus, whose helical nucleocapsids become enveloped not in a coiled but in an extended configuration, which causes the enveloped virus particles also to be long and filamentous (see Fig. 2-4).

Nucleocapsids

Viral nucleocapsids are constructed according to a small number of basic patterns. Two of these have been studied in great detail at both the structural and the molecular level: in one the nucleic acid is extended, in the other it is

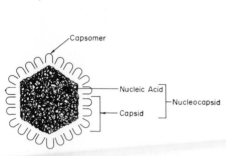

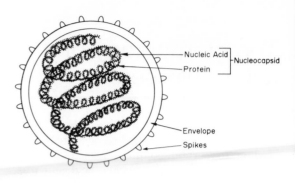

NAKED ICOSAHEDRAL NUCLEOCAPSID   ENVELOPED HELICAL NUCLEOCAPSID

FIG. 2-1. The two basic patterns of animal virus structure. Left: the condensed genome is enclosed by a shell of capsomers arranged so as to display 5:3:2 rotational symmetry. Right: the extended genome is enclosed by protein molecules arranged so as to display helical symmetry. The resultant structure, the nucleocapsid, is enclosed in an envelope to whose outer surface glycoprotein spikes are attached.

FIG. 2-2. Schematic representation of tobacco mosaic virus. As can be seen in the cutaway section, the ribonucleic acid helix is associated with protein molecules in the ratio of three nucleotides per protein molecule. (From Klug and Caspar: Adv Virus Res 7:225, 1960.)

condensed (Fig. 2-1). Superimposed on these two patterns are variations dictated both by the size of the genome and by the nature of the capsid polypeptides.

Nucleocapsids with Helical Symmetry The prototype of nucleocapsids in which the nucleic acid occurs as an extended filament is a plant virus, tobacco mosaic virus (TMV), whose structure has been studied extensively by x-ray diffraction. In this virus, the extended nucleic acid molecule is surrounded by protein molecules arranged helically so as to yield a structure with a single rotational axis (Fig. 2-2). The ortho- and paramyxoviruses and rhabdoviruses possess nucleocapsids constructed in this manner, each with its own characteristic length, width, periodicity, flexibility, and stability (Fig. 2-3). It should be noted that these nucleocapsids are not the complete virions: the virions of these virus families consist of the nucleocapsids coiled more or less tightly inside envelopes (Figs. 2-4 and 2-5).

Nucleocapsids with Icosahedral Symmetry In the second pattern of virus structure, the nucleic acid is condensed and forms the central portion of a quasispherical nucleocapsid. Here the capsid consists of a shell of protein molecules that are clustered into small groups called capsomers, with the bonds between molecules within capsomers being stronger than those between capsomers. Capsomers are morpho-

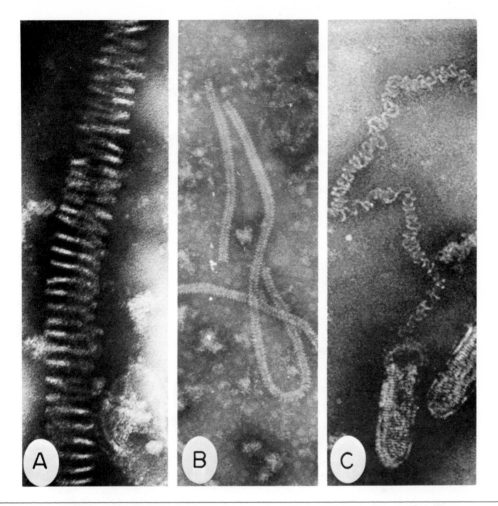

FIG. 2-3. The nucleocapsids of (**A**) influenza virus strain PR8 ($\times 225,000$); (**B**) measles virus ($\times 150,000$); and (**C**) vesicular stomatitis virus (VSV) ($\times 160,000$). The latter is emerging from a damaged virion. (From: (A) Almeida and Waterson: In Barry and Mahy (eds): The Biology of Large RNA Viruses, 1970. Courtesy of Academic Press; (B) Finch and Gibbs: J Gen Virol 6:144, 1970; (C) Simpson and Hauser: Virol 29:660, 1966.)

logic units which can often be seen with the electron microscope; they vary in size and shape from virus to virus.

X-ray diffraction analysis indicates that in this type of nucleocapsid the capsomers are arranged very precisely according to icosahedral patterns characterized by 5:3:2-fold rotational symmetry (Fig. 2-6). Two such patterns are found among animal viruses. The first is exhibited most clearly by adenoviruses. The adenovirus capsid is constructed in the shape of an icosahedron with 6 capsomers along each edge and 252 capsomers altogether (Fig. 2-7). Of these, 240 are spherical and are situated along the edges and on the faces of the icosahedron; each has 6 nearest neighbors and is known as a hexon or hexamer. The remaining 12 are situated at the 12 vertices of the icosahedron and have 5 nearest neighbors; these are known as pentons or pentamers. They have a highly characteristic shape, consisting of a spherical base and a long fiber which may serve as the cell attachment organ.

The capsids of four other virus families are constructed similarly. Iridovirus capsids possess 10 capsomers along each edge, and there are 1,112 capsomers altogether, 1,100 hexons and 12 pentons; herpesvirus capsids possess 5 capso-

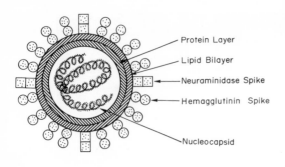

Protein Layer

Lipid Bilayer

Neuraminidase Spike

Hemagglutinin Spike

Nucleocapsid

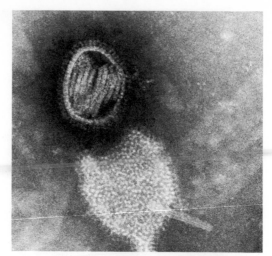

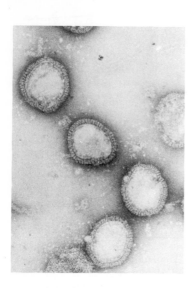

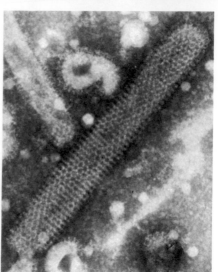

FIG. 2-4. The structure of influenza virus. Top Left: Model of influenza virus particle. The nucleocapsid is loosely coiled inside an envelope that consists of both protein and lipid, and to which two different types of glycoprotein spikes, each consisting of two protein subunits, are attached. Hemagglutinin spikes are about five times as numerous as neuraminidase spikes. Top right: Influenza virus A_2, stained with phosphotungstate. One particle is penetrated by the stain, thereby revealing the arrangement of the internal nucleocapsid. ×155,000 (Courtesy of Dr. M. V. Nermut.) Bottom left: Influenza virus A_0/WSN, stained with phosphotungstate, revealing the characteristic arrangement of spikes on the particle surface. ×135,000 (Courtesy of Dr. I. T. Schulze.) Bottom right: A filamentous particle of an influenza C strain. Note regular subunit surface pattern. ×115,000. (From Apostolov and Flewett: J Gen Virol 4:366, 1969.)

mers along each edge and 162 capsomers altogether, 150 prism-shaped hexons and 12 pentons (Fig. 2-8); papovavirus capsids consist of 72 capsomers, 60 hexons and 12 pentons (Fig. 2-9); and parvovirus capsids are probably made up of 32 capsomers, 20 hexons and 12 pentons (Fig. 2-9).

The second pattern is exhibited most clearly by picornaviruses. Here sixty identical capsomers, each composed of 4 different proteins,

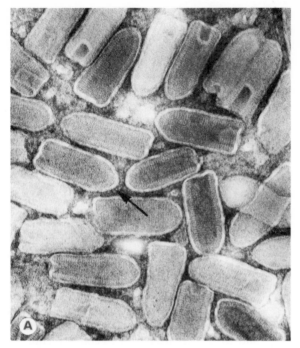

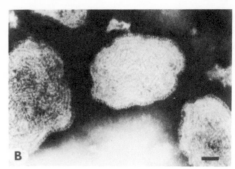

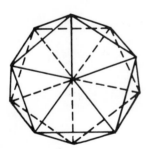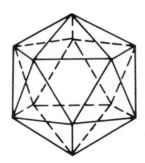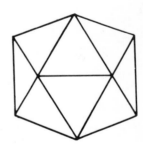

FIG. 2-5. **A.** Highly characteristic bullet-shaped vesicular stomatitis virus (VSV) particles, some penetrated by stain, revealing the tightly coiled nucleocapsid. Note glycoprotein spikes (arrow) (Courtesy of Dr. Erskine Palmer). **B.** Sendai virus, a paramyxovirus. Note tightly coiled nucleocapsid. ×73,500. (From Maeno et al: J Virol 6:492, 1970.)

are situated equidistantly from a common center, which results in a spherical capsid (Fig. 2-10); but instead of being bonded equally strongly to each other, these capsomers are bonded into groups of five, twelve of which make up the capsid. Reoviruses are unique in possessing two capsids shells (Fig. 2-11). Both possess icosahedral symmetry, but it has so far proved impossible to discern either the total number of capsomers or the precise manner in which they are arranged. The capsid shell of some members of this family, such as the orbiviruses and the rotaviruses, appears to be composed of 32 large ring-shaped capsomers. It is more likely however that the capsid is composed of numerous small subunits arranged in ring-shaped (or hexagonal) patterns, and that many of these subunits are shared by adjacent

FIG. 2-6. The icosahedron viewed normal to 5-, 3-, and 2-fold rotational axes. Edges of the upper and lower surfaces are drawn in solid and broken lines respectively. The 5-fold rotational axes pass through the vertices (left); the 3-fold rotational axes pass through the centers of the triangular faces (center); and the 2-fold rotational axes pass through the edges (right). In this view, the edges on the upper and lower surfaces coincide. Note that the icosahedron possesses 12 vertices, 20 triangular faces, and 30 edges.

rings, so that what is visible is 32 holes, rather than 32 capsomers. Reovirus capsids are probably structured similarly, although here the rings are much less prominent.

In the case of the picornaviruses, caliciviruses, adenoviruses, papovaviruses, parvoviruses and reoviruses, the virions are the na-

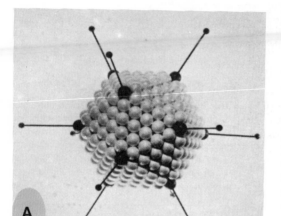

ked nucleocapsids. In the case of the herpesviruses and the sole iridovirus that is a mammalian virus, however, the naked nucleocapsids themselves are relatively noninfectious; here the virions consist of enveloped icosahedral nucleocapsids.

THE STRUCTURE OF TOGAVIRUSES, BUNYAVIRUSES, ARENAVIRUSES AND CORONAVIRUSES

These viruses are all enveloped particles to whose outer surface glycoprotein spikes are attached; the spikes are particularly prominent in coronaviruses (Fig. 2-12), where they are large and club-shaped and surround the virus particles like a halo, hence the name. The nucleocapsids of these viruses exhibit a variety of morphologic patterns. In the case of the togaviruses, the alphaviruses—which comprise the old group A arboviruses—contain condensed

FIG. 2-7. **A.** Model of adenovirus particle constructed by R. C. Valentine. **B.** Adenovirus freeze-dried and shadowed with platinum. ×400,000. (Courtesy of Dr. M. V. Nermut.)

RNA molecules intimately associated with protein to form nucleocapsids that possess distinct icosahedral symmetry elements (Fig. 2-13); but the nucleocapsids of the slightly smaller flavoviruses—the old group B arboviruses, the name deriving from the prototype strain *yel-* *low* fever virus strain—possess no obvious symmetry elements. The bunyaviruses have a diameter that is about twice that of togaviruses and possess three coiled circular nucleocapsids that possess helical symmetry (Fig. 2-14). The arenaviruses (Fig. 2-15) are slightly larger than

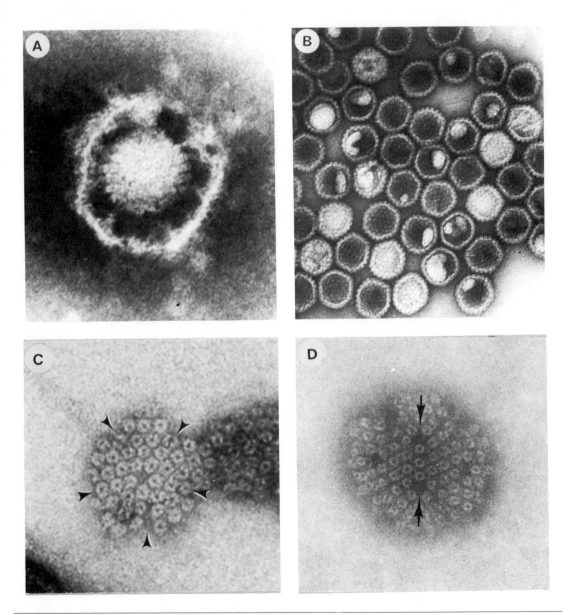

FIG. 2-8. The morphology of herpesviruses. **A.** Enveloped equine abortion virus (EAV) particle. ×125,000. **B.** EAV particles from which the envelope has been removed by treatment with detergent. ×75,000. **C** and **D.** Ultrastructure of partially disrupted herpesvirus nucleocapsids exhibiting 5- and 2-fold symmetry axes respectively. ×320,000. (A and B, from Abodeely, Lawson, and Randall: J Virol 5:513, 1970; C and D, courtesy of Dr. Erskine Palmer, Center for Disease Control, Atlanta.)

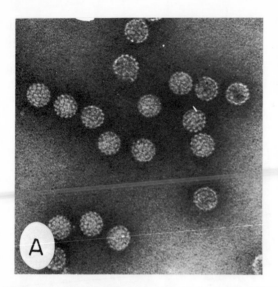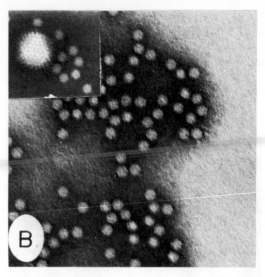

FIG. 2-9. **A.** The papovavirus SV40 ($\times$160,000) and **B,** the parvovirus Adeno-Associated Virus (AAV) type 4 ($\times$150,000). Insert: a virion of the simian adenovirus SV15, which enabled the AAV to multiply (see Chap. 6). (Courtesy of Dr. Heather Mayor.)

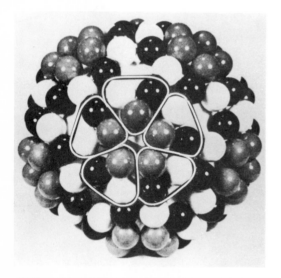

FIG. 2-10. Model of the picornavirus capsid. There are 60 capsomers, each consisting of three protein molecules, represented here by white, grey, and black balls. (In fact, in the mature virion, one of these protein molecules is cleaved into a larger and a smaller fragment, both of which remain in position.) Groups of 5 capsomers form units that are intermediates during morphogenesis; the capsid is composed of 12 of these "groups of five." (Adapted from Johnston and Martin: J Gen Virol 11:77, 1971.)

bunyaviruses and contain two circular coiled nucleocapsids; they also contain several highly characteristic granules (Latin arenosus, sandy) that have been shown to be ribosomes (which, however, are not required for virus multiplication). Finally, the coronaviruses (Fig. 2-12). which are about the same size as bunyaviruses, possess linear helical nucleocapsids that resemble those of myxoviruses.

THE STRUCTURE OF RNA TUMOR VIRUSES, POXVIRUSES AND SOME MISCELLANEOUS VIRUSES

RNA tumor viruses have a more complex structure; they consist of a concentrically coiled RNA-containing filament enclosed in a nucleocapsid that possesses icosahedral symmetry and is closely associated with an "inner coat" that is itself bounded by a membrane that bears more or less prominent glycoprotein spikes (Fig. 2-16) (see Chap. 9).

Poxviruses are the largest and most complex of all animal viruses. Morphologically, there are two classes of poxviruses. Most poxvirus particles are prolate ellipsoids that are often brick-shaped when fixed for electron microscopy. They are covered on their outer surface with

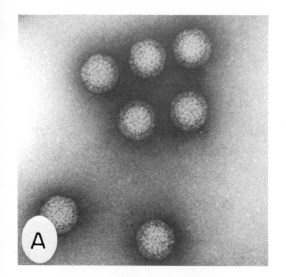

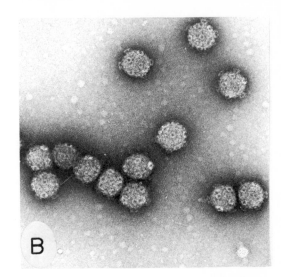

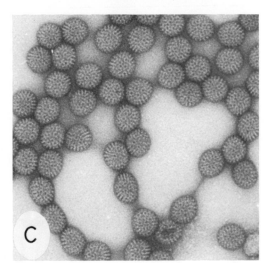

FIG. 2-11.   Morphology of reovirus and human rotavirus. **A.** Reovirus. Note double capsid shell. The arrangement of capsomers is clearly discernible only at the periphery. ×120,000. **B.** Reovirus cores. Cores are derived from reovirions by digesting the outer capsid shell with chymotrypsin. Note the large spikes; there are 12, located as if situated on the 12 vertices of an icosahedron. **C.** Human rotavirus. Its structure is similar to, but not identical with, that of orbiviruses. ×136,000. (A and B, Courtesy of Drs. R. B. Luftig and W. K. Joklik; C, courtesy of Dr. Erskine Palmer, Center for Disease Control, Atlanta.)

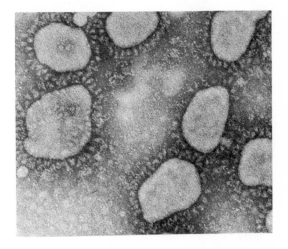

tubules or filaments arranged in a characteristic whorled or mulberry pattern (Fig. 2-17). Within this outer layer there is a protein coat that contains two lateral bodies of unknown composition and function, and a DNA-containing nucleoid or core bounded by a layer of well-defined protein subunits.

Parapoxviruses are somewhat smaller, ovoid rather than brick-shaped, and covered on their outer surface by tubules or filaments similar to those that cover the poxvirus particles just described, except that they are arranged in a highly regular, crisscross pattern, which is in all probability caused by one continuous filament

FIG. 2-12.   Coronavirus particles. Note pleomorphic envelopes studded with characteristic widely-spaced club- or pear-shaped surface projections. ×144,000. (From Kapikian, In Lennette EH, Schmidt NJ (eds): Diagnostic Procedures for Viral and Rickettsial Infections, 4th ed. American Public Health Assoc, 1969.)

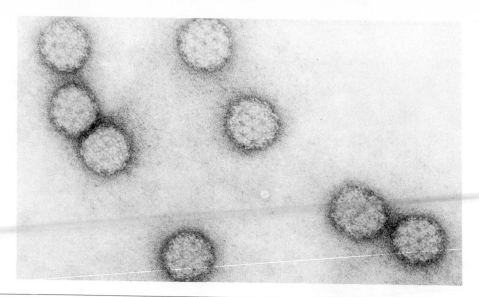

FIG. 2-13.   The togavirus Sindbis virus stained with uranyl acetate. ×240,000. Courtesy of Dr. P. J. Enzmann.

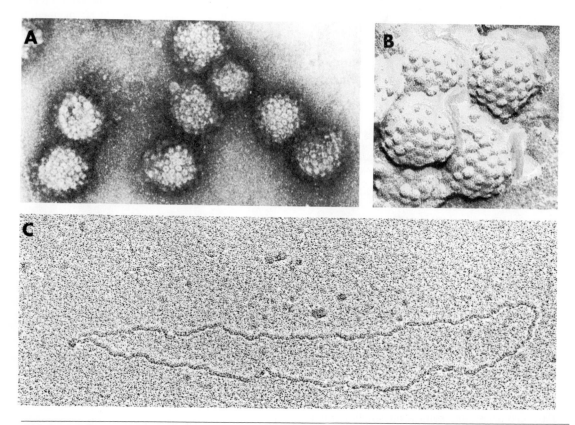

FIG. 2-14.   The structure of Uukuniemi virus, a bunyavirus. **A.** Virus particles fixed with glutaraldehyde and negatively stained with uranyl acetate. ×100,000. **B.** Freeze-etched, glutaraldehyde-fixed virus particles: a group of cleaved particles showing the icosahedral arrangement of surface projections. ×180,000. **C.** The circular nucleocapsid of this virus, shadowed with platinum. Nucleocapsids are released from virus particles by treatment with the nonionic detergent Triton X-100. ×60,000. (All photos courtesy of Dr. C. H. von Bonsdorff. J Virol 16:1296, 1975.)

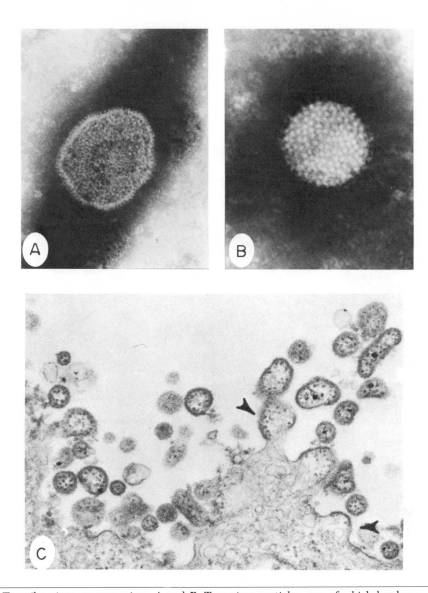

FIG. 2-15.   Tacaribe virus, an arenavirus. **A** and **B.** Two virus particles, one of which has been partially penetrated by negative contrast medium, showing the glycoprotein spikes that cover the surface. ×135,000 and ×235,000 respectively. **C.** Thin section of Parana virus particles budding from the plasma membrane of Vero African Green Monkey kidney cells. Note the characteristic dense granules. ×45,000. (From Murphy et al: J Virol 6:507, 1970.)

wound round each particle in 12 to 15 left-handed turns (Fig. 2-18). The internal components of parapoxvirus particles are similar to those of the poxvirus particles described above.

Finally, there are several viruses that have not yet been assigned to any virus family. Among these are hepatitis A virus (infectious hepatitis virus) which has a diameter of 27 nm and may be a picornavirus; hepatitis B virus (serum hepatitis virus) which has a diameter of 42 nm (Dane particles, see Chap. 23); and the chronic infectious neuropathic agents (see Table 2-4).

The relative sizes of some important animal viruses are illustrated in Fig. 2-19. Their morphology is summarized in Table 2-1.

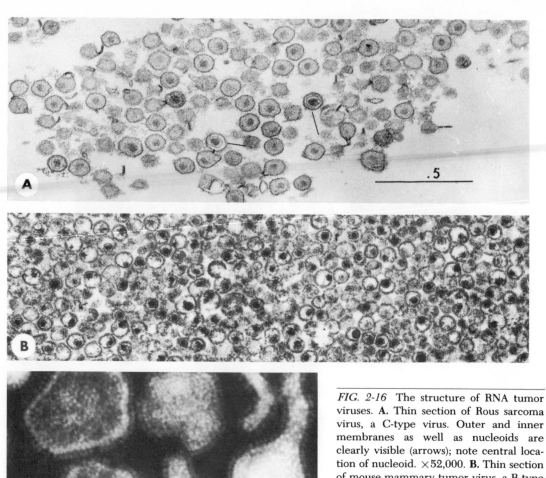

FIG. 2-16 The structure of RNA tumor viruses. **A.** Thin section of Rous sarcoma virus, a C-type virus. Outer and inner membranes as well as nucleoids are clearly visible (arrows); note central location of nucleoid. ×52,000. **B.** Thin section of mouse mammary tumor virus, a B-type virus. Note eccentric location of nucleoid. **C.** Mouse mammary tumor virus stained with phosphotungstate. Note prominent glycoprotein spikes. (A, from Courington and Vogt: J Virol 1:400, 1967; B and C, courtesy of Dr. D. Moore.)

The nature of the components of animal viruses

The Purification of Viruses

Little serious work on the properties, composition, and molecular biology of viruses is possible unless pure virus is available. The starting material for virus purification may be any material that contains a sufficiently large amount of virus. This may be cellular material such as infected organs or cultured cells, or it may be extracellular material such as plasma, allantoic fluid, or cell culture medium. If the virus concentration in such material is not high enough, an initial concentration step that employs either precipitation or centrifugation may be necessary. The next step is then generally designed to achieve a preliminary purification by removing the bulk of nonviral material. This may be achieved by treatment with detergents or emulsification with organic solvents followed by centrifugation or by adsorption to and elution from red blood cells (in the case of myxovi-

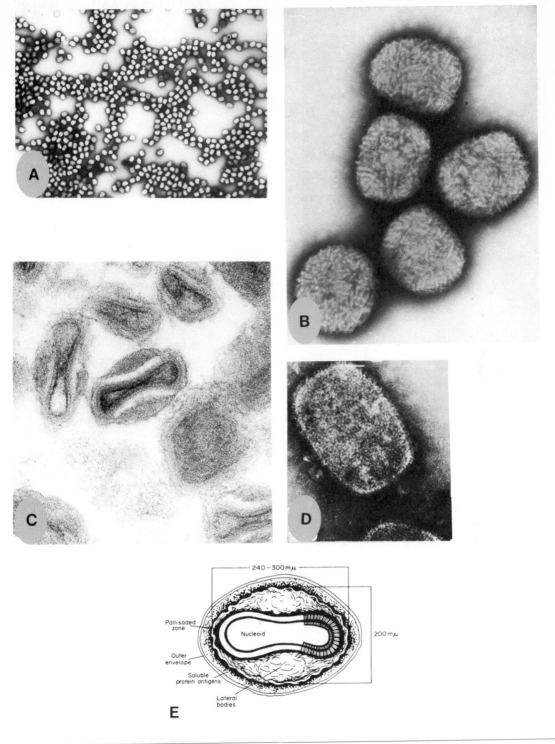

FIG. 2-17. The structure of vaccinia virus. **A.** A purified virus preparation. Note characteristic brick shape. ×6,000. **B.** Vaccinia virus particles stained with phosphotungstate to reveal surface structure. Note characteristic arrangement of rodlets or tubules. ×60,000. **C.** Cross-section of vaccinia virus particle. ×70,000. **D.** An isolated core, showing regular surface elements. ×90,000. **E.** A model of the vaccinia virus particle. (A, B and C, courtesy of Dr. Samuel Dales; D, from Easterbrook: J Ultrastr Res. 14:484, 1966; E, adapted from Westwood et al: J Gen Microbiol 34:67, 1964.)

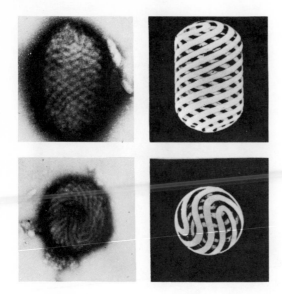

FIG. 2-18. Parapoxvirus particles. These are particles of contagious pustular dermatitis virus (ORF), negatively stained so as to reveal the crisscross arrangement of surface strands or tubules. ×90,000. (From Büttner, Giese, and Peters: Arch Ges Virusforsch 14:657, 1964.)

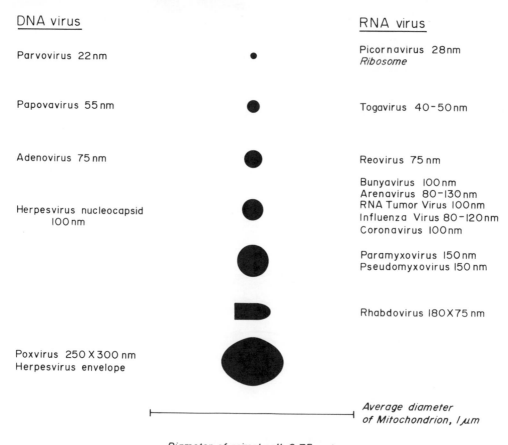

DNA virus

Parvovirus 22 nm

Papovavirus 55 nm

Adenovirus 75 nm

Herpesvirus nucleocapsid
100 nm

Poxvirus 250 X 300 nm
Herpesvirus envelope

RNA virus

Picornavirus 28 nm
Ribosome

Togavirus 40-50 nm

Reovirus 75 nm

Bunyavirus 100 nm
Arenavirus 80-130 nm
RNA Tumor Virus 100 nm
Influenza Virus 80-120 nm
Coronavirus 100 nm

Paramyxovirus 150 nm
Pseudomyxovirus 150 nm

Rhabdovirus 180 X 75 nm

*Average diameter
of Mitochondrion, 1 μm*

Diameter of animal cell, 0.75 meter

Length of DNA in poxvirus particle, 7.5 meters

FIG. 2-19. The relative sizes of the principal families of animal viruses. Unless otherwise indicated, the scale is the same for all.

TABLE 2-1   THE MORPHOLOGY OF
             ANIMAL VIRUSES

Virus	Morphology
DNA VIRUSES	
Poxvirus	Complex
Iridovirus	Enveloped icosahedral nucleocapsid
Herpesvirus	Enveloped icosahedral nucleocapsid
Adenovirus	Naked icosahedral nucleocapsid
Papovavirus	Naked icosahedral nucleocapsid
Parvovirus	Naked icosahedral nucleocapsid
RNA VIRUSES	
Picornavirus	Naked icosahedral nucleocapsid
Togavirus	Enveloped icosahedral nucleocapsid
Bunyavirus	Enveloped coiled circular nucleocapsids
Reovirus	Naked double-shelled icosahedral nucleocapsid
Orthomyxovirus	Enveloped helical nucleocapsids
Paramyxovirus	Enveloped helical nucleocapsid
Rhabdovirus	Enveloped helical nucleocapsid
RNA tumor virus	Enveloped icosahedral nucleocapsid
Arenavirus	Enveloped coiled nucleocapsid
Coronavirus	Enveloped helical nucleocapsid

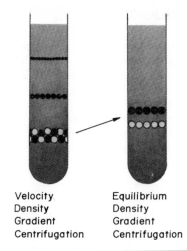

Velocity   Equilibrium
Density   Density
Gradient   Gradient
Centrifugation   Centrifugation

FIG. 2-20. Application of the technique of density gradient centrifugation to virus purification. A partially purified virus preparation is layered onto the left-hand density gradient, which is centrifuged so that particles with the same sedimentation coefficient form distinct bands. Three such bands are shown, one of which contains virus (open spheres). This band is then centrifuged to equilibrium in the right-hand density gradient, in which particles are separated according to their buoyant density.

ruses) or by passage through columns that contain materials capable of separating viral and cellular components. For the final purification step fractionation by means of density gradient centrifugation is almost always used. There are two modes of employing this technique (Fig. 2-20). In velocity density gradient or rate zonal centrifugation, the virus suspension is layered onto a density gradient, that is, a solution of either sucrose or glycerol or some salt of gradually increasing density, the maximum density being such that virus particles would migrate to the bottom of the tube if centrifuged long enough. If centrifuged for shorter periods, particles with the same sedimentation coefficient, which depends on size, density, and shape, sediment as homogeneous bands which may be collected. This step eliminates all impurities except those with the same sedimentation coefficient as virus particles. In order to eliminate

the remaining impurities, the particles in the virus-containing band recovered from the first density gradient are then centrifuged to equilibrium in a second density gradient composed of solutions of higher density. Here particles form bands where the density of the medium is identical to their own buoyant density. Since contaminants with both the same sedimentation coefficient and the same density as virus particles are rare, virus purified by two such density gradient centrifugation steps is generally considered to be essentially pure (that is, at least 97-99 percent free of nonviral material).

The purification of icosahedral viruses generally presents no difficulties. Some enveloped viruses, however, are not easily purified because the amount of envelope per virus particle may be variable, which causes them to be heterogeneous with respect to both size and density.

It is almost impossible to establish absolute criteria of purity for viruses, mainly because small amounts of cellular constituents tend to adsorb to them. Absence of particulate impurities is best assessed by electron microscopic examination.

VIRAL NUCLEIC ACIDS

The nucleic acids of animal viruses are astonishingly diverse. Some are DNA, others RNA; some are double-stranded, others single-stranded; some are linear, others circular; some have plus polarity, others minus polarity. Information concerning these and other properties of viral nucleic acids is essential for an understanding of the key reactions during virus multiplication cycles.

The Size of Viral Nucleic Acids

The nucleic acid content of animal viruses varies within wide limits. At the lower end of the scale, only about 2 percent of influenza virus particles is RNA, and only about 5 percent of poxvirus particles is DNA; at the upper end, about 25 percent of picornavirus particles is RNA. However, the proportion of nucleic acid in virus particles is not as significant as its absolute amount, which is the factor that determines the amount of genetic information that it contains. The smallest animal virus genomes are those of the picornaviruses and parvoviruses, whose molecular weights range from 1.5 to 3×10^6. Since the coding ratio (the ratio of the molecular weight of single-stranded nucleic acid to the molecular weight of the protein for which it can code) is about 9, these viral genomes can code for protein with a total molecular weight of from 150,000 to 300,000, which is equivalent to 4 to 8 average-size proteins. The largest animal virus genomes, those of the poxviruses and herpesviruses, are about 100 times larger; their molecular weights range from 100 to almost 200×10^6 (Fig. 2-21). The molecular weights of viral genomes are listed in Table 2-2.

The Structure of Viral Nucleic Acids

Strandedness Both double-stranded and single-stranded DNA as well as RNA can act as the genome of animal viruses (Table 2-2).

Terminal Redundancy The nucleic acids of several animal viruses are terminally redundant or repetitious, that is, their base composition may be represented as A, B, C . . . X, Y, Z, A, where A, B, C, and so on are nucleotide sequences. This is most readily demonstrated, in the case of double-stranded DNAs, by treating them briefly with bacteriophage lambda exonuclease, which digests DNA strands from their 5' phosphate-containing termini. In this way one end of one strand and the other end of the other strand are digested. On melting and reannealing, DNA digested in this manner circularizes, indicating that the two single-stranded regions at the two ends are complementary, hence that their sequences must be repetitious (Fig. 2-22). In the case of herpesvirus DNA, the length of the repeated sequence is about 400 nucleotide base pairs, which is about 0.25% of its total length. Terminal redundancy of this type is also exhibited by the RNA of RNA tumor viruses.

Adenovirus, parvovirus and poxvirus DNAs are also terminally redundant, but here the repeated sequences are inverted or reversed. This is demonstrated by the fact that here single-stranded DNA circles form upon melting and reannealing, without first having to be digested with nuclease (Fig. 2-23). The length of the repeated sequence in adenovirus DNA ranges from 100 to about 6000 base pairs (0.4-15 percent of its length) in different strains. For adeno-associated virus and poxvirus DNAs the corresponding figures are 1.5 and about 6 percent, respectively, the latter being variable in uncloned virus populations.

Yet a different type of terminal redundancy is exhibited by molecules of bunyavirus and arenavirus RNA which exist in circular form in the native state, but not following denaturation. The reason for this appears to be the presence of inverted complementary base sequences some 10-30 nucleotides long at their ends, which causes them to be "sticky" or cohesive, that is able to hybridize and therefore circularize.

The significance of the various forms of terminal redundancy is no doubt related to the mode of replication and expression of these nucleic acids. Examples of where the reasons for terminal redundancy seem clear are described in Chapters 9 and 10.

Cross-linking The DNAs of poxviruses are unique in being cross-linked covalently at or close to their ends (within 50 base pairs). The nature of the cross-links, which can be hydro-

FIG. 2-21. An electron micrograph of an intact herpesvirus DNA molecule. It is about 44 μm long. The small circular molecules are intact DNA molecules of the bacteriophage φX174 (see Chap. 10), which were added to provide a size marker. (Courtesy of Dr. Edward K. Wagner.)

lized by single-strand specific nucleases and are known to be broken shortly after infection, is unknown. Their presence is revealed by melting the DNA, when single-stranded circles are generated whose circumference is twice the length of the original linear double-stranded poxvirus DNA molecules (Fig. 2-24).

Covalent linkage with protein Several viral nucleic acids are linked covalently to protein molecules. Thus the RNAs of picornaviruses like poliovirus are linked at their 5'-termini to a protein with a molecular weight of about 5,000 (through a tyrosine residue), and each 5'-

terminus of the double-stranded DNA of adenoviruses is linked to a protein with a molecular weight of about 55,000. It is conceivable that these proteins function in the replication of these nucleic acids.

Circularity and Supercoiling Most viral nucleic acids are linear (Fig. 2-21), but papovavirus DNA exists in the form of double-stranded supercoiled circles (Fig. 2-25). The reason this DNA is supercoiled is a deficiency of 15-20 turns in the double helix, introduced by the action of a DNA gyrase that catalyzes negative supercoiling at the expense of hydrolyzing ATP

TABLE 2-2   CHARACTERISTICS OF VIRAL NUCLEIC ACIDS

Virus	Nature of Nucleic Acid	MW $\times$ 10^6	Strandedness	Structure	Number of Segments	Polarity	Infectivity of Naked Nucleic Acid
Poxvirus	DNA	125 – 185	Double	Linear	1		+
Herpesvirus	DNA	100,150*	Double	Linear	1		+
Adenovirus	DNA	23	Double	Linear	1		+
Papovavirus	DNA	3.5†	Double	Supercoiled circular	1		+
Parvovirus	DNA	2	Single	Linear	1	(a) + and – (b) –	+
Picornavirus	RNA	2 – 3	Single	Linear	1	+	+
Calicivirus	RNA	2 – 3	Single	Linear	1	+	+
Togavirus	RNA	4.5	Single	Linear	1	+	+
Bunyavirus	RNA	5	Single	Linear, cohesive ends	3	–	–
Reovirus	RNA	15	Double	Linear	10		–
Orthomyxovirus	RNA	4	Single	Linear	8	–	–
Paramyxovirus	RNA	6	Single	Linear	1	–	–
Rhabdovirus	RNA	3 – 4	Single	Linear	1	–	–
RNA tumor virus	RNA	5 – 7	Single	Linear	2‡	+	–
Arenavirus	RNA	5	Single	Linear, cohesive ends	2	–	–
Coronavirus	RNA	5.5	Single	Linear	1	+	+

*100 $\times$ 10^6 for alpha- and gammaherpesviruses; 150 $\times$ 10^6 for betaherpesviruses.
†5 $\times$ 10^6 for papilloma virus; 3 $\times$ 10^6 for all others.
‡The genome is not really segmented, as the two molecules are identical.

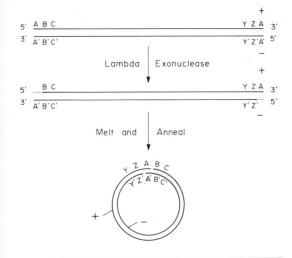

FIG. 2-22. Demonstration of terminal redundancy in double-stranded DNA. Note that the A-A' sequences that are paired in the circularized molecule were originally present at opposite ends of the linear molecule.

by stepwise nicking and reforming of phosphodiester bonds. The significance of supercoiling is not clear; however it is known that most circular double-stranded DNA molecules that occur in nature, such as plasmids in bacteria, chloroplast DNA in plants and mitochondrial DNA in animal cells, also exist in supercoiled form. Supercoiling is relieved by the action of wide-spread untwisting enzymes that remove superhelical turns by providing a swivel through a sequence of successive nicking and closing events, and finally yield perfect circular molecules (that is, no nicks in either strand). It

is also relieved by introducing single-strand breaks or nicks into one of the two strands of supercoiled DNA by the action of deoxyribonuclease; it is also relieved by intercalation of substances such as ethidium bromide (Fig. 2-26).

The significance of circularity is not known. It is not essential for infectivity, since DNA in poxviruses, herpesviruses and adenoviruses is linear. Conceivably, circularity is a prerequisite for integration into the host genome (see Chap. 9), and the DNAs of adenoviruses and herpesviruses, which, like papovaviruses, can transform cells (see Chap. 9), circularize before being integrated.

Circular permutation   As discussed in Chap. 10, bacteriophage DNAs are frequently circularly permuted. The DNAs of poxviruses, herpesviruses, and adenoviruses are not circularly permuted.

Segmentation   For a long time it was assumed that viral genomes consist of unbroken strands of nucleic acid. However, this is not always so. The genomes of reoviruses and rotaviruses consist of 10 and 11 segments of double-stranded RNA, respectively (Fig. 2-27); the genomes of influenza viruses, bunyaviruses and arenaviruses consist of 8, 3 and 2 segments of single-stranded RNA molecules, respectively; and the genome of RNA tumor viruses consists of two single-stranded RNA molecules which are identical [so that the genome is not really segmented (see Chap. 10)]. One of the conse-

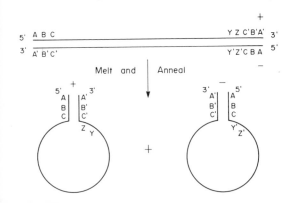

FIG. 2-23. Demonstration of the presence of inverted terminally redundant sequences in double-stranded DNA.

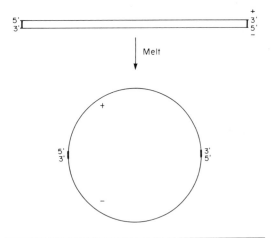

FIG. 2-24. Demonstration of cross-links at the ends of linear double-stranded vaccinia DNA.

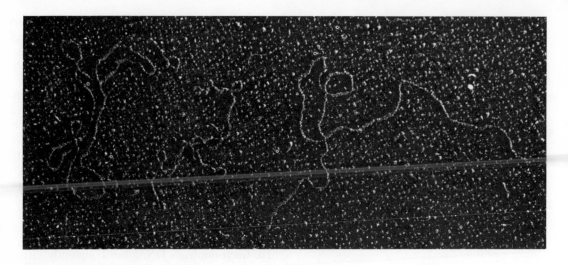

FIG. 2-25. The three forms of bovine papilloma virus DNA. Center, a supercoiled twisted circular molecule, which is the form in which the DNA exists within the virus. Left, a "relaxed" circular molecule, in which one strand has been "nicked" or broken by treatment with deoxyribonuclease, thereby relieving the supercoiling by permitting free rotation of the remaining intact strand. Right, a linear molecule, generated by the introduction of nicks close to one another in both strands. ×66,000. (Courtesy of Dr. H. J. Bujard.)

quences of segmentation is highly efficient genetic recombination caused by random reassortment of segments in multiply infected cells (see Chap. 8).

The Polarity of Viral Nucleic Acids

The single-stranded RNA molecules present in picornavirus, calicivirus, togavirus, coronavirus and RNA tumor virus particles can combine with ribosomes and serve as messenger RNA, and all the genetic information necessary for the formation of progeny virus is translated directly from them. This is not the case for the RNA molecules present in ortho- and paramyxovirus, rhabdovirus, bunyavirus and arenavirus particles. These RNAs must first be transcribed into RNA strands of opposite polarity, and it is these transcripts that are then translated by ribosomes. Since the polarity of messenger RNA is generally designated as plus, the polarity of the RNA in the former group of viruses is plus, and that of the RNA in the latter group is minus.

The adeno-associated satellite viruses, which belong to the parvovirus family and contain single-stranded DNA, present a unique situation. There are two kinds of these virus particles: one contains plus strands, the other minus strands. These two kinds of particles are produced in equal amounts. When the DNA is extracted from them it hybridizes rapidly, thus giving the illusion that it is double-stranded. It is only when the DNA is extracted under conditions when the plus and the minus strands are prevented from hybridizing with each other that the true situation is revealed. All other parvoviruses (except the insect Densonucleosis viruses) contain single-stranded DNA of one polarity only, namely minus.

The Base Composition of Viral Nucleic Acids

The base composition of viral genomes varies within wide limits: thus the guanine plus cytosine (G+C) content of many poxvirus DNAs is only about 36%, while that of many herpesvirus DNAs is as high as 68%. Base composition is indicative neither of ability to multiply in any given cell, nor of taxonomy in general. However, it is true that among the adenoviruses the DNAs of nononcogenic, weakly oncogenic and highly oncogenic strains contain about 58, 52 and 42% G+C respectively, the latter being

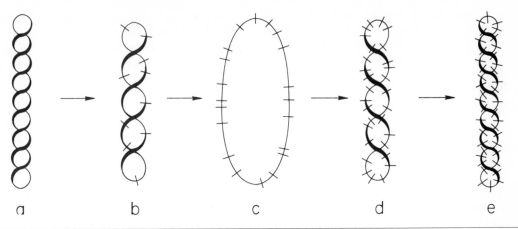

a   b   c   d   e

FIG. 2-26.   Diagrammatic representation of the removal and reversal of supercoiling turns by intercalation of ethidium bromide. The Watson-Crick helix is here represented as a single continuous line. The number of supercoiling turns in the original molecule (**a**) decreases as ethidium bromide molecules (represented by bars perpendicular to the helix axis) bind to the DNA and intercalate between base pairs (**b**). At equivalence, the accumulated untwisting due to the number of intercalated drug molecules balances the initial number of supercoiling turns (**c**). Intercalation of further drug molecules (**d**) leads to the introduction of supercoiling turns in the opposite sense (**e**). (From Crawford and Waring: J Gen Virol 1:387, 1967.)

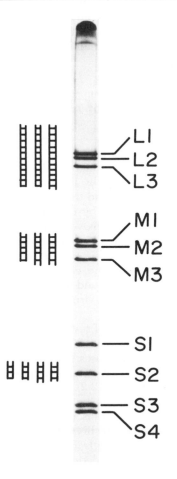

L1
L2
L3

M1
M2
M3

S1

S2

S3
S4

FIG. 2-27.   The structure of the reovirus genome. RNA extracted from reovirions by treatment with the detergent sodium dodecyl sulfate (SDS) was subjected to electrophoresis in a polyacrylamide gel. In such gels molecules of various sizes migrate as discrete bands, the smallest ones moving fastest. The direction of migration in the gel shown here was from top to bottom. Bands of RNA were visualized by autoradiography.

The reovirus genome is seen to consist of 10 molecular species that fall into 3 size classes, designated L, M and S. Since the rate of migration is inversely proportional to the square of the molecular weight, estimates of relative molecular weights can be made by measuring the distances traveled by the various bands. The relative sizes of the 10 reovirus RNA segments are indicated at the left of the gel.

closest to the G+C content of the DNA of mammalian cells (about 45% G+C); but although the DNAs of some oncogenic herpesviruses have a G+C content of 42%, the DNAs of others possess a G+C content that is as high as that of the DNAs of many nononcogenic ones.

The Genetic Relatedness of Viral Nucleic Acids

The genetic relatedness of animal virus nucleic acids is of interest for its taxonomic and evolutionary significance. For example, there are many viruses that are poxviruses, according to a variety of criteria (see Table 2-4). The question naturally arises as to how closely related they are.

The most definitive measure of genetic relatedness is determination of similarity of nucleic acid-base sequence by direct sequence analysis. Although very efficient techniques for sequencing both RNA and DNA have now been perfected, the nucleic acids of animal viruses are so large—comprising from about 5000 to 150,000 bases or base pairs—that the entire sequence has so far been established only for the DNAs of the two papovaviruses SV40 and polyoma. However, there is another, much less laborious way to assess genetic relatedness, which is still sensitive enough to be very useful taxonomically, and that is measurement of how extensively nucleic acids can hybridize with each other. If the nucleic acids under investigation are double-stranded, the two molecules whose relatedness is to be determined are denatured to the single-stranded state, mixed and allowed to reanneal. Single strands derived from the two genomes are thus presented with the opportunity of pairing with each other, and conditions are readily arranged so that such pairing can be quantitated. If the two genomes are very closely related, pairing will occur extensively; if there is no relatedness, no pairing will occur. For example, the genomes of the highly oncogenic adenoviruses referred to above hybridize with each other to the extent of 80 percent or more; in other words, they share over 80 percent of their base sequences. However, they share only about 25 percent of base sequences with the genomes of the nononcogenic adenoviruses, which, on the other hand, share over 80 percent of base sequences

among themselves. Another example is provided by strains of the three serotypes of reovirus. The RNAs of strains of serotypes 1 and 3 hybridize with each other to the extent of about 90%, but they both hybridize only to the extent of about 10% with the RNAs of reovirus strains of serotype 2.

If the viruses that are to be compared possess single-stranded nucleic acid, the strategy is to hybridize the viral nucleic acid in question to the double-stranded, or replicative form (see Chap. 5) of the nucleic acid of the virus strain whose genetic relatedness is to be assessed.

Another measure of genetic relatedness employs comparison of proteins rather than of nucleic acids. Similarity of nucleic acid-base sequences signifies similarity of the amino acid sequences of the proteins coded by them. This in turn implies antigenic similarity, that is, ability of proteins to react with each other's antibodies. Ability of nucleic acids to hybridize, and of the proteins coded by them to cross-react immunologically are thus measures of the same parameter. As an example of this approach, Fig. 2-28 illustrates the relatedness of five herpesviruses, based upon analysis of their antigenic similarity.

The Infectivity of Animal Virus Nucleic Acids

Viral nucleic acids contain all the information necessary for the formation of virus particles. This was first shown by Hershey and Chase in 1952, when they found that infection by bacteriophage is initiated by the injection into the host cell of viral DNA (along with a small amount of protein, which has no genetic function). Later, in 1956, Gierer and Schramm showed that the same was true for plant viruses, when they found that RNA extracted from tobacco mosaic virus particles was infectious.

The nucleic acids of several groups of animal viruses, such as that of the picornaviruses, caliciviruses, togaviruses, coronaviruses, papovaviruses, adenoviruses, and herpesviruses are also infectious (Table 2-2). These are all nucleic acids that either can act as messenger RNA themselves or are transcribed into messenger RNA by host-coded RNA polymerases. Viral nucleic acids that are transcribed into messenger RNA by virus-coded polymerases, such as

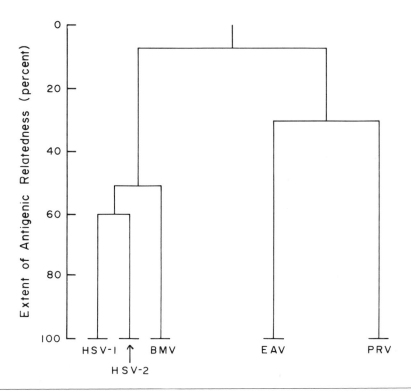

FIG. 2-28. Dendrogram illustrating the serological relatedness of five herpesviruses, herpes simplex virus type 1 (HSV-1), herpes simplex virus type 2 (HSV-2), bovine mammillitis virus (BMV), equine abortion virus (EAV) and pseudorabies virus (PRV). The extent of antigenic relatedness has been assessed by counting the number of precipitin bands formed in agar gel immunodiffusion reactions, using antisera to the various herpesviruses and antigens of cells infected with homologous and heterologous herpesviruses. (Adapted from Honess and Watson: J Gen Virol 37:15, 1977.)

the minus-stranded RNAs of ortho- and paramyxoviruses, rhabdoviruses, bunyaviruses and arenaviruses, or the double-stranded RNA of reovirus, would not be expected to be infectious, as they could not possibly express themselves in the cell.

In all cases the naked nucleic acids are less infectious by factors varying from 10^3 to 10^6 than the virus particles from which they were extracted. There are two principal reasons for this. First, naked viral nucleic acids are quickly degraded by nucleases which are generally present in extracellular fluids, as well as on outer cell membranes; second, naked nucleic acids are taken up very poorly by cells. Uptake can be increased significantly by (1) treating cells briefly with concentrated salt solutions, which promotes pinocytosis, thereby facilitating nucleic acid entry; or (2) by complexing the naked nucleic acids with polycations such as protamine or DEAE-dextran; or (3) by adsorbing them to precipitated calcium phosphate, which is taken up well by cells. The inefficiency with which naked viral nucleic acids penetrate into cells emphasizes the role of the viral capsid in this vital function.

The host range of naked viral nucleic acids is very much broader than that of the respective virus particles. This stems from the fact that the host range of a virus particle is restricted by the specificity of the interaction between capsid and cell surface receptors (see Chap. 5). This is not the case for naked viral nucleic acids, which can apparently multiply in any cell into which they can penetrate without being degraded. For example, whereas poliovirus can infect and therefore multiply only in cells of human or primate origin, poliovirus RNA can also infect chick cells and mouse cells.

Infectious viral nucleic acid can be extracted

not only from infectious virus particles but also from all virus particles that contain undamaged nucleic acid. Among such are virus particles inactivated by heat, proteolytic enzymes, and detergents (see Chap. 3).

The Presence of Host Cell Nucleic Acids in Virus Particles

As a rule, capsids enclose viral nucleic acids. However, sometimes segments of host nucleic acid become encapsidated instead. For example, particles exist that contain, within a papovavirus capsid, a linear piece of host DNA roughly the same size as papovavirus DNA. As far as is known, each such particle, known as a pseudovirion, contains a different segment of host DNA. Pseudovirions usually make up only a small fraction of the yield, but in some cell lines the majority of the particles formed as a result of infection with polyoma virus are pseudovirions. Another example is provided by virus particles that are formed when papovaviruses are passaged repeatedly at high multiplicity, that is, in cells infected with many virus particles. These virus particles contain DNA molecules that consist partly of virus and partly of host cell sequences. They are formed when the viral genomes, which become integrated into host cell DNA prior to replication (see Chaps. 9 and 10), are imperfectly excised (cut out of the host genome).

Virus particles that contain host DNA sequences have attracted attention because of their potential ability to transduce host genes from one cell to another; this ability could conceivably be exploited for correcting inborn errors of metabolism.

VIRAL PROTEINS

The principal constituent of all animal viruses is protein. Proteins are the sole component of capsids; they are the major component of envelopes; and they are also intimately associated, as internal or core proteins, with the nucleic acids of many icosahedral viruses. All these proteins are referred to as structural proteins, since their primary function is to serve as virion building blocks. They are almost always coded by the viral genome.

Viral proteins vary widely in size, from less than 10,000 to more than 150,000 daltons.

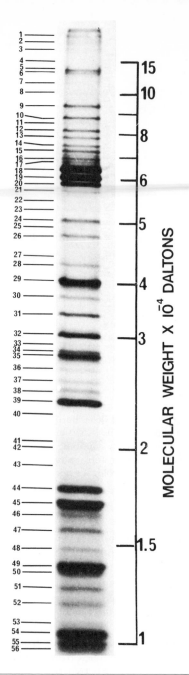

FIG. 2-29. Polyacrylamide gel of vaccinia virus structural polypeptides. Vaccinia virus labeled with ^{35}S-methionine was dissolved in a buffer containing SDS, and the resulting solution was electrophoresed in a 11 percent polyacrylamide gel. An autoradiogram was then prepared from it by exposing X-ray film to it. The direction of electrophoresis was from top to bottom. Over 50 polypeptides are clearly visible. There is evidence that the vaccinia virus particle may comprise as many as 100 different species of polypeptides. (Courtesy of Dr. S. Dales. Virology 95:355, 1979.)

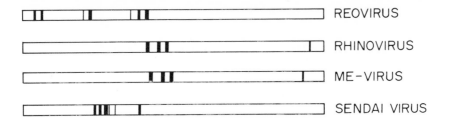

FIG. 2-30. Drawings of polyacrylamide gel patterns of the proteins of reovirus, rhinovirus, ME-virus and Sendai virus. Highly purified preparations of these viruses were dissolved in buffer containing sodium dodecyl sulfate; the illustration shows gels in which the SDS complexes of the proteins of these four viruses had been electrophoresed from left to right. The location of each band is a measure of its molecular weight; thus the largest protein shown is the left-most protein of reovirus (MW 150,000), the smallest the right-most protein of rhinovirus (MW about 8000). The relative amount of each protein is indicated by the thickness of the band representing it. It is clear from the gel patterns shown that the protein complements of rhinovirus and ME-virus, both picornaviruses, are very similar, and that they are quite different from those of reovirus and Sendai virus.

They also vary in number, some virus particles containing as few as 3 species, others more than 50. Viral proteins are characterized most conveniently by dissociating highly purified virus preparations and subjecting the resulting polypeptide mixtures to electrophoresis in polyacrylamide gels. The most commonly used dissociating agent is the detergent sodium dodecyl sulfate (SDS), which not only destroys the secondary structure of proteins but also forms complexes with them. These complexes carry numerous strong negative charges, so that upon electrophoresis in polyacrylamide gels they migrate strictly according to size. As a result, SDS-polyacrylamide gel electrophoresis provides a very convenient method for determining not only the number of different polypeptide species that make up virus particles, but also their sizes. The polyacrylamide gel electrophoresis profile of vaccinia virus polypeptides is shown in Figure 2-29.

All members of the same virus family display the same or almost the same highly characteristic electrophoretic polypeptide patterns. The patterns exhibited by reovirus, rhinovirus, ME-virus, and Sendai virus (a paramyxovirus) are illustrated in Figure 2-30.

Glycoproteins

Viral envelopes often contain glycoproteins in the form of spikes or projections. Their carbohydrate moieties consist of oligosaccharides comprising 10 to 15 monosaccharide units, which are linked to the polypeptide backbone through N— and O— glycosidic bonds involving asparagine and serine or threonine respectively. Their principal components are generally galactose and galactosamine, glucose and glucosamine, fucose, mannose, and neuraminic acid, which always occupies a terminal position.

It was long thought that the structure of the oligosaccharides of virus glycoproteins is specified solely by the nature and relative abundance of the various host cell glycosyl transferases that assemble them from their monosaccharide components. This is certainly a very important factor, but it now appears that the nature of the protein that is being glycosylated also influences oligosaccharide structure, particularly near the chain termini. As a result, the oligosaccharides of viral glycoproteins probably differ from those of cells, the oligosaccharides of glycoproteins of different viruses grown in the same cells are probably not identical, and the oligosaccharides of the glycoproteins of the same virus grown in different cells, although closely related, are probably different. The structure of the oligosaccharide of the spike glycoprotein of VSV is shown in Figure 2-31.

Viral Proteins with Specialized Functions

Some viral proteins have specialized properties and functions. Among them are the following.

Hemagglutinins Many animal viruses, both naked and enveloped (such as picornaviruses,

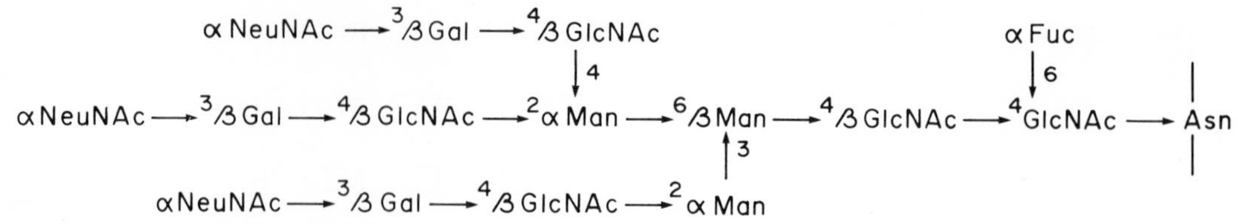

$$\alpha\,\text{NeuNAc} \xrightarrow{\;3\;} \beta\,\text{Gal} \xrightarrow{\;4\;} \beta\,\text{GlcNAc} \qquad\qquad\qquad \alpha\,\text{Fuc}$$

$$\downarrow 4 \qquad\qquad\qquad\qquad\qquad\qquad \downarrow 6$$

$$\alpha\,\text{NeuNAc} \xrightarrow{\;3\;} \beta\,\text{Gal} \xrightarrow{\;4\;} \beta\,\text{GlcNAc} \xrightarrow{\;2\;} \alpha\,\text{Man} \xrightarrow{\;6\;} \beta\,\text{Man} \xrightarrow{\;4\;} \beta\,\text{GlcNAc} \xrightarrow{\;4\;} \text{GlcNAc} \longrightarrow \text{Asn}$$

$$\uparrow 3$$

$$\alpha\,\text{NeuNAc} \xrightarrow{\;3\;} \beta\,\text{Gal} \xrightarrow{\;4\;} \beta\,\text{GlcNAc} \xrightarrow{\;2\;} \alpha\,\text{Man}$$

FIG. 2-31. The structure of the oligosaccharide unit of the VSV spike glycoprotein. There are two of these units per protein molecule. Asn, asparagine; GlcNAc, N-acetylglucosamine; Man, mannose; Gal, galactose; NeuNAc, N-acetyl-neuraminic acid; Fuc, fucose. (From Reading, Penhoet and Ballou: J Biol Chem: 253:5600, 1978.)

togaviruses, reoviruses, ortho- and paramyxoviruses, adenoviruses, and papovaviruses), agglutinate the red blood cells of certain animal species. This property, which is called hemagglutination, reflects the fact that these red blood cells possess receptors for certain surface components of virus particles. In the case of ortho- and paramyxoviruses, the hemagglutinins are glycoprotein spikes that are located on the virus particle surface (Fig. 2-4). The ability

TABLE 2-3   ENZYMES IN ANIMAL VIRUSES

Virus	Enzyme	Coded by
DNA VIRUSES		
Poxvirus	DNA-dependent RNA polymerase	Virus
	Nucleotide phosphohydrolase	Probably Virus
	Poly A polymerase	Virus
	Capping enzymes	Virus
	Protein kinase	Virus
	Nucleases	Virus
Herpesvirus	ATPase*	Cell
Adenovirus	Protein kinase	?
Papovavirus	None	
Parvovirus	DNA polymerase	?
Hepatitis virus	DNA polymerase	Probably Virus
RNA VIRUSES		
Picornavirus	None	
Calicivirus	None	
Togavirus	None	
Bunyavirus	RNA polymerase	Virus
Reovirus	RNA-dependent RNA polymerase	Virus
	Nucleotide phosphohydrolase	?
	Capping enzymes	?
Orthomyxovirus	Neuraminidase	Virus
	RNA polymerase	Virus
Paramyxovirus	Neuraminidase	Virus
	RNA polymerase	Virus
Rhabdovirus	RNA polymerase	Virus
	Poly A polymerase	?
	Protein kinase	?
	Capping enzymes	?
RNA tumor virus	RNA-dependent DNA polymerase (reverse transcriptase)	Virus
	Ribonuclease H	Virus
	Protein-cleaving enzyme	Probably Virus
	Protein kinase	Probably Virus
	ATPase*	Cell
Arenavirus	RNA polymerase	Virus
	Poly A polymerase	?
Coronavirus	None	

* This enzyme is present only if it is a component of the plasma membrane of the cells in which these viruses multiplied.

to hemagglutinate can be used to quantitate virus (see Chap. 1).

Enzymes Virus particles often contain enzymes (Table 2-3). Among them are the following.

(1) Orthomyxovirus and paramyxovirus particles contain an enzyme, neuraminidase, that hydrolyzes the galactose-N-acetylneuraminic acid bond at the ends of the oligosaccharide chains of glycoproteins and glycolipids, thereby liberating N-acetylneuraminic acid. Like the hemagglutinin, this enzyme is located on glycoprotein spikes: orthomyxoviruses possess two types of spikes, one with hemagglutinin and the other with neuraminidase activity (Fig. 2-4), while paramyxoviruses also possess two types of spikes, on one of which both activities are located. The primary function of neuraminidase appears to be to release virus particles from the cells in which they were formed. The reason the release of these viruses requires a special enzyme is that their hemagglutinins possess strong affinity for the galactose-N-acetylneuraminic acid (sialic acid) residues at the ends of oligosaccharide chains; by hydrolyzing them the enzyme permits the liberation of virus particles that would otherwise not be released or that would readsorb immediately after being liberated.

(2) Many virus particles contain RNA polymerases. The necessity for such enzymes stems from the fact that the nucleic acid present in virus particles must be able to express itself once it has gained access to the interior of host cells. If this nucleic acid itself can act as messenger RNA, like picornavirus or togavirus RNA, there is no problem. If it cannot, either because it is double-stranded or because it is RNA with minus polarity, RNA with plus polarity must first be synthesized. There are two possible sources for the enzymes necessary for this purpose. The first is enzymes existing in the host cell. Herpesvirus, adenovirus, and papovavirus DNAs are probably transcribed into messenger RNA by DNA-dependent RNA polymerases specified by the host cell. The alternative is that the virus particle itself contains the enzyme. Examples of this type are the poxviruses which possess a DNA-dependent RNA polymerase in their cores, reoviruses which possess an RNA-dependent RNA polymerase in their cores, and negative RNA-stranded viruses which contain RNA polymerases that synthesize plus RNA strands from the minus RNA strands present within them. These enzymes are not active in intact virus particles but are activated when the envelope or capsid is partially degraded, which generally occurs very soon after infection.

(3) RNA tumor viruses possess a DNA polymerase, the so-called reverse transcriptase, which transcribes their single-stranded RNA into double-stranded DNA (which is then integrated into the genome of the host cell, see Chap. 9).

All these enzymes are virus-specified. The evidence for this conclusion is provided by the existence of virus strains that contain variant enzymes (as is the case for neuraminidase) and the fact that the enzymes in question often simply do not exist in uninfected cells (as is the case for many of the polymerases).

Viruses also often contain other enzymes, and for many it is not quite clear whether they are virus-coded or host-coded. Among them are nucleases, nucleotide phosphohydrolases, protein kinases, and enzymes that modify both ends of the messenger RNA molecules synthesized by their polymerases (capping enzymes, that is, enzymes that add GMP to the 5'-termini and methylate both it as well as the ribose residue that was N-terminal in the uncapped molecule, and poly A polymerases, which add sequences of poly A ranging in length from about 50 to 300 residues to the 3'-termini). Some of these enzymes are virus-coded; others may be cellular enzymes that become associated with virus particles by as yet unknown mechanisms.

VIRAL LIPIDS

Viral envelopes contain complex mixtures of neutral lipids, phospholipids, and glycolipids. As a rule, the composition of these mixtures resembles that of the membranes of the host cells in which the virus multiplied. Since the lipid composition of membranes varies markedly from one cell strain to another, and even for the same cell strain depending on the composition of the medium, this means that the same virus grown in different cell strains may have widely differing neutral lipid, phospholipid, and glycolipid complements, even though its biologic properties, including its infectivity, are identical.

The nature and extent of the differences in composition that are encountered are best illustrated by the example of the paramyxovirus SV5. On the average, this virus contains about

20 percent total lipid, of which roughly 25 percent and 5 percent, respectively, are cholesterol and triglyceride, 55 percent is phospholipid, and 15 percent is glycolipid. If grown in cells the membranes of which have a molar ratio of cholesterol to phospholipid of 0.81, the ratio for the viral envelope is 0.89; if the cellular ratio is 0.51, that for the viral envelope is 0.60. The same holds for the relative amounts of several phospholipids. If the ratio of phosphatidylcholine to phosphatidylethanolamine in host cell membranes is 3.55, the ratio in the viral envelope is 2.55; if the ratio in host cell membranes is 0.8, the ratio in the viral envelope is 0.6. However, for one cell the ratio is 1.6, while that in the viral envelope derived from it is 0.6, which suggests that a limited degree of lipid selection is possible.

The classification of animal viruses

Attempts to devise a system of classification for animal viruses began almost as soon as their importance as pathogens became apparent. However, the criteria on which to base such a system have changed with advances in our knowledge of their nature and properties. For example, host range cannot be such a criterion; not only is each animal species subject to infection by a wide variety of viral agents, but also numerous viruses infect several different animal species. Similarly, the pattern of pathogenesis of the disease that is caused is an unreliable foundation on which to base a system of classification.

It has become more and more apparent that a system of classification must be based on the physical and chemical properties of the virus particles themselves.

Criteria for Classification

Morphology The primary criterion for classification is morphology. It is easy to apply because it does not require purified virus. Virus particles can be examined either within cells,

that is, in thin sections of infected tissue, or in their extracellular state. Morphologic detail is usually brought out by shadowing with thin films of some heavy metal, such as uranium or tungsten, by staining with osmic acid or uranyl acetate, or by negative staining with phosphotungstic acid.

Physical and Chemical Nature of Virion Components Morphologic similarity correlates closely with similarity of virion components. For examples, all viruses with the morphology of adenoviruses contain double-stranded DNA genomes with a molecular weight of about 23 million; all papovaviruses contain circular supercoiled DNA genomes with molecular weights of 3 to 5 million, and all reoviruses contain segmented double-stranded RNA genomes. In fact, a system of virus classification based on the structure and size of viral genomes yields the same grouping as one based on morphology. Similarly, viruses with similar morphology are composed of similar populations of proteins, and a classification system based on polyacrylamide gel patterns, such as illustrated in Figure 2-26, again yields the same grouping.

Genetic Relatedness Although the members of the major virus families presumably derived from a common ancestor, genetic relationships among them are often no longer discernible today. For example, several groups of mammalian poxviruses exist (Table 2-4). whose members are unrelated genetically, as judged either by nucleic acid hybridization or by immunologic cross-reactivity. The same applies to many members of the herpesvirus family, the human, simian, canine, and avian adenoviruses, and so on. These groups form the basis for subdividing the viruses within families into genera.

THE MAJOR FAMILIES OF ANIMAL VIRUSES

Table 2-4 presents a summary of the distinguishing characteristics of the major families of animal viruses, together with a list of the most important animal and human pathogens in each. It is based in large part upon a series of recommendations that have recently been made by the International Committee on Taxonomy of Viruses.

TABLE 2-4   THE MAJOR FAMILIES OF ANIMAL VIRUSES

DNA-containing Viruses

Classification	Description
I. POXVIRIDAE	All genera except *Parapoxvirus:* Brick-shaped complex particles, whorled surface filament pattern Dimensions: 225 × 300 nm *Parapoxvirus:* Ovoid complex particles, regular surface filament pattern Dimensions: 150 × 200 nm

Strains pathogenic for many animal species exist. There are seven genera; the members of each are closely related antigenically but share only one antigen with members of the others. Poxviruses multiply in the cytoplasm.

Subfamily	Host	Symptoms in humans
Chordopoxvirinae (poxviruses of vertebrates)		
A. Genus *Orthopoxvirus*		
Variola major	Humans	Smallpox
Variola minor	Humans	Alastrim
Monkeypox	Monkey, humans	Smallpox-like disease
Vaccinia	Cattle	–
Cowpox	Cattle, humans	Vesicular eruption of the skin
Buffalopox	Buffalo	–
Rabbitpox	Rabbit	–
Ectromelia (mousepox)	Mouse	–
B. Genus *Leporipoxvirus*		
Myxoma	Rabbit	–
Fibroma	Rabbit	–
C. Genus *Avipoxvirus*		
Fowlpox	Chicken	–
Other birdpox viruses	Various birds	–
D. Genus *Capripoxvirus*		
Sheeppox	Sheep	–
Goatpox	Goat	–
E. Genus *Suipoxvirus*		
Swinepox	Pig	–
F. Genus *Parapoxvirus*		
Orf (contagious postular dermatitis [CPD])	Sheep, goats humans	Nodules on hands
Pseudocowpox (milker's nodule virus)	Cattle, humans	Nodules on hands
Bovine papular stomatitis	Cattle	–
G. Ungrouped		
Molluscum contagiosum	Humans	Benign epidermal tumors
Yaba monkey tumor virus	Monkey, humans	Benign subcutaneous tumors
Entomopoxvirinae		
Various strains	Arthropods	

(continued)

TABLE 2-4 (cont.)

DNA-containing Viruses

Classification

II. IRIDOVIRIDAE

These are the icosahedral cytoplasmic deoxyviruses. Most iridoviruses are arthropod viruses and nonenveloped. The sole mammalian iridovirus is enveloped. The size of its DNA is about 150×10^6.

	Host	*Symptoms in humans*
African swine fever virus	Swine	–
Frog virus 2	Amphibia	–
Frog virus 3		
Lymphocystis virus of fish	Fish	–
Tipula iridescent viruses	Arthropods	–
Iridescent viruses of	Molluscs, Annelids, Protozoa	–

Classification	*Description*
III. HERPESVIRIDAE	Enveloped icosahedral nucleocapsids Diameter of enveloped virions: 180 – 250 nm Diameter of naked nucleocapsids: 100 nm

The most logical classification of herpesviruses is based on a combination of biological properties and genome structure. Alphaherpesviruses have a variable host range (from very wide to very narrow); most, but not all, have a relatively short replication cycle; most, but not all, are highly cytopathic in cultured cells; they frequently establish latent infections in sensory ganglia; and their DNA has a molecular weight of about 100×10^6. Betaherpesviruses have a narrow host range and a relatively long replication cycle; are less cytopathic; frequently establish latent infections in the salivary gland and in other tissues; and their DNA has a molecular weight of about 150×10^6. Gammaherpesviruses have a narrow host range; a predilection for lymphoblastoid cells which they can transform (that is, they can cause tumors); and their DNA has a molecular weight of about 100×10^6. Few herpesviruses are closely related to each other as judged by DNA-DNA hybridization, but most show some relatedness (2-10 percent cross-hybridization). Almost all herpesviruses possess some common antigenic determinants. Herpesviruses cause type A nuclear inclusions (single large acidophilic inclusion bodies separated by a nonstaining halo from basophilic marginated chromatin).

	Host	*Symptoms in humans*
A. *Alphaherpesviruses*		
Herpes simplex virus type 1 (human (alpha)herpesvirus 1)	Humans	Stomatitis, upper respiratory infections, generalized systemic disease, severe and generally fatal encephalitis
Herpes simplex virus type 2 (human (alpha)herpesvirus 2)	Humans	Genital infections
Varicella-zoster (human (alpha)herpesvirus 3)	Humans	Chickenpox, herpes zoster
B virus	Monkey, humans	Fatal encephalitis in humans
Equine abortion virus (EAV) (equine herpesvirus 1)	Horse	–
LK virus (equine herpesvirus 2)	Horse	–
Coital exanthema virus (equine herpesvirus 3)	Horse	–
Pseudorabies virus (suid herpesvirus)	Pig	–
Infectious bovine rhinotracheitis (IBR)	Cattle	–
Infectious bovine kerato-conjunctivitis (IBKC)	Cattle	–
Feline rhinotracheitis virus	Cat	–
Infectious laryngotracheitis virus (ILT)	Chicken	–
B. *Betaherpesviruses*		
Human Cytomegalovirus (human (beta)herpesvirus 5)	Humans	Jaundice, hepatosplenomegaly, brain damage, death

(continued)

TABLE 2-4 (cont.)

DNA-containing Viruses

	Host	Symptoms in humans
Other Cytomegaloviruses	Monkey	–
	Rodents	–
	Swine	–

C. *Gammaherpesviruses*—All these viruses either cause, or are strongly suspected of causing, tumors

	Host	Symptoms in humans
Epstein-Barr virus (human (gamma)herpesvirus 4)	Humans	Burkitt lymphoma Nasopharyngeal carcinoma Infectious mononucleosis
Marek's disease virus	Chicken	–
Herpesvirus saimiri	Squirrel monkey	–
Herpesvirus ateles	Spider monkey	–
Herpesvirus sylvilagus	Rabbit	–
Guinea pig herpesvirus	Guinea pig	–
Lucké tumor virus	Frog	–

Classification	Description
IV. ADENOVIRIDAE	Naked icosahedral nucleocapsids Diameter: 75 nm

Strains pathogenic for many animal species exist. All adenoviruses except avian ones share common group-specific complement-fixing antigenic determinants (on hexons) and in addition possess type-specific determinants (on pentons and fibers). They can be grouped into subgroups on the basis of antigenic cross-reactivity, DNA hybridization characteristics, ability to transform cells of various animal species, and ability to agglutinate rhesus monkey and rat erythrocytes. Some human and simian strains are very tumorigenic in certain rodents; all can transform cultured mammalian cells more or less efficiently (Chap. 9). Adenoviruses produce intranuclear Type B inclusions (basophilic masses sometimes connected to the nuclear periphery by strands of chromatin).

	Host	Symptoms in humans
A. Genus *Mastadenovirus*		
Human adenoviruses 31 serotypes	Humans	Primarily respiratory (serotypes 4, 7, 14, and 21) and conjunctival (serotype 8) infections. Serotypes 1, 2, 5, and 6 cause acute febrile pharyngitis in infants and young children.
Simian adenoviruses 23 serotypes	Monkey	–
Canine adenoviruses (infectious canine hepatitis [ICH])	Dog	–
B. Genus *Aviadenovirus*		
Avian adenoviruses (CELO, chicken-embryo-lethal-orphan; GAL, gallus-adenolike)	Chicken, quail, other birds	–

Classification	Description
V. PAPOVAVIRIDAE	Naked icosahedral nucleocapsids Diameter: 55 nm (papilloma viruses) 45 nm (all other papovaviruses)

Papovaviruses (**pa**pilloma-**po**lyoma-simian **va**cuolating agent) fall into two distinct groups on the basis of size; members of the genus papillomavirus possess both larger capsids and larger genomes than members of the genus polyomavirus. All except K virus have oncogenic potential; all produce latent and chronic infections in their natural hosts. No members of this family are antigenically related except SV40 and the viruses associated with PML in man. Polyoma and K virus agglutinate erythrocytes of certain animal species.

(continued)

TABLE 2-4 (cont.)

DNA-containing Viruses

	Host	Symptoms in humans
A. Genus *Papillomavirus*		
Human papilloma viruses 1-5	Humans	Benign tumors (warts)
Shope rabbit papilloma virus	Rabbit	
Various viruses pathogenic for other animal species	Cattle	–
B. Genus *Polyomavirus*		
Polyoma virus	Mouse	–
Simian vacuolating agent (SV40)	Monkey kidney cell cultures	–
Rabbit vacuolating agent (RKV)	Rabbit	–
K virus	Mouse	–
BK virus	Humans	Isolated from urine of renal transplant patients. Its DNA is 20-30% homologous with SV40 DNA. Transforms cultured cells
DAR virus	Humans	Isolated from brains of patients with progressive multifocal leukoencephalopathy (PML). Its DNA is over 90% homologous with SV40 DNA.
JC virus	Humans	Isolated from brains of patients with PML. Its DNA is 10-20% homologous with SV40 DNA. [Highly oncogenic in hamsters]

Classification	Description
VI. PARVOVIRIDAE	Naked icosahedral nucleocapsids Diameter: 22 nm

These viruses are grouped together on the basis of morphology and nucleic acid structure (single-stranded DNA). All multiply in the nucleus. Most are antigenically unrelated. AAVs multiply only in cells simultaneously infected with adenoviruses (Chap. 8); RV multiplies only in cells that are themselves multiplying actively. Aleutian mink disease virus causes a slow disease characterized by hypergamma-globulinemia, systemic proliferation of plasma cells, glomerulonephritis and hepatitis, and is invariably fatal. The disease condition is caused by activation of the immune response, since it is exacerbated by administration of inactivated virus or passive antibody. Although it causes a slow disease, Aleutian mink disease virus replicates as rapidly in cultured cells as viruses that cause acute disease.

	Host	Symptoms in humans
A. Genus *Adeno-associated virus*		
Human adeno-associated virus (AAV) 4 serotypes	Humans	No known symptoms
Other adeno-associated viruses	Cattle, dog, chicken	
B. Genus *Parvovirus*		
Rat virus (Kilham)	Rat	Latent viruses isolated from various hosts including human tumors; none is oncogenic or capable of transforming cultured cells. [They produce a mongoloid osteolytic deformity in newborn hamsters.]
Hamster osteolytic viruses (H-1, H-3, X-14, etc.)		
Minute virus of mice (MVM)	Mouse	–
Other parvoviruses (porcine parvovirus, feline panleuko-penia virus, bovine parvo-virus, etc.)	Pig, cattle, cat	–
Aleutian mink disease virus	Mink	–
Norwalk agent (3 serotypes)	Humans	Gastroenteritis

(continued)

TABLE 2-4 (cont.)

DNA-containing Viruses

	Host	Symptoms in humans
C. Genus *Densovirus*		
Densonucleosis viruses	Arthropods	–

RNA-containing Viruses

Classification	Description
I. PICORNAVIRIDAE	Naked icosahedral nucleocapsids Diameter: 25 – 30 nm

Picornaviruses comprise a large number of virus strains pathogenic for many animal species. They are subdivided into four genera: Enterovirus and Cardiovirus, whose members are acid-stable, and Rhinovirus and Aphthovirus, whose members are acid-labile.

	Host	Symptoms in humans
A. Genus *Enterovirus*		
Human enteroviruses		
Poliovirus 3 serotypes	Humans, monkey	Poliomyelitis
Coxsackie virus A 24 serotypes	Humans, mouse	Differentiated from Group B Coxsackie viruses primarily on the basis of selective tissue damage: Group A, primarily general striated muscle damage; Group B, primarily fatty tissue and central nervous tissue damage. Group A viruses are associated with herpangina, aseptic meningitis, paralysis, and the common cold syndrome
Coxsackie virus B 6 serotypes	Humans, mouse	Pleurodynia (Bornholm disease), aseptic meningitis, paralysis, severe systemic illness of newborns
ECHO viruses (*E*nteric *C*ytopathogenic *H*uman *O*rphan) 33 serotypes	Humans	Paralysis, diarrhea, aseptic meningitis
Simian enteroviruses Multiple serotypes	Monkey	–
Murine encephalomyelitis viruses Poliovirus muris (Theiler's virus) GDVII strain and others	Mouse	–
Bovine enteroviruses	Cattle	–
Porcine enteroviruses Teschen virus 1 Teschen virus 2	Swine	–
B. Genus *Cardiovirus*		
Encephalomyocarditis virus (EMC) (mengovirus)	Various species, including humans	Mild febrile illness
C. Genus *Rhinovirus*		
Human rhinoviruses Multiple serotypes	Humans	Common cold, bronchitis, croup, bronchopneumonia
Other rhinoviruses	Strains pathogenic for horses, cattle, etc.	–

(continued)

TABLE 2-4 (cont.)

RNA-containing Viruses

	Host	Symptoms in humans
D. Genus *Aphthovirus*		
Foot-and-mouth disease virus (FMDV) 7 serotypes	Cattle, swine, sheep, goats	–
E. Probable Picornavirus		
Hepatitis A virus	Humans	Infectious Hepatitis

Classification	Description
II. CALICIVIRIDAE	Naked icosahedral nucleocapsids Diameter: 35 – 40 nm

Caliciviruses differ significantly from Picornaviruses in size and structure; but the primary reason they have been constituted a separate family is that the strategy of genome expression during the multiplication cycle of picornaviruses and caliciviruses is quite different. VE virus and SMSV are very closely related; feline picornaviruses are related to them to the extent of about 10% as judged by RNA hybridization analysis.

	Host	Symptoms in humans
Vesicular exanthema of swine virus (VE)	Swine	–
San Miguel sea lion virus (SMSV)	Seals	–
Feline picornaviruses	Cat	–

Classification	Description
III. TOGAVIRIDAE	Enveloped nucleocapsids Diameter: 50 – 70 nm (alphaviruses and rubiviruses) 40 – 50 nm (flaviviruses and pestiviruses)

Togaviruses include many of the viruses previously known as arboviruses (**arthropodborne**). They multiply in bloodsucking insects as well as in vertebrates; in their natural environment they alternate between an insect vector (usually a mosquito or tick) and a vertebrate reservoir, rarely producing disease in either. Many cause subclinical infections in humans, particularly in the tropics, but several are among the most virulent and lethal of all viruses. They are commonly named for the geographic site where they were isolated. They are divided into four genera, primarily on the basis of antigenic relationships (neutralization, complement fixation and hemagglutination inhibition). The alphaviruses and flaviviruses are the old arbovirus Groups A and B.

	Reservoir	Symptoms in humans
A. Genus *Alphavirus* (mosquitoborne)		
Eastern equine encephalitis (EEE)	Birds	Encephalitis: frequently fatal
Semliki forest virus	Monkey	Undifferentiated febrile illness
Sindbis	Monkey	None
Chikungunya	Monkey	Myositis-arthritis
O'Nyong-Nyong	?	Fever, arthralgia, rash
Ross river virus	Mammals	Fever, rash, arthralgia
Venezuelan equine encephalitis (VEE)	Rodents	Encephalitis
Western equine encephalitis (WEE)	Birds	Encephalitis
B. Genus *Flavivirus*		
1. Mosquitoborne Yellow fever	Monkey	Hemorrhagic fever, hepatitis, nephritis, often fatal
Dengue (4 serotypes)	Man	Fever, arthralgia, rash
Japanese encephalitis	Birds	Encephalitis: frequently fatal
St. Louis encephalitis	Birds	Encephalitis

(continued)

TABLE 2-4 (cont.)

RNA-containing Viruses

	Host	Symptoms in humans
Murray Valley encephalitis	Birds	Encephalitis
West Nile	Birds	Fever, arthralgia, rash
Kunjin	Birds	–
Kyasanur forest	Rodents	Hemorrhagic fever
2. Tickborne		
Central European tickborne encephalitis (biphasic meningoencephalitis)	Rodents, hedgehog	Encephalitis
Far Eastern tickborne encephalitis [Russian spring-summer encephalitis (RSSE)]	Rodents	Encephalitis
Louping Ill	Sheep	Encephalitis
Powassan	Rodents	Encephalitis
Omsk hemorrhagic fever	Mammals	Hemorrhagic fever

C. Genus *Rubivirus*

	Host	Symptoms in humans
Rubella virus	Humans	Severe deformities of fetuses in first trimester of pregnancy. [Like most togaviruses, it multiplies in the brains of newborn mice.]

D. Genus *Pestivirus*

	Host	Symptoms in humans
Mucosal disease-bovine virus diarrhea virus	Cattle	–
Hog cholera virus (European swine fever)	Pig	–
Border disease virus	Sheep	–

E. Also included among the togaviruses are several other as yet unclassified viruses. Among them are:

	Host	Symptoms in humans
Riley's lactic dehydrogenase elevating virus (LDHV)	Mouse	[Produces lifelong chronic viremia in mice; elevates LDH levels by decreasing the rate of enzyme clearance]
Equine arteritis virus	Horse	–

Classification	Description
IV. BUNYAVIRIDAE	Enveloped nucleocapsids Diameter: about 100 nm

Bunyaviruses include all former arbovirus Group C viruses as well as several previously ungrouped arboviruses. They comprise a family separate from the togaviruses, since they are larger and possess a fundamentally different structure: Their nucleocapsids possess helical rather than icosahedral symmetry and are circular, as well as segmented.

	Host	Symptoms in humans
Bunyamwera and related viruses	Mammals	–
California encephalitis	Mammals	Encephalitis
Uukuniemi and related viruses	Birds	–
Congo-Crimean hemorrhagic fever viruses	Mammals	Hemorrhagic fever
Phlebotomus fever virus	Sandfly	Facial erythema
Various strains including Rift Valley fever	Sheep, cattle	Fever, arthralgia, retinitis

Classification	Description
V. REOVIRIDAE	Naked nucleocapsids possessing two capsid shells (except genus *Cytoplasmic Polyhedrosis Virus*), each with icosahedral symmetry Diameter: 75 nm

(continued)

TABLE 2-4 (cont.)

RNA-containing Viruses

The name is an acronym based on respiratory-enteric-orphan (because of lack of association with any disease in humans). The primary criterion for inclusion in this family is possession of a genome consisting of 10, 11 or 12 segments of double-stranded RNA. There are six genera with widely differing host ranges and somewhat differing morphologies. The vertebrate reoviruses possess two clearly defined capsid shells; the orbiviruses (many of which are transmitted by arthropods and are functionally arboviruses) possess a structurally featureless outer shell and an inner shell composed of 32 large ring-shaped capsomers (hence the name, Latin orbis, ring; however, see p. 21). The cytoplasmic polyhedrosis viruses possess only one capsid shell with clearly defined icosahedral symmetry. Phytoreoviruses closely resemble the vertebrate reoviruses, while Fijiviruses possess a structure more reminiscent of cytoplasmic polyhedrosis viruses. Rotaviruses present a wheel-like appearance (see Fig. 2-11), hence the name. The members of the six genera are not related antigenically.

	Host	*Symptoms in humans*
A. Genus *Orthoreovirus*		
Mammalian reoviruses 3 serotypes	Humans, other mammals	Pathogenicity not established
Avian reoviruses 5 serotypes	Chicken, duck	–
B. Genus *Orbivirus*		
Bluetongue virus	Culicoides, sheep	–
Eubenangee virus	Mosquitoes	–
Komarovo	Ticks	–
African horse sickness virus	Culicoides, horse	–
Colorado tick fever virus	Ticks, humans	Encephalitis
C. Genus *Cypovirus*		
Cytoplasmic Polyhedrosis Virus Numerous strains	*Bombyx mori* (silkworm) and other Lepidoptera, Diptera, and Hymenoptera	–
D. Genus *Phytoreovirus*		
Wound tumor virus	Plants, leaf hoppers	–
Rice dwarf virus	Plants, leaf hoppers	–
E. Genus *Fijivirus*		
Maize rough dwarf virus	Plants, leaf hoppers	–
Fiji disease virus	Plants, leaf hoppers	–
F. Genus *Rotavirus*		
Human rotavirus (2 serotypes)	Humans	Diarrhea in infants
Calf rotavirus (Nebraska calf diarrhea virus)	Calf	–
Murine rotavirus (Epizootic diarrhea of infant mice [EDIM]	Mouse	–
Simian rotavirus (SA11)	Monkey	–
Bovine or ovine rotavirus ("O" agent)	Cattle or sheep	–
Numerous other rotaviruses	Guinea pig, goat, horse, deer, antelope, rabbit, dog, duck	–

Classification	*Description*
VI. ORTHOMYXOVIRIDAE	Enveloped helical nucleocapsids Diameter: 80–120 nm

(continued)

TABLE 2-4 (cont.)

RNA-containing Viruses

The term myxovirus was coined to denote the unique affinity of influenza viruses for glycoproteins. Nowadays members of this family are characterized by possession of nucleocapsids with helical symmetry that reside within lipid-containing envelopes, to whose outer surface are attached glycoprotein spikes of two types: One is the hemagglutinin, the other the neuraminidase. Influenza virus strains of type C differ from those of type A and type B in that (1) their buoyant density is lower, (2) the receptors for their hemagglutinins do not appear to contain sialic acid, and (3) their receptor-destroying enzyme seems to cleave some bond other than the -gal-neuraminic acid bond.

	Host	Symptoms in humans
A. Genus *Influenza virus*		
Influenza virus type A		
Human subtypes	Human	Acute respiratory disease
A_0 1933-1947*		"
A_1 1947-1957		"
A_2 1957-1964 (Asian)		"
A_2 1968- (Hong Kong)		"
Swine influenza virus	Swine	"
Avian subtypes		
Fowl plague virus and numerous other strains	Chicken, duck, turkey and others	–
Equine subtypes	Horse	–
Influenza virus type B		
Human subtypes		
B_0 1940-1945	Human	Acute respiratory disease
B_1 1945-1955		"
B_2 1962-1964		"
B_3 1962 (Taiwan)		"
Influenza virus type C (possible separate genus)	Humans	Acute respiratory disease

*In 1971 a WHO Study Group adopted a new nomenclature for influenza type A viruses. In this system virus strains are described in terms of both the hemagglutinin (HA) and neuraminidase (NA) antigens. A_0 strains are now designated as H0N1, A_1 strains as H1N1, the Asian type A_2 of 1957 as H2N2, and the Hong Kong type A_2 of 1968 as H3N2.

Classification	Description
VII. PARAMYXOVIRIDAE	Enveloped helical nucleocapsids Diameter: about 150 nm

Members of this family were until recently grouped with the orthomyxoviruses in the family Myxoviridae. They have been placed in a separate family because they differ from orthomyxoviruses in the following characteristics: Their genomes are not segmented, and their hemagglutinin and neuraminidase are located on the same glycoprotein spike, the other type of spike being responsible for the cell-fusing and hemolyzing activities (genus Paramyxovirus), or they possess no neuraminidase (the other two genera). Members of the genus Pneumovirus do not possess a hemagglutinin.

	Host	Symptoms in humans
A. Genus *Paramyxovirus*		
Parainfluenza virus type 1		
Sendai virus (hemagglutinating virus of Japan [HVJ])	Human, pig, mouse	Croup, common cold syndrome
HA-2 (hemadsorption virus)	Human	Mild respiratory disease
Parainfluenza viruses types 2 to 5		
Numerous strains including HA-1, SV5	Human and other animals	Respiratory tract infections
Newcastle disease virus (NDV)	Chicken	–
Mumps	Human	Parotitis, orchitis, meningoencephalitis
B. Genus *Morbillivirus*		
Measles	Human	Measles

(continued)

TABLE 2-4 (cont.)

RNA-containing Viruses

	Host	Symptoms in humans
Subacute sclerosing panencephalitis (SSPE)	Human	Chronic degeneration of the central nervous system
Distemper	Dog	–
Rinderpest	Cattle	–

C. Genus *Pneumovirus*

	Host	Symptoms in humans
Respiratory syncytial virus (RSV)	Human	Pneumonia and bronchiolitis in infants and children, common cold syndrome
Bovine respiratory syncytial virus	Cattle	–
Pneumonia virus of mice (PVM)	Mouse	–

Classification	Description
VIII. RHABDOVIRIDAE	Bullet-shaped, enveloped helical nucleocapsids Dimensions: 180×75 nm

This group comprises all viruses with the unique bullet-shaped morphology, as well as some that are bacilliform (rounded at both ends). It includes VSV, rabies, and some viruses isolated from insects that do not appear to cause disease in vertebrates, but antibodies to them are found in birds and mammals, including humans. Some members of this family multiply in arthropods as well as in vertebrates.

	Host	Symptoms in humans
A. Genus *Vesiculovirus*		
Vesicular stomatitis virus (VSV)	Cattle, horse, swine	–
Chandipura virus	Isolated from humans	–
Flanders-Hart Park virus	Mosquitoes, birds	–
Kern Canyon virus	Bats	–
B. Genus *Lyssavirus*		
Rabies	All warm-blooded animals	Encephalitis, almost invariably fatal
C. Two Genera of Fish *Rhabdoviruses*	Fish	–
D. Other *Rhabdoviruses*		
Drosophila Sigma virus	Drosophila	–
Lettuce necrotic yellow virus	Plants	–
Other plant rhabdoviruses		–
E. Possible *Rhabdoviruses*		
Marburg Virus	Monkey	Hemorrhagic fever, frequently fatal
Ebola hemorrhagic fever	?	Acute hemorrhagic fever, almost 90% case mortality

Classification	Description
IX. RETROVIRIDAE (RNA tumor viruses)	Enveloped particles containing a coiled nucleocapsid with an icosahedral core shell Diameter: about 150 nm

The RNA tumor virus family comprises a large group of viruses characterized by possession of an RNA genome that comprises two identical molecules, a common morphology and reverse transcriptase. There are three subfamilies. The first, the Oncovirinae, comprises the C-, B-, and D-type RNA tumor viruses, which include both viruses that exist in proviral form in the genomes of animals (endogenous viruses) and those that do not (exogenous viruses). Type A virus particles may also belong here. The second subfamily, the Lentivirinae, comprises the Visna group of viruses. They resemble the Oncovirinae with respect to morphology, nature of the genome and possession of a DNA polymerase, and can transform cultured cells in vitro, but have not yet been shown to possess oncogenic potential. The third subfamily, the Spumavirinae, comprises the foamy viruses, which are found in spontaneously degenerating kidney (and

(continued)

TABLE 2-4 (cont.)

RNA-containing Viruses

---

other) cell cultures, causing the formation of multinucleated vacuolated giant cells that have a highly characteristic appearance. Spumavirinae resemble Oncovirinae in morphology and in certain key characteristics of their mode of replication.

	Host	Symptoms in humans
Subfamily Oncovirinae		
A. Genus *Oncornavirus C*		
Subgenus *Oncornavirus C Avian*		
Chicken sarcoma and leukosis group (Taxonomic criterion: Shared antigen on p27. Subgroups A-G, defined primarily by the properties of the major envelope glycoprotein gp85).	Chicken	–
Avian Sarcoma Viruses (exogenous) Rous sarcoma virus (RSV), several strains B77 virus Fujinami sarcoma virus		
Nondefective Avian Leukosis Viruses (endogenous) Rous-associated viruses, several strains, such as RAV-0, RAV-1, RAV-2, etc.		
Defective Avian Leukosis Viruses (exogenous) Avian myeloblastosis virus (AMV) Avian erythroblastosis virus (AEV) Avian myelocytomatosis virus (MC29) Avian lymphomatosis virus— RPL-12 and other strains Avian Carcinoma Virus (Mill Hill 2) Avian Reticuloendotheliosis virus		
Duck infectious anemia virus Duck spleen necosis virus	Duck	
Subgenus *Oncornavirus C Mammalian* (Shared interspecies antigen on p30)		
Murine Sarcoma and Leukemia Viruses	Mouse	–
Numerous exogenous virus strains Murine sarcoma viruses (MSV) Moloney sarcoma virus Harvey sarcoma virus Kirsten sarcoma virus	May be recombinants with rat type C oncovirus	
Murine leukemia viruses (MLV) Rauscher Moloney Friend, etc.		

(continued)

TABLE 2-4 (cont.)

RNA-containing Viruses		
	Host	*Symptoms in humans*
Numerous endogenous virus strains		
Ecotropic viruses		
Xenotropic viruses		
Amphotropic viruses		
Feline Sarcoma and Leukemia Viruses	Cat	–
Sarcoma and leukemia viruses, primarily exogenous (MAH, F4, FL 237, etc)		
Endogenous viruses (RD 114 and CCC)		
Primate and Monkey Sarcoma and Leukemia Viruses	Primates, Monkeys	?
Simian sarcoma virus [SSV]		
Gibbon ape leukemia virus		
Primate endogenous viruses (baboon endogenous type C virus, M 7)		
Sarcoma and Leukemia Viruses of numerous other species including cattle, rat, hamster, etc.	Cattle, Rat, Hamster	–
Subgenus *oncornavirus C Reptilian*	Reptiles	–
Reptilian Type C oncornaviruses		

B. Genus *Oncornavirus B*

Mouse mammary tumor virus (Bittner virus [milk factor]	Mouse	–
Viruses of guinea pigs, baboons, and possible other mammals	Mammals	?

C. Genus *Oncornavirus D*

Mason-Pfizer monkey virus (MPMV)	Rhesus monkey	–
Viruses from primates	Primates	?
Guinea pig virus	Guinea pig	–

Subfamily Lentivirinae

Visna	Sheep	–
Maedi	Sheep	–
Progressive pneumonia virus	Mice	–

Subfamily Spumavirinae

Human foamy virus	Human cells	–
Simian foamy viruses (9 serotypes)	Monkey kidney cells	–
Canine foamy virus	Dog kidney cells	–
Bovine syncytial virus	Bovine kidney cells	–
Feline syncytial virus	Feline cells	–
Hamster syncytial virus	Hamster cells	–

Probable Retrovirus

Equine Infectious Anemia Virus	Horse	

Classification	*Description*
X. ARENAVIRIDAE	Enveloped coiled nucleocapsids Diameter: 80-130 nm

(continued)

TABLE 2-4 (cont.)

<hr>

RNA-containing Viruses

This family comprises viruses characterized by well-defined envelopes that bear closely spaced projections and enclose an unstructured interior containing a variable number of characteristic electron-dense granules about 25 nm in diameter which have been shown to be ribosomes. They share a group-specific antigen, but antisera do not cross-neutralize.

	Host	Symptoms in humans
Lymphocytic choriomeningitis virus (LCM)	Mouse	Latent infection in mice, may produce fatal meningitis in many other species, including humans
Tacaribe virus complex		
Several viruses including Argentinian (Junin) and Bolivian (Machupo) hemorrhagic fever	Isolated from insects and rodents	Hemorrhagic fever, Machupo frequently fatal
Lassa virus	?	Hemorrhagic fever, frequently fatal

Classification	Description
XI. CORONAVIRIDAE	Enveloped helical nucleocapsids Diameter: about 100 nm

Nucleocapsids helical, characteristic large club-shaped projections (spikes). There is some antigenic relationship between certain human and murine strains.

	Host	Symptoms in humans
Infectious bronchitis virus (IBV)		
Avian strains	Chicken	–
Human strains	Human	Acute upper respiratory disease
Mouse hepatitis virus	Mouse	–

XII. MISCELLANEOUS VIRUSES

Several viruses do not fit into any of the families listed so far. The most important are:

A. Hepatitis Virus. Clinically, two types of viral hepatitis are distinguished. One is characterized by a short incubation (infectious hepatitis, or epidemic jaundice); the other is characterized by a long incubation and usually requires parenteral transmission (serum hepatitis). The etiologic agent of the former, hepatitis virus A, appears to be a particle about 27 nm in diameter which is probably a picornavirus; that of the latter (hepatitis virus B) is probably a particle about 40 nm in diameter that contains DNA and a DNA polymerase (Dane particle).

B. Chronic infectious neuropathic agents (CHINA viruses). These agents have a preclinical period lasting months to several years, succeeded by a slowly progressing, usually fatal disease. Most of them affect the central nervous system. Degenerative diseases of other organs and tissues may be caused by similar agents. They include the agents that cause Kuru and Creutzfeldt-Jakob disease in man, scrapie in sheep, and transmissible mink encephalopathy in mink. All are slow degenerative disorders of the central nervous system, marked by ataxia and wasting, and end in death. The etiologic agents have been transmitted, but not yet isolated or even visualized.

<hr>

FURTHER READING

Books and Reviews

Aposhian HV: Pseudovirions in animals, plants and bacteria. In Fraenkel-Conrat H, Wagner RR (eds): Comprehensive Virology, Vol 5, p 155. New York and London, Plenum, 1975

Baer G (ed): The Natural History of Rabies. New York, Academic Press, 1975

Barry RD, Armond JW, McGeoch DJ, Inglis SC, Mahy BWJ: Structure and function of the influenza virus genome. In Mahy BWJ, Barry RD (eds): Negative Strand Viruses and the Host Cell, p 1. London, Academic Press, 1978

Bishop DHL: Virion polymerases. In Fraenkel-Conrat H, Wagner RR (eds): Comprehensive Virology, Vol

10, p 117. New York and London, Plenum, 1977

Bishop DHL (ed): The Rhabdoviruses. CRC Press, 1979

Bishop JM: Retroviruses. Ann Rev Biochem 47:35, 1978

Caspar DLD, Klug A: Physical principles in the construction of regular viruses. Cold Spring Harbor Symp Quant Biol 27:1, 1962

Choppin PW, Compans RW: Reproduction of paramyxoviruses. In Fraenkel-Conrat H, Wagner RR (eds): Comprehensive Virology, Vol 4, p 179. New York and London, Plenum, 1975

Dalton AJ, Hagenau F (eds): Ultrastructure of Animal Viruses and Bacteriophages: An Atlas. New York, Academic Press, 1973

Diener TO, Hadidi A: Viroids. In Fraenkel-Conrat H, Wagner RR (eds): Comprehensive Virology, Vol 11, p 285. New York and London, Plenum, 1977

Fareed GC, Davoli D: Molecular biology of papovaviruses. Ann Rev Biochem 46:471, 1977

Fenner F: Portraits of viruses: The poxviruses. Intervirol 11:137, 1979

Finch JT, Crawford LV: Structure of small DNA-containing animal viruses. In Fraenkel-Conrat H, Wagner RR (eds): Comprehensive Virology, Vol 5, p 119. New York and London, Plenum, 1975

Gajdusek DC, Gibbs CJ, Jr: Slow virus infections of the nervous system and the laboratories of slow, latent and temperate virus infections. In: The Nervous System, Vol 2: The clinical neurosciences. New York, Raven, 1975

Honess RW, Watson DH: Unity and diversity in the herpesviruses. J Gen Virol 37:15, 1977

Hooks JJ, Gibbs CJ, Jr: The foamy viruses. Bacteriol Rev 39:169, 1975

Joklik WK: Reproduction of reoviridae. In Fraenkel-Conrat H, Wagner RR (eds): Comprehensive Virology, Vol 2, p 231. New York and London, Plenum, 1974

Kapikian AZ: The coronaviruses. Develop Biol Stand 28:42, 1975

Kaplan AS (ed): The Herpesviruses. New York and London, Academic Press, 1973

Kilbourne ED (ed): The Influenza Viruses and Influenza. New York, San Francisco, London, Academic Press, 1975

Kingsbury DW: Paramyxoviruses. In Nayak DP (ed): The Molecular Biology of Animal Viruses, Vol 1, p 349. New York, Marcel Dekker, 1977

Knight CA: Chemistry of Viruses, 2nd ed. New York and Vienna, Springer-Verlag, 1975

Lehmann-Grube F (ed): Lymphocytic Choriomeningitis Virus and Other Arenaviruses. Berlin, Heidelberg and London, Springer-Verlag, 1973

Levintow L: Reproduction of picornaviruses. In Fraenkel-Conrat H, Wagner RR (eds): Comprehensive Virology, Vol 2, p 109. New York and London, Plenum, 1974

McIntosh K: Coronaviruses: A comprehensive review. Curr Top Microbiol Immunol 63:86, 1974

McNulty NS: Rotaviruses. J Gen Virol 40:1, 1978

Mahy BWJ, Barry RD (eds): Negative Strand Viruses and the Host Cell. London, Academic Press, 1978

Mattern CFT: Symmetry in virus architecture. In Nayak DP (ed): The Molecular Biology of Animal Viruses, Vol 1:1. New York, Marcel Dekker, 1977

Melnick JL (ed): Slow virus diseases. Progr Med Virol, Vol 18, 1974

Montelaro RC, Bolognesi DP: Structure and morphogenesis of the type C retroviruses. Adv Cancer Res 28:63, 1978

Moss B: Reproduction of poxviruses. In Fraenkel-Conrat H, Wagner RR (eds): Comprehensive Virology, Vol 3, p 405. New York and London, Plenum, 1974

Murphy FA: Arboviruses: Value of the new taxonomy. In: Proc US Animal Health Assoc, p 425, 1974

Pfau CJ: Biochemical and biophysical properties of the arenaviruses. Progr Med Virol 18:64, 1974

Pfefferkorn ER, Shapiro D: Reproduction of togaviruses. In Fraenkel-Conrat H, Wagner RR (eds): Comprehensive Virology, Vol 2, p 171. New York and London, Plenum, 1974

Philipson L, Lindberg U: Reproduction of adenoviruses. In Fraenkel-Conrat H, Wagner RR (eds): Comprehensive Virology, Vol 3, p 143. New York and London, Plenum, 1974

Raghow R, Kingsbury DW: Endogenous viral enzymes involved in messenger RNA production. Ann Rev Microbiol 30:21, 1976

Rekosh DMK: The molecular biology of picornaviruses. In Nayak DP (ed): Molecular Biology of Animal Viruses, Vol 1, p 63. New York, Marcel Dekker, 1977

Robb JA, Bond CW: Coronaviridae. In Fraenkel-Conrat H and Wagner RR (eds): Comprehensive Virology, Vol 14. New York, Plenum, 1978

Robinson WS: The genome of hepatitis B virus. Ann Rev Microbiol 31:357, 1977

Roizman B: The structure and isomerization of herpes simplex virus genomes. Cell 16:481, 1979

Roizman B, Furlong D: The replication of herpesviruses. In Fraenkel-Conrat H, Wagner RR (eds). Comprehensive Virology, Vol 3, p 229. New York and London, Plenum, 1974

Rowson KEK, Mahy BWJ: Lactic Dehydrogenase Virus. Virol Monographs, Vol 13. Vienna, New York, Springer-Verlag, 1975

Schlesinger RW: Dengue Viruses. Virol Monographs, Vol 16. Vienna, New York, Springer-Verlag, 1977

Siegl G: The Parvoviruses. Virol Monographs, Vol 15. Vienna, New York, Springer-Verlag, 1976

Silverstein SC: The reovirus replicative cycle. Ann Rev Biochem 45:376, 1976

Stevens JG, Todaro, GJ, Fox CF (eds): Persistent Viruses. New York, San Francisco, London, Academic Press, 1978

Theiler M, Downs WG: The Arthropod-borne Viruses of Vertebrates. New Haven and London, Yale Univ Press, 1973

Wagner RR: Reproduction of rhabdoviruses. In Fraenkel-Conrat H, Wagner RR (eds): Comprehensive Virology, Vol 4, p 1. New York and London, Plenum, 1975

Selected Papers

POXVIRUSES

Essani K, Dales S: Biogenesis of vaccinia: Evidence for more than 100 polypeptides in the virion. Virol 95:385, 1979

Geshelin P, Berns KI: Characterization and localization of the naturally occurring cross-links in vaccinia virus DNA. J Mol Biol 88:785, 1974

Kates JR, McAuslan BR: Messenger RNA synthesis by a "coated" viral genome. Proc Natl Acad Sci USA 57:314, 1967

Monroy G, Spencer E, Hurwitz J: Purification of mRNA guanylyltransferase from vaccinia virions. J Biol Chem 253:4481, 1978

Nevins JR, Joklik WK: Isolation and properties of the vaccinia virus DNA-dependent RNA polymerase. J Biol Chem 252:6930, 1977

Paoletti E, Rosemond-Hornbeak H, Moss B: Two nucleic acid nucleoside triphosphate phosphohydrolases from vaccinia virus. J Biol Chem 249:3273, 3281, 1974

Preston VG, Davison AJ, Garon CF, Barbosa E, Moss B: Visualization of an inverted terminal repetition in vaccinia virus DNA. Proc Natl Acad Sci USA 75:4863, 1978

Sarov I, Joklik WK: Studies on the nature and location of the capsid polypeptides of vaccinia virions. Virol 50:579, 1972

Wittek R, Menna A, Schumperli B, Stoffel S, Mueller HK, Wyler R: Hind III and SstI restriction sites mapped on rabbitpox virus and vaccinia virus DNA. J Virol 23:669, 1977

HERPESVIRUSES

Dolyniuk M, Pritchett R, Kieff E: Proteins of Epstein-Barr virus. J Virol 17:935, 1976

Graham FL, Veldhuisen G, Wilkie NM: Infectious herpesvirus DNA. Nature [New Biol] 245:265, 1973

Heine JW, Honess RW, Cassai E, Roizman B: Proteins specified by herpes simplex virus. XII. The virion polypeptides of type 1 strains. J Virol 14:640, 1974

Kudler L, Hyman RW: Exonuclease III digestion of herpes simplex virus DNA. Virol 92:68, 1979

Palmer EL, Martin ML, Gary GW: The ultrastructure of disrupted herpesvirus nucleocapsids. Virol 65:260, 1975

Wagner MJ, Summers WC: Structure of the joint region and the termini of the DNA of the herpes simplex virus type 1. J Virol 27:374, 1978

ADENOVIRUSES

Brown DP, Westphal M, Burlingham BT, Winterhoff U, Doerfler W: Structure and composition of the adenovirus type 2 core. J Virol 16:366, 1975

Everitt E, Lutter L, Philipson L: Structural proteins of adenoviruses. Virol 67:197, 1975

Maizel JV, White DO, Scharff MD: The polypeptides of adenovirus. II. Soluble proteins, cores, top components and the structure of the virion. Virol 36:126, 1968

Nermut MV: Fine structure of adenovirus type 5. I. Virus capsid. Virol 65:480, 1975

Rekosh DMK, Russell WC, Bellett AJD: Identification of a protein linked to the ends of adenovirus DNA. Cell 11:283, 1977

Roberts RJ, Arrand JR, Keller W: The length of the terminal repetition in adenovirus-2 DNA. Proc Natl Acad Sci USA 71:3829, 1974

Wadell G: Classification of human adenoviruses by SDS polyacrylamide gel electrophoresis of structural polypeptides. Intervirol 11:47, 1979

Wadell G, Varsanyi TM: Demonstration of three different subtypes of adenovirus type 7 by DNA restriction site mapping. Inf Imm 21:238, 1978

PAPOVAVIRUSES

Fiers W, Contreras R, Haegeman G, Rogier R, Van de Voorde A, Van Heuverswyn H, Van Herreweghe J, Volckaert G, Ysebaert M: Complete nucleotide sequence of SV40 DNA. Nature 273:113, 1978

Finch JT: The surface structure of polyoma virus. J Gen Virol 24:359, 1974

Finch JT, Klug A: The structure of viruses of the papilloma-polyoma type. J Mol Biol 13:1, 1968

Gibson W: Polyoma virus proteins: A description of the structural proteins of the virion based on polyacrylamide gel electrophoresis and peptide analysis. Virol 62:319, 1974

Reddy VB, Thimmappaya B, Dhar R, Subramanian KN, Zain BS, Pan J, Ghosh PK, Celma ML, Weissman SM: The genome of simian virus 40. Science 200:494, 1978

PARVOVIRUSES

Berns KI, Kelley TJ: Visualization of the inverted terminal repetition of adeno-associated DNA. J Mol Biol 82:267, 1974

Chesebro B, Bloom M, Hadlow W, Race R: Purification and ultrastructure of Aleutian disease virus of mink. Nature 254:456, 1975

de la Maza LM, Carter BJ: Adeno-associated virus DNA structure: Restriction endonuclease maps and arrangement of terminal sequences. Virol 82:409, 1977

Mayor HD, Torikai K, Melnick JL: Plus and minus single-stranded DNA separately encapsidated in adeno-associated satellite virions. Science 166:1280, 1969

Rose JA, Maizel JV, Inman JK, Shatkin AJ: Structural proteins of adenovirus-associated viruses. J Virol 8:766, 1971

Spear IS, Fife KH, Hauswirth WW, Jones CJ, Berns

KI: Evidence for two nucleotide sequence orientations within the terminal repetition of adeno-associated virus DNA. J Virol 24:627, 1977

Türler H: Interactions of polyoma and mouse DNAs. III. Mechanism of polyoma pseudovirion formation. J Virol 15:1158, 1975

PICORNAVIRUSES

Ambros V, Baltimore D: Protein is linked to the 5′ end of poliovirus RNA by a phosphodiester linkage to tyrosine. J Biol Chem 253:5263, 1978

Golini F, Nomoto A, Wimmer E: The genome-linked protein of picornaviruses. Virol 89:112, 1978

Lund GA, Ziola BR, Salmi A, Scraba DG: Structure of the mengo virion. V. Distribution of the capsid polypeptides with respect to the surface of the virus particle. Virol 78:35, 1977

Medappa KC, McLean C, Rueckert RR: On the structure of rhinovirus 1A. Virol 44:259, 1971

Young, NA: Polioviruses, coxackieviruses, and echoviruses: Comparison of the genomes by RNA hybridization. J Virol 11:832, 1973

CALICIVIRUSES

Black DN, Brown F: A major difference in the strategy of the calici- and picornaviruses and its significance in classification. Intervirol 6:57, 1975/1976

Burroughs JN, Brown F: Physico-chemical evidence for the reclassification of the caliciviruses. J Gen Virol 22:281, 1974

Burroughs JN, Doel TR, Smale CJ, Brown F: A model for vesicular exanthema virus, the prototype of the calicivirus group. J Gen Virol 40:161, 1978

TOGAVIRUSES

Aliperti G, Schlesinger MJ: Evidence for an auto-protease activity of Sindbis virus capsid protein. Virol 90:366, 1978

Burke D, Keegstra K: Carbohydrate structure of Sindbis virus glycoprotein E2 from virus grown in hamster and chicken cells. J Virol 29:546, 1979

Clewley JP, Kennedy SIT: Purification and polypeptide composition of Semliki Forest virus RNA polymerase. J Gen Virol 32:395, 1976

Enzmann PJ, Weiland F: Studies on the morphology of the alphaviruses. Virol 95:501, 1979

de Madrid AT, Porterfield JS: The flaviviruses (group B arboviruses): A cross-neutralization study. J Gen Virol 23:91, 1974

Pedersen CE, Eddy GA: Separation, isolation, and immunological studies of the structural proteins of Venezuelan equine encephalomyelitis virus. J Virol 14:740, 1974

von Bonsdorff CH, Harrison SC: Hexagonal glycoprotein arrays from Sindbis virus membranes. J Virol 28:578, 1978

BUNYAVIRUSES

Dahlberg JE, Obijeski JS, Korb J: Electron microscopy of the segmented RNA genome of La Crosse virus: Absence of circular molecules. J Virol 22:203, 1977

Hewlett MJ, Petterson RF, Baltimore B: Circular forms of Uukuniemi virion RNA: An electron microscopic study. J Virol 21:1085, 1977

Kascsak RJ, Lyons MJ: Bunyamwera virus. 1. The molecular complexity of the virion RNA. Virol 92:37, 1977

Murphy FA, Harrison AK, Whitefield SG: Bunyaviridae: Morphologic and morphogenetic similarities of Bunyamwera serologic supergroup viruses and several other arthropod-borne viruses. Intervirol 1:297, 1973

Petterson RF, von Bornsdorff CH: Ribonucleoproteins of Uukuniemi virus are circular. J Virol 15:386, 1975

REOVIRUSES

Esparza J, Gil F: A study on the ultrastructure of human rotavirus. Virol 91:141, 1978

Furuichi Y, Morgan M, Muthukrishnan S, Shatkin AJ: Reovirus mRNA contains a methylated, blocked 5′-terminal structure: $m^7G(5′)ppp(5′)G^mpCp-$ Proc Natl Acad Sci USA 72:262, 1975

Luftig RB, Kilham SS, Hay AJ, Zweerink HJ, Joklik WK: An ultrastructural study of virions and cores of reovirus type 3. Virol 48:170, 1972

Martin SA, Zweerink HJ: Isolation and characterization of two types of bluetongue virus particles. Virol 50:495, 1972

Palmer EL, Martin ML: The fine structure of the capsid of reovirus type 3. Virol 76:109, 1977

Shatkin AJ, Sipe JD, Loh P: Separation of ten reovirus genome segments by polyacrylamide gel electrophoresis. J Virol 2:986, 1968

Smith RE, Zweerink HJ, Joklik WK: Polypeptide components of virions, top component and cores of reovirus type 3. Virol 39:791, 1969

Woode G, Bridger JC, Jones JM, Flewett TH, Bryden AS, Davies HA, White GBB: Morphological and antigenic relationships between viruses (rotaviruses) from acute gastroenteritis of children, calves, piglets, mice and foals. Inf Imm 14:804, 1976

ORTHOMYXOVIRUSES

Collins JK, Knight CA: Purification of the influenza hemagglutinin glycoprotein and characterization of its carbohydrate components. J Virol 26:457, 1978

Desselderger U, Palese P: Molecular weights of RNA segments of influenza A and B viruses. Virol 88:394, 1978

Inglis SC, MeGeoch DJ, Mahy BWJ: Polypeptides specified by the influenza virus genome. II. Assignment of protein coding functions to individual genome segments by in vitro translation. Virol 78:522, 1977

Racaniello VR, Palese P: Influenza B virus genome: Assignment of viral polypeptides to RNA segments. J Virol 29:361, 1979

Scholtissek C, von Hoyningen V, Rott R: Genetic relatedness between the new 1977 epidemic strains (H1N1) of influenza and human influenza strains isolated between 1947 and 1957 (H1N1). Virol 89:613, 1978

PARAMYXOVIRUSES

Gething MJ, White JM, Waterfield MD: Purification of the fusion protein of Sendai virus: analysis of NH₂-terminal sequence generated during precursor activation. Proc Natl Acad Sci USA 75:2737, 1978

Moore PME, Hayes EC, Miller SE, Wright LL, Machamer CE, Zweerink HJ: Measles virus nucleocapsids: Large-scale purification and use in radioimmunoassay. Inf and Imm 20:842, 1978

Orvell C: Structural polypeptides of mumps virus. J Gen Virol 41:527, 1978

Scheid A, Choppin PW: Identification of biological activities of paramyxovirus glycoproteins. Activation of cell fusion, hemolysis, and infectivity by proteolytic cleavage of an inactive precursor protein of Sendai virus. Virol 57:475, 1974

Seifried AS, Albrecht P, Milstein JB: Characterization of an RNA-dependent RNA polymerase activity associated with measles virus. J Virol 25:781, 1978

Shimizu K, Shimizu YK, Kohama T, Ishida N: Isolation and characterization of two distinct types of HVJ (Sendai virus) spikes. Virol 62:90, 1974

RHABDOVIRUSES

Bishop DHL, Repik P, Obijeski JF, Moore NF, Wagner RR: Reconstitution of infectivity to spikeless vesicular stomatitis virus by solubilized viral components. J Virol 16:75, 1975

Emerson SU, Wagner RR: L protein requirement for in vitro RNA synthesis by vesicular stomatitis virus. J Virol 12:1325, 1973

Etchison JR, Holland JJ: Carbohydrate composition of the membrane glycoprotein of vesicular stomatitis virus grown in four mammalian cell lines. Proc Natl Acad Sci USA 71:4011, 1974

Imblum RL, Wagner RR: Protein kinase and phosphoproteins of vesicular stomatitis virus. J Virol 13:113, 1974

Reading CL, Penhoet EE, Ballou CE: Carbohydrate structure of vesicular stomatitis virus glycoprotein. J Biol Chem 253:5600, 1978

Szilagyi JF, Uryvayev L: Isolation of an infectious ribonucleoprotein from vesicular stomatitis virus containing an active RNA transcriptase. J Virol 11:279, 1973

Tabas I, Schlesinger S, Kornfeld S: Processing of high mannose oligosaccharides to form complex type oligosaccharides on the newly synthesized polypeptides of the vesicular stomatitis virus G protein and the IgG heavy chain. J Biol Chem 253:716, 1978

Tan KB: Comparative study of the protein kinase associated with animal viruses. Virol 64:566, 1975

Wagner RR, Prevec L, Brown F, et al: Classification of rhabdovirus proteins: A proposal. J Virol 10:1228, 1972

ARENAVIRUSES

Palmer EL, Obijeski JF, Webb PA, Johnson KN: The circular, segmented nucleocapsid of an arenavirus-Tacaribe virus. J Gen Virol 36:541, 1977

CORONAVIRUSES

Guy JS, Brian DA: Bovine coronavirus genome. J Virol 29:293, 1979

MacNaughton MR, Davies HA, Nermut MV: Ribonucleoprotein-like structures from coronavirus particles. J Gen Virol 39:545, 1978

MacNaughton MR, Madge MH: The genome of human coronavirus strain 229 E. J Gen Virol 39:497, 1978

Tannock GA, Hierholzer JC: Presence of genomic polyadenylate and absence of detectable virion transcriptase in human coronavirus OC-43. J Gen Virol 39:29, 1978

Wege H, Wege H, Nagashima K, ter Meulen V: Structural polypeptides of the murine coronavirus JHM. J Gen Virol 42:37, 1979

HEPATITIS VIRUSES

Barker LF, Almeida JD, Hoofnagle JG, et al: Hepatitis B core antigen: Immunology and electron microscopy. J Virol 14:1552, 1974

Krugman S, et al: Nomenclature of antigens associated with viral hepatitis B. Intervirol 2:134, 1974

Robinson WS, Greenman RL: DNA polymerase in the core of human hepatitis B virus. J Virol 13:1231, 1974

VIROIDS

Dickson E, Diener TO, Robertson HD: Potato spindle tuber and citrus exocortis viroids undergo no major sequence changes during replication in two different hosts. Proc Natl Acad Sci USA 75:951, 1978

Owens RA, Erbe E, Hadidi A, Steere RL, Diener TO: Separation and infectivity of circular and linear forms of potato spindle tuber viroids. Proc Natl Acad Sci USA 74:3859, 1977

CHAPTER 3

The Inactivation of Viruses

Knowledge of the nature of the interaction of viruses with the chemical and physical components of the extracellular environment is of significance for three reasons: it is important to know (1) how to inactivate viruses when the object is to eliminate them and (2) how to preserve them when the object is to avoid loss of infectivity. (3) Analysis of the mode of action of specific reagents on virions often throws light on the nature of viral capsids and of their association with nucleic acids.

Enzymes

All viral genomes are protected from nucleases by virtue of the capsids that enclose them. Animal viruses are generally resistant to attack by the proteases of higher animals, such as pepsin, trypsin, and chymotrypsin. Some, such as the enteroviruses, are completely resistant, while others, such as the poxviruses and reoviruses, possess susceptible outer shells but resistant cores. Glycoprotein spikes of enveloped viruses can generally be removed by treatment with proteolytic enzymes, and the resulting "bald" particles lack infectivity. Phospholipases usually inactivate enveloped viruses by hydrolyzing their phospholipids.

Chemical reagents

Viruses are inactivated by many types of chemical compounds. Among them are oxidizing agents, salts of heavy metals, and most reagents that interact with proteins chemically. Many of these reagents are of little importance for the selective destruction of viral infectivity, and the study of their interaction with viruses has not contributed significantly to our knowledge of virion structure. There are several chemical reagents, however, whose action on viruses is of great interest.

Detergents

1. Nonionic detergents (such as Nonidet P-40, Triton X-100, and the like), which are usually polyoxyethylene ethers or sorbitans, solubilize lipid components of viral envelopes, thereby releasing undenatured internal components and glycoprotein spikes, which can then be examined further with respect to morphology, antigenic constitution, and enzymatic activity.
2. Anionic detergents, the most important of which is sodium dodecyl sulfate, not only solubilize viral envelopes but also dissociate capsids into their constitutent polypeptides. The use of sodium dodecyl sulfate for the characterization of these polypeptides by means of polyacrylamide gel electrophoresis has already been discussed (Chap. 2).
3. Cationic detergents have so far found little application in virology.

Protein Solvents

Guanidine, urea, and phenol are powerful protein solvents that are used extensively to dissociate viral capsids into their component polypeptide chains. In contrast to sodium dodecyl sulfate, they do not form complexes with polypeptides but act by minimizing the formation of hydrogen bonds on which protein structure is largely dependent. Phenol is the most commonly used reagent for liberating viral nucleic acids.

Formaldehyde

Formaldehyde destroys infectivity without significantly affecting antigenicity and has therefore been used extensively for preparing inactivated virus vaccines. It destroys infectivity primarily by reacting with those amino groups of adenine, guanine, and cytosine that are not involved in hydrogen bond formation. Viruses that contain single-stranded nucleic acid are therefore inactivated readily, while those that contain double-stranded nucleic acid are resistant to formaldehyde. It also reacts with amino groups of proteins, forming addition compounds of the Schiff's base type and cross-linking polypeptide chains without, however, significantly disturbing protein conformation. Reaction with protein is most probably responsible for the occasional generation of a formaldehyde-resistant infectious virus fraction, which appears to be caused by such extensive cross-linking of capsid proteins that formaldehyde cannot reach and inactivate the viral nucleic acid. Careful control of reaction conditions and rigorous checks for residual infectious virus are mandatory for the preparation of formaldehyde-inactivated virus vaccines.

pH

Viruses differ greatly in their resistance to acidity. For example, enteroviruses are very resistant to acid conditions, while rhinoviruses are very susceptible. All viruses are disrupted under alkaline conditions.

Physical agents

Heat

Viruses differ enormously in heat stability. In general, icosahedral viruses, such as enteroviruses, reoviruses, papovaviruses, and adenoviruses, as well as poxviruses, are reasonably heat-stable; their infectivity decreases by no more than two- to fourfold during six hours at 37C. By contrast, many enveloped viruses, especially myxoviruses and RNA tumor viruses, are very heat-labile; their half-life at 37C is sometimes no more than one hour. The infectivity titer of the former group of viruses remains stable for months at 4C; but viruses of the latter group must be stored at $-70C$, the temperature of dry ice, or in liquid nitrogen ($-196C$).

The initial rate of heat inactivation is exponential, that is, a constant fraction of virus is inactivated in each unit of time. This is most readily explained in terms of an energy distribution, with those molecules that possess more than a certain minimal energy of activation having a significant probability of undergoing an inactivating change. As the temperature increases, so does the proportion of molecules with this minimal amount of energy.

Heat stability is strongly influenced by environmental conditions. Proteins stabilize all viruses to greater or lesser extent, as do metal ions, especially diavalent cations such as Mg^{2+} and Ca^{2+}. Measurement of heat inactivation therefore requires careful standardization of suspension media.

Radiation

All viruses are inactivated by electromagnetic radiation, especially X-radiation or gamma-radiation and ultraviolet (UV) radiation. X-rays inactivate viruses primarily by causing scissions (breaks) in nucleic acid strands. If the viral genome consists of single-stranded nucleic acid, every scission is lethal; if the genome consists of double-stranded nucleic acid, scissions in both strands located near to each other are required for inactivation. X-rays therefore inactivate viruses that contain single-stranded nucleic acid much more efficiently (about tenfold) than viruses that contain double-stranded nucleic acid.

Ultraviolet radiation also damages nucleic acids. In particular, it causes the formation of covalent bonds between adjacent pyrimidine molecules, thereby giving rise to cyclobutane derivatives. In DNA, the most commonly formed pyrimidine dimers are those between adjacent thymine rings; in RNA, dimers are formed between any adjacent uracil and cytosine rings. Dimer formation inactivates viral genomes by preventing replication and probably also transcription and translation.

Dimers are removed from DNA by several mechanisms. These have been studied mainly in bacteria, but there is evidence that they operate in animal cells as well. One involves an enzyme system that utilizes radiation of longer wavelengths, particularly those of visible light, for dissociating dimers; this is the so-called photoreactivating repair system. Another mechanism (the dark or excision repair mechanism) involves nucleases that recognize dimers and excise them, the gaps then being repaired by a DNA polymerase acting in conjunction with polynucleotide ligase; this is the host-cell reactivating system.

Ultraviolet radiation also induces cross-linking of the two strands of double-stranded

DNA by a mechanism that is not clear. No doubt this also contributes to virus inactivation.

Ultraviolet radiation also causes the addition of water molecules across the C5-C6 double bond of pyrimidines in both DNA and RNA, which results in the formation of photohydrates (6-hydroxy-5,6-dihydro derivatives). These photohydrates represent a major portion of the lethal damage caused by ultraviolet light in many RNA-containing viruses.

The most radiation-sensitive property of a virion is its infectivity. The reason is that infectivity requires expression of the genome's entire information content and thus presents the largest target. Sometimes virus particles that have lost the ability to reproduce can still express some special function or group of functions that originate from cistrons that have not sustained radiation damage. Examples of such functions are the ability to synthesize early enzymes (Chap. 5) and the ability to transform cells (Chap. 9). At very high radiation doses damage to capsid proteins becomes important. This causes loss of the ability to interfere with the multiplication of related viruses, loss of ability to hemagglutinate, and loss of antigenicity.

Photodynamic Inactivation

Virions interact with certain organic dyes in such a manner that illumination with visible light inactivates them. One such dye is methylene blue. When methylene blue is added to vaccinia virus and the mixture is kept dark, infectivity is not affected. However, if it is illuminated with white light in the presence of oxygen, the virus is inactivated. The mechanism of this inactivation is not well understood. Presumably the dye must penetrate into the virus and become associated with the nucleic acid.

The acridine dye, neutral red, acts in a similar manner. Acridines characteristically intercalate between the stacked hydrogen-bonded base pairs of double-stranded nucleic acids, and they also intercalate, although more weakly,

between base pairs of double-helical regions of single-stranded nucleic acids. If a virus, such as poliovirus, is allowed to multiply in the presence of neutral red, the viral RNA together with intercalated dye becomes enclosed in capsids, forming virions that are fully infectious unless they are illuminated with white light. This phenomenon is of practical importance for two reasons. First, neutral red is a vital dye that is commonly used for the enumeration of plaques (Chap. 1). Plates on which plaque titrations are performed must be protected from bright white light once neutral red is added to them, since otherwise progeny virus is inactivated, and visible plaques do not form. Second, the fact that neutral red-containing virus is photosensitive provides a very useful method of inactivating infectious virus, which finds application in several types of studies.

FURTHER READING

Books and Reviews

Bachrach HL: Reactivity of viruses in vitro. Prog Med Virol 8:214, 1966

Selected Papers

Cooper PD: Studies on the structure and formation of the polioviruses: Effect of concentrated urea solutions. Virol 16:485, 1962

Gard S: Theoretical considerations in the inactivation of viruses by chemical means. Ann NY Acad Sci 83:638, 1960

Oster G, McLaren AD: The ultraviolet light and photosensitized inactivation of tobacco mosaic virus. J Gen Physiol 33:215, 1950

Salk JE: Considerations in the preparation and use of poliomyelitis virus vaccine. JAMA 158:1239, 1955

Shapiro AL, Viñuela E, Maizel JV: Molecular weight estimation of polypeptide chains by electrophoresis in SDS-polyacrylamide gels. Biochem Biophys Res Commun 28:815, 1967

CHAPTER 4

Viruses and Viral Proteins as Antigens

Many proteins coded by viruses are good antigens. This is of vital significance both medically and scientifically. Although great strides have been made during the last decade in defining the biochemistry and molecular biology of the multiplication cycles of animal viruses, the chemotherapeutic control of virus diseases is not yet a practical proposition. In fact, with very few exceptions, there is no way in which virus infections can be controlled; in almost all cases one relies on the natural ability of the host to form antibody to the invading virus. When the spread of viral infection to essential organs is too rapid, or when for some reason antibody formation does not take place early enough, the patient may succumb. Gamma globulin from hyperimmune sera is sometimes administered as a last resort: even then one must rely on antibodies, not on drugs. By the same token, the only presently practical form of antiviral prophylaxis is provided by antibodies produced in response to vaccines (Chap. 8).

Structural viral proteins stimulate the formation of antibodies not only as components of virions but also as components of virion subunits, such as capsomers or nucleocapsids, and also in the free state. The principal antigenic determinants are often the same in all three forms, but extra antigenic sites are sometimes generated as individual proteins become part of more complex structures; for example, adenovirus hexons exhibit antigenic determinants not expressed by free hexon proteins.

The range of antiviral antibodies formed under conditions when viruses can and cannot multiply differs greatly. If a virus cannot multiply, either because it has been inactivated or because the host is not susceptible, only antibodies to surface components of the virus particle are usually formed. However, if the virus can multiply, not only is far more antibody formed because progeny virus will also act as antigen, but also the range of antibodies produced is much wider, since antibodies are then also formed to the unassembled and partially assembled virion components that are synthesized as a result of virus multiplication, as well as to nonstructural virus-coded proteins. For example, antisera to inactivated vaccinia virus contain only a few species of antibody directed against its surface components; but antisera from animals in which vaccinia virus has multiplied contain antibodies against at least 25 different viral proteins in readily detectable amounts.

The most specific, or individual, components of a virion's antigenic complement are its surface antigens, which usually vary from strain to strain and are known as the "type-specific antigens." Internal proteins of virus strains that belong to the same subgroup (which often infect the same host species) usually possess common antigenic determinants; these are therefore known as "group-specific (gs) antigens" or "species-specific antigens." In the case of mammalian RNA tumor viruses (Chap. 9), some of the proteins that possess group-specific determinants also possess antigenic specificities common to several groups of viruses that infect different animal species; these are said to be "interspecies-specific" antigenic determinants.

Possession of group- and type-specific antigenic determinants is of great taxonomic significance and provides an important tool for epidemiology. On the one hand, newly isolated virus strains usually receive their preliminary characterization as a result of tests against a variety of antisera of known specificity. On the other hand, it is possible to determine whether a given human or animal population has been exposed to a particular virus strain by testing serum samples for antiviral antibody.

The interaction of viruses and viral proteins with antibodies can be recognized and measured in several ways. The three most important follow.

The interaction of virus with neutralizing antibody

Antibodies to viral surface components neutralize infectivity; these are the neutralizing antibodies that protect against disease. They usually persist in the body for many years, and even when their level drops, a secondary or anamnestic response to virus generally boosts their titers to very high levels, so that no second cycle of infection ensues. This explains the fact that animals generally contract any particular virus disease only once. Exceptions, such as the common cold and influenza, are due to special circumstances. The reason for the frequent recurrence of the common cold syndrome is that it is elicited by a very large group of viruses, among which are rhinoviruses, enteroviruses, adenoviruses, ortho- and paramyxoviruses, and coronaviruses. The reason for recurrent epidemics caused by the influenza virus is its genetic and, therefore, immunologic variability (Chap. 7).

The reaction between neutralizing antibody and virus follows first order kinetics, which indicates that one antibody molecule can inactivate one virus particle. It does so by interfering with one of the initial events of the virus multiplication cycle, most probably uncoating. As we shall see, under conditions of normal infection, viral genomes are liberated into the interior of the cell, ready to start multiplication; but virus-antibody complexes are apparently engulfed and inactivated by phagocytic vacuoles, so that no intact viral genomes are able to reach the interior of the cell (Chap. 5).

In practice, a small fraction of virus that may vary from 0.1 to 10 percent generally remains infectious even in the presence of a large excess of antibody. The reason for this is that the virus-antibody union is reversible. Simple dilution causes dissociation, and so does contact with cells. Different cell types differ in their ability to reactivate virus-antibody complexes, so that the titers of virus-antibody mixtures tend to vary with the type of cell used for assay. A contributing factor to the formation of a persistent non-neutralizable virus fraction is the fact that antisera generally contain antibody molecules with diverse avidities which compete with each other for virus, so that some virus particles are usually combined with low avidity antibody. The lower the avidity, the less perfect is the fit between the antigenic site on the virus and the antibody-combining site, and the more readily reversible is the antigen-antibody union.

Complement-fixing antigen and antibody

The virus-antibody interaction can also be measured by taking advantage of the fact that complexes of viral protein and antibody often fix complement. Since sensitive methods for titrating complement are available, this provides a convenient and accurate method of measuring either the amount of viral antigens (complement-fixing antigens or CFA) or of antibody to such antigens. The chief advantage of this method of detecting viral antigens is that *any* virus-coded protein may be a complement-fixing antigen; both structural and nonstructural virus-coded proteins, such as the T antigens formed in cells infected with adenoviruses and papovaviruses (Chap. 5), as well as viral subunits and virus particles themselves, may form complexes with antibody that fix complement. This method of quantitating viral proteins is particularly useful for detecting abortive virus infections when only part of the genetic information present in the viral genome is expressed and no virus particles are produced (Chap. 6). It is also of great importance in epidemiology, since it is often far easier to classify newly isolated virus strains by determining whether extracts of cells infected with them fix complement with antisera of known specificity than by measuring the ability of such antisera to neutralize infectivity.

Gel immunodiffusion and immuno-electrophoresis

Under appropriate conditions of antigen-antibody equivalence, antigen-antibody complexes are insoluble. This property is used in gel immunodiffusion and gel immunoelectrophoresis, techniques that are widely employed for resolving mixtures of viral antigens such as occur in extracts of infected cells.

The most widely used modification of gel immunodiffusion is the Ouchterlony method, which employs petri plates containing agar into which are cut a number of wells, one being situated centrally and the others equidistantly from it and from each other. Antiserum or antibody is placed into the center well, antigen is placed into the outer ones, and diffusion is then allowed to proceed. Where the concentration of antigen-antibody complexes exceeds their solubility product, precipitin lines form, the location of which depends on the relative diffusion rates, and therefore on the relative sizes,

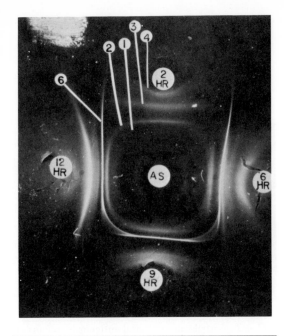

FIG. 4-1. The use of gel diffusion analysis to detect virus-specified proteins in extracts of infected cells. Antiserum to vaccinia virus was placed into the center well (AS) of a petri dish containing a layer of agar. Extracts of HeLa cells infected for 2, 6, 9, and 12 hours with vaccinia virus were placed into the other four wells, and the antibodies and antigens were allowed to diffuse toward each other. Precipitin lines, formed as described in the text, were then stained with Poinceau S. The pattern becomes increasingly complex with increasing time after infection, as more virus-specified proteins are synthesized. The advantage of this method is that virus-specified proteins are revealed without the necessity for purification. (From Salzman and Sebring: J Virol 1:16, 1967.)

of antigen and antibody. Identity of antigens is revealed by the familiar fusion of precipitin lines (Fig. 4-1).

In the gel immunoelectrophoresis technique, antigens are not separated by free diffusion but by electrophoresis in agar slabs, after which antiserum is applied in a trough cut parallel to the direction of electrophoresis. After diffusion, precipitin lines form as above (Fig. 4-2). The concentrations of antibody and antigen used in both these techniques are generally adjusted so that the precipitin lines are very thin, thus permitting great resolution.

A modification of this technique is "rocket" immunoelectrophoresis, in which proteins are electrophoresed into agarose gels that contain antibodies to them. If the pH of such gels is adjusted so that antibody molecules remain stationary (about pH 8.6), there result rocket-shaped precipitates, demonstrable with protein stains, the area of which is proportional to the quantity of antigen that is applied. This technique is very versatile since it permits not only the quantitation of single antigens in complex mixtures, provided that the appropriate monospecific antibody is available, but is also applicable to mixtures of antigens. The technique is then known as "crossed immunoelectrophoresis," and the antigens are first separated by electrophoresis in a standard agarose gel before being electrophoresed at right angles into the gel that contains the appropriate mixture of antibodies. An example of this technique is as follows: the mixture of virus-coded antigens is an extract of infected cells, and the antibody mixture is antiserum from an infected

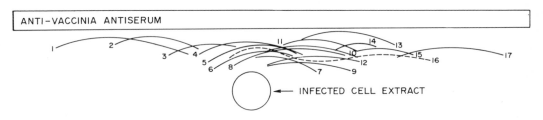

FIG. 4-2. Diagrammatic representation of an immunoelectrophoretic pattern of the virus-specified proteins present in an extract of chick cells infected with vaccinia virus. In this technique the cell extract is placed into a circular well cut into an agar slab and an electric field is applied, causing antigens to migrate at rates governed by their charge and size. After electrophoresis, antiserum to vaccinia virus is placed into a trough cut parallel to the direction of electrophoresis and allowed to diffuse toward the separated cell extract components. Virus-specified proteins able to react with antibodies in the antiserum form precipitin lines. Seventeen such proteins can be detected in the extract shown here. (From Rodriguez-Burgos et al: Virol 30:569, 1966.)

rabbit. This combination would yield a profile of the type which would permit the simultaneous quantitation of numerous virus-coded proteins.

The visualization of viral antigens using tagged antibody

There are many occasions when direct visual localization of viral antigens is desired. This can be achieved by using antibody that is tagged, conjugated, or labeled with some material that can be visualized with either the light microscope or the electron microscope.

Antibody labeled with the dye fluorescein fluoresces brightly when viewed with a microscope equipped with a source of ultraviolet light (Fig. 4-3). Such antibody is a sensitive research tool in viral pathogenesis, that is, in studies of the route of infection and the spread of virus within the organism, since it can reveal a small number of infected cells in large populations of uninfected ones. It is also useful for measuring the proportion of infected cells in a variety of experimental situations, and it can serve as a rapid diagnostic tool, since minute amounts of infected biopsy material can be treated with fluorescein-labeled antibodies to several suspected viruses, one of which will cause the infected cells to fluoresce. Finally, since fluorescein-labeled antibody can also reveal the pattern of viral antigen distribution within infected cells, and since this pattern is often highly characteristic (nuclear or cytoplasmic, diffuse or highly localized), it can also serve in this way as a useful adjunct to virus identification.

Antibody molecules can also be tagged with large molecules or particles that can be seen with the electron microscope. Among these are ferritin, a large iron-containing protein, bacteriophage or virus particles with characteristic shapes, and latex spheres. Use of such antibody permits exquisitely detailed observation of the distribution of viral antigens in infected cells. It is therefore invaluable in studies aimed at establishing the exact location of the sites of synthesis, accumulation, and assembly of viral protein components (Fig. 4-4).

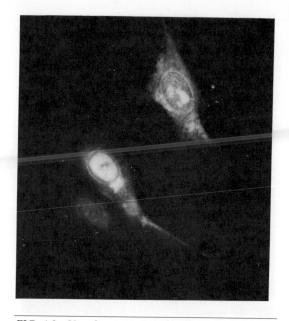

FIG. 4-3. Visualization of viral antigens by means of immunofluorescence. Cells infected with herpes simplex virus were washed, fixed in acetone and allowed to react either with herpesvirus antibody conjugated with fluorescein isothiocyanate (direct immunofluorescence), or with herpesvirus antibody prepared in rabbits, followed by anti-rabbit globulin conjugated with fluorescein isothiocyanate (indirect immunofluorescence). Cells were then examined under ultraviolet light. The cells shown here were stained by the indirect method. The top cell shows fluorescent nuclear patches as well as fluorescence at the nuclear membrane and some diffuse cytoplasmic fluorescence. The other cell shows bright fluorescence of practically the whole nucleus, as well as cytoplasmic fluorescence. ×550. (From Ross, Watson, and Wildy: J Gen Virol 2:115, 1968.)

The detection of minute amounts of viral antigens and antibodies

Appropriate patient management frequently depends on early detection of virus infections or on the recognition of persistent virus infections. In both situations diagnosis requires the correct and rapid identification of minute amounts of viral antigens and antibodies; and in

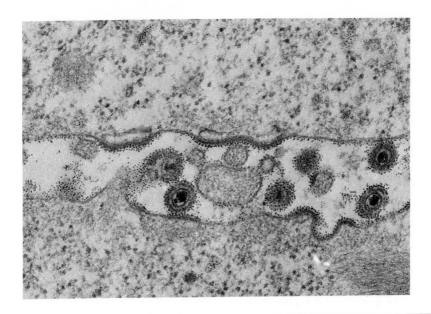

FIG. 4-4. Visualization of viral antigens by means of ferritin-conjugated antibody. This electron micrograph reveals the localization of virus-specified proteins on the surfaces of cells infected with herpes simplex virus. Infected cells were allowed to react with herpesvirus antibody conjugated with ferritin; the cells were then washed, fixed, embedded and sectioned. The surfaces of two adjacent cells are seen, both with intensely labeled patches. Budding virus particles and detached cytoplasmic fragments are also labeled. ×48,000. (From Nii et al: J Virol 2:1172, 1968.)

recent years great strides have been made in the development of extremely sensitive techniques for this purpose. Two such techniques are as follows. The first is radioimmunoassay. Not only is this technique capable of detecting extremely small amounts of virus-coded proteins, but it is also very useful for assessing serological relatedness, since one can measure the efficiency with which unknown antigens can prevent antibody from precipitating the standard labeled test antigen: the closer the serologic relationship, the more efficient the competition.

The second technique is the enzyme-linked immunosorbent assay (ELISA), in which antigen is immobilized on some surface, such as the wells of a plastic microtiter plate (see Fig. 1-8); and specific antiserum coupled with alkaline phosphatase is then added. Following extensive washing, p-nitrophenyl phosphate is added and the p-nitrophenol that is liberated by the antibody-linked enzyme bound to the antigen is then measured colorimetrically. This technique can be made to be over 100 times more sensitive than complement fixation assays.

A further dimension has recently been added to these techniques by the use of the "protein A" that is present in the cell walls of certain strains of *Staphylococcus*, which possesses high affinity for the Fc portion of most mammalian IgGs. Cells of such strains, or isolated A protein coupled to inert particles such as agarose, can be used to adsorb minute amounts of antigen-antibody complexes from complex mixtures.

Virus-coded cell surface antigens

As described in Chapters 5 and 9, new antigenic determinants frequently appear on the surfaces (outer plasma membranes) of infected cells. These new determinants are on virus-coded polypeptides. In the case of enveloped viruses these polypeptides become, in due course, part of the viral envelope; in the case of nonenveloped viruses they are nonstructural polypeptides that are synthesized early during the infection cycle. In either case, virus-coded cell surface polypeptides provide a clear signal to the immune mechanism that a cell is infected, and antibodies are formed against them.

The immune response to viral infection

During viral infection antibodies are formed against all classes of virus-coded antigens. Those that are most important for eliminating virus from the body are those that are directed against virion components and virus-coded cell surface antigens. The former include the neutralizing and complement-fixing antibodies that prevent virus particles from infecting cells (p. 67). As for the latter, their combination with virus-coded antigens on cell surfaces renders cells subject to destruction by at least two mechanisms: combination with complement followed by lysis, and attack by cytolytic T lymphocytes. In either case the infected cell is eliminated as a source of progeny virus.

Until recently, it was thought that T lymphocytes recognize only the virus-coded antigen(s) on infected cells; recent work has shown, however, that cytolytic T lymphocytes generated during viral infection can only lyse infected cells that share at least part of the major histocompatibility region (HLA in humans) with the cells that originally stimulated their development. Two hypotheses have been advanced as the reason for this genetic restriction. The first theory postulates that the viral antigens and the histocompatibility antigens are separate entities on the surface of infected cells and that the T cells possess independent receptors for these two types of molecules (dual recognition). The second theory suggests that the viral and histocompatibility antigens form a molecular complex that is recognized by only one receptor on T lymphocytes (altered self). Evidence in favor of both hypotheses has accumulated, the balance perhaps favoring the altered self model.

Usually the mechanisms for destroying infectious virus and infected cells are beneficial to the host. However, it is now recognized that sometimes these mechanisms may be very harmful. Let us consider first the destruction of infected cells. As a rule the number of cells that are destroyed is not large enough to cause serious problems for the host organism, but there are exceptions. A good example is provided by lymphocytic choriomeningitis virus (LCM), which causes encephalitis in mice and also in humans. LCM, an enveloped virus, is not a very "lytic" virus; cells infected with it are not severely damaged and may survive for long periods of time. In mice, LCM produces no overt disease if the immune mechanism is not operative (in immunosuppressed or tolerant animals). However, in immunologically competent mice, LCM causes a fatal meningitis within a week, that is, as soon as antibody begins to be formed; death being due to the destruction of infected cells by activated macrophages. The disease is thus not caused by the destruction of the host's cells by the virus, but by the destruction of infected cells by the host's immune mechanism. A similar interaction between immune lymphocytes and virus-coded cell surface antigens may account for the symptoms associated with some viral diseases of humans such as hepatitis.

The same is true for the second mechanism for destroying infected cells—namely, combination with antibody and complement, which will destroy infected cells long before cells break down as a direct result of viral infection. This mechanism also, though no doubt generally very valuable as a defense against infection, may sometimes cause severe damage to the host, for it appears that the sometimes fatal hemorrhagic shock syndrome associated with dengue fever is caused by sudden increases in vascular permeability that may be triggered by the interaction of immune complexes with the complement and clotting systems.

Although virus-antibody complexes are usually eliminated from the body without difficulty either before or after combination with complement, they may cause diseases quite unrelated to those caused by viruses alone. This realization has come from studies of several virus infections in animals, particularly LCM and lactic dehydrogenase elevating virus (LDHV). Infection with both these viruses results in the presence in the bloodstream of large amounts of virus-antibody complexes; it is also characterized by the development of glomerulonephritis and the presence in kidney capillaries of large amounts of virus-antibody-complement complexes. Similar observations have been made with respect to Aleutian mink disease and equine infectious anemia, in which the inflammatory changes are not confined to the kidneys but also involve the blood vessels (with the development of arteritis) and other parts of the body. Clearly, human glomerulonephritis may also be caused by virus-antibody complexes. Further, it is known that some virus-antibody complexes, such as complexes between adenovirus and antibody to hexon (but not to penton or fiber), are cytotoxic even in the absence of complement, and although this has so far been demonstrated only in vitro, the possibility exists that such complexes also cause tissue damage in the body.

Finally, it is now suspected that autoimmune diseases, such as rheumatoid arthritis and lupus erythematosus, are also caused by the interaction of viruses and the immune mechanism.

FURTHER READING

Books and Reviews

Almeida JD, Waterson AP: The morphology of virus-antibody interactions. Adv Virus Res 15:307, 1969

Casals J: Immunological techniques for animal viruses. In Maramorosch K, Koprowski H (eds): Methods in Virology. Vol. 3 pp 113-198. New York and London, Academic Press, 1967

Della-Porta AJ, Westaway EG: A multi-hit model for the neutralization of animal viruses. J Gen Virol 38:1, 1977

Dent PB: Immunodepression by oncogenic viruses. Prog Med Virol 14:1, 1972

Howe C, Morgan C, Hsu KC: Recent virologic application of ferritin conjugates. Prog Med Virol 11:307, 1969

Notkins A (ed): Viral Immunology and Immunopathology. New York and London, Academic Press, 1976

Oldstone MBA: Virus neutralization and virus-induced immune complex disease. Prog Med Virol 19:85, 1975

Porter DD: Destruction of virus-infected cells by immunological mechanisms. Ann Rev Microbiol 25:283, 1971

Sommerville RG: Rapid diagnosis of viral infections by immunofluorescent staining of viral antigens in leukocytes and macrophages. Prog Med Virol 10:398, 1968

Svehag S: Formation and dissociation of virus-antibody complexes with special reference to the neutralization process. Prog Med Virol 10:1, 1968

Selected Papers

Babiuk LA, Acres SD, Rouse BT: Solid-phase radioimmunoassay for detecting bovine (neonatal calf diarrhea) rotavirus antibody. J Clin Microbiol 6:10, 1977

Bidwell DE, Buck AA, Diesfeld HJ, Enders B, Haworth J, Huldt G, Kent MJ, Kirsten C, Mattern P, Ruitenberg EJ, Voller A: The enzyme-linked immunosorbent assay (ELISA). Bull World Health Org 54:129, 1976

Brown F, Smale CJ: Demonstration of three specific sites on the surface of FMDV by antibody complexing. J Gen Virol 7:115, 1970

Engvall E, Perlmann P: Enzyme-linked immunosorbent assay (ELISA). Quantitative assay of immunoglobulin G. Immunochem 8:871, 1971

Fazekas de St Groth S, Webster RG: Disquisitions on original antigenic sin. J Exp Med 124:331, 347, 1966

Ghose LH, Schnagl RD, Holmes IH: Comparison of an enzyme-linked immunosorbent assay for quantitation of rotavirus antibodies with complement fixation in an epidemiological survey. J Clin Microbiol 8:268, 1978

Halstead SB, O'Rourke EJ: Antibody enhanced dengue virus infection in primate leukocytes. Nature 265:739, 1977

Kapikian AZ, Cline WL, Mebus CA, Wyatt RG, et al: New complement-fixation test for the human reovirus-like agent in infantile gastroenteritis. Lancet 1:1056, 1975

Nakane PK, Kawaoi A: Peroxidase-labeled antibody: A new method of conjugation. J Histochem Cytochem 22:1084, 1974

Norrby E: The relationship between the soluble antigen and the virion of adenovirus type 3. 4. Immunological complexity of soluble components. Virol 37:565, 1969

Oldstone MBA: Immune complexes in cancer: Demonstration of complexes in mice bearing neuroblastomas. J Natl Cancer Inst 54:223, 1975

Pettersson U: Structural proteins of adenoviruses. 6. On the antigenic determinants of the hexon. Virol 43:123, 1971

Radwan AL, Burger D: The complement-requiring neutralization of equine arteritis virus by late antisera. Virol 51:71, 1973

Rodda SJ, White DO: Cytotoxic macrophages: A rapid nonspecific response to viral infection. J Immunol 117:2067, 1976

Saunders GC, Clinard EH: Rapid micromethod of screening for antibodies to disease agents using the indirect enzyme-labeled antibody test. J Clin Microbiol 3:604, 1976

Stollar V: Immune lysis of Sindbis virus. Virol 66:620, 1975

Wallis C, Melnick JL: A persistent fraction of herpesvirus caused by insufficient antibody. Virol 42:128, 1970

Welsh RM Jr, Lampert PW, Burner PA, Oldstone

MBA: Antibody-complement interactions with purified lymphocytic choriomeningitis virus. Virol 73:59, 1976

Woodroofe GM, Fenner F: Serological relationships within the poxvirus group: An antigen common to all members of the group. Virol 16:334, 1962

Yoshiki T, Mellors RC, Strand M, August JT: The viral envelope glycoproteins of murine leukemia virus and the pathogenesis of immune complex glomerulonephritis of New Zealand mice. J Exp Med 140:1011, 1974

Zinkernagel RM, Oldstone MBA: Cells that express viral antigens but lack H-2 determinants are not lysed by immune thymus-derived lymphocytes but are lysed by other antiviral immune attack mechanisms. Proc Natl Acad Sci USA 72:3666, 1976

Zweerink HJ, Courtneidge SA, Skehel JJ, Crumpton MJ, Askonas BA: Cytotoxic T cells kill influenza-virus infected cells but do not distinguish between serologically distinct type A viruses. Nature 267:354, 1977

CHAPTER 5

The Virus Multiplication Cycle

Virions represent the static or inert form of viruses. The very existence of viruses is recognizable only in terms of their interaction with cells, which is the central theme of virology.

The interaction of virus and cell generates a novel entity, the virus-cell complex, the fate of which varies widely, since it depends both on the nature of the cell and on the nature of the virus. The two most commonly observed virus-cell interactions are (1) the lytic interaction, which results in virus multiplication and lysis of the host cell, and (2) the transforming interaction, which results in the integration of the viral genome into the host genome and the permanent transformation or alteration of the host cell with respect to morphology, growth habit, and the manner in which it interacts with cells with which it comes into contact.

In studying the virus-host interaction, one can focus primarily either on the fate and functioning of the invading virus particle and on the production of virus progeny or on the reaction of the host cell to virus infection. Both approaches are of fundamental importance to the medical practitioner. The former is particularly relevant to the development of a rational approach to antiviral chemotherapy, the latter to an understanding of chronic virus infection and cancer. In this chapter we will focus on the invading virus particle; in the next, on the response of the cell.

are the proteins for which it codes, and what are their functions? How are mature virus particles assembled? What is the fate of the 200 parental protein molecules? What effect do they have on the host cell? What are the reactions that cause the host cell to die?

It is impossible to answer these and many other questions by studies in the intact organism. Instead, simple experimental systems are required that can be manipulated at will. Such systems are provided by cloned strains of cultured animal cells that grow in vitro and can be infected under any desired set of conditions with pure (plaque-purified) strains of virus. Among their many advantages is the fact that they permit focusing on one multiplication cycle rather than on many repeated cycles, which is achieved by infecting all cells at the same time. In fact, one of the major conceptual breakthroughs in virology occurred when Ellis and Delbrück demonstrated, almost 40 years ago, how very much simpler the analysis of the one-step growth cycle is than that of numerous successive unsynchronized cycles. In populations of cultured cells infected at high multiplicity—that is, with many virus particles per cell—so as to ensure that infection commences at the same time in all cells, the various reactions that together comprise virus multiplication proceed synchronously according to a regulated progressive pattern that is amenable to study by the techniques of biochemistry, biophysics, and cellular and molecular biology.

The lytic virus-cell interaction

The lytic virus-cell interaction is best thought of in terms of a cycle, the infection or multiplication cycle, during which the virus enters cells, multiplies, and is released. This cycle is repeated many times when a virion infects an organism, until, for one reason or another, further multiplication is arrested, or the host dies.

One of the principal goals of virology is the definition in molecular terms of all the various reactions that proceed during the virus multiplication cycle. As an example, when a poliovirion infects a cell, one RNA molecule and about 200 protein molecules are introduced into it. How does this RNA molecule replicate? What

The one-step growth cycle: general aspects

The virus multiplication cycle can be divided into several phases, using events of critical importance as markers (Fig. 5-1). As we shall see, infectivity of virus particles is destroyed or eclipsed when they adsorb; the initial phase of the cycle is therefore often referred to as the eclipse period. This phase ends with the formation of the first mature progeny virus particle, which marks the beginning of the rise period. Alternatively, the synthesis of the first progeny genome is often taken to divide the multiplication cycle into the early and late periods. The

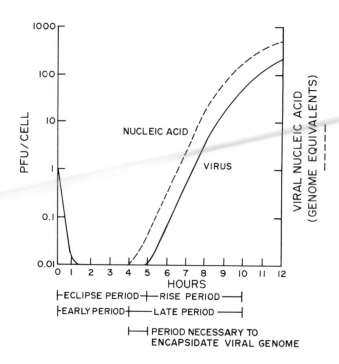

FIG. 5-1. The one-step growth cycle. Its essential features are: following adsorption, infectivity is abolished or "eclipsed"; this is caused by uncoating of the infecting virus particles. Then follows the eclipse or early period, which can last from a few minutes to many hours, during which the stage is set for viral nucleic acid replication. The appearance of the first progeny genome marks the beginning of the late period, and the beginning of the first mature progeny virus particle the beginning of the rise period. It should be noted that the interval between the beginning of the late and rise periods represents the average time necessary for virus maturation, that is, the incorporation of a free nucleic acid molecule into a mature virus particle. The lengths of all periods, as well as the extent of virus multiplication, vary greatly for different viruses and cells.

eclipse and early periods and the rise and late periods overlap substantially, with the interval between the beginning of the late and rise periods representing the time necessary for the incorporation of a viral genome into a mature virion.

THE ECLIPSE PERIOD

Adsorption

The first step of the virus-cell interaction is adsorption, which can itself be separated into several stages. The first of these is ionic attraction. Both cells and virus particles are negatively charged at pH 7, and positive ions are therefore required as counter-ions. As a rule this requirement is met most efficiently by magnesium ions. The second stage involves an accurate aligning of virus particles with the cell surface by virtue of interaction with specific receptor molecules, whose existence is particularly apparent in the case of enteroviruses. There are two lines of evidence. First, poliovirus adsorbs only to cells of human or primate origin. In fact, in the body, poliovirus adsorbs only to cells of the central nervous system and to cells lining the intestinal tract. Other human and primate cells develop the ability to adsorb

poliovirus only after being cultured in vitro, which causes unmasking of receptors. Second, the receptors for different enteroviruses, particularly for certain strains of ECHO virus and Coxsackie virus, can be either removed or inactivated differentially. Futhermore, virus particles that do not combine with the same species of receptor molecule can nevertheless, by steric hindrance, inhibit each other's ability to adsorb.

A further example of virus receptors is provided by the oligosaccharide groups of the glycoproteins of the outer cell membranes of mammalian cells, which function as the receptors for all ortho- and many paramyxoviruses, since their hemagglutinins possess strong affinity for them. It has also been reported that components of the histocompatibility complex function as receptors for alphaviruses.

The time course of virus adsorption follows first order kinetics. The rate of adsorption is independent of temperature if suitable corrections are made for changes in the viscosity of the medium, but the rate is directly proportional to the amount of surface to which virus can adsorb, that is, to the cell concentration. The kinetics of adsorption are described by the relation

$$\frac{V_t}{V_0} = e^{-Ktc}$$

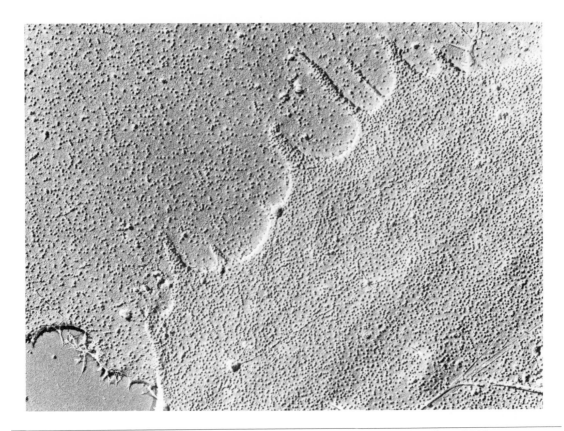

FIG. 5-2. Surface replica of Sindbis virus adsorbed to the surface of two chick cells. ×9,300. (From Birdwell and Strauss: J Virol 14:672, 1974.)

where V_0 and V_t are the concentrations of free virus at time 0 and after t minutes, respectively, c is the cell concentration, t is the time in minutes, and K is the adsorption rate constant.

The number of virus particles or infectious units adsorbed per cell is referred to as the "multiplicity of infection" (moi). Animal cells are generally capable of adsorbing very large amounts of virus; it has been shown, for example, that cells contain about 100,000 receptor sites for Sindbis virus (Fig. 5-2).

Penetration and Uncoating

The second stage of the virus multiplication cycle involves penetration and uncoating, which are considered together because although they are separated both temporally and spatially for some viruses, they occur simultaneously for others. Penetration, sometimes referred to as viropexis, concerns the entry into the cytoplasm of either the whole virus particle or that part of it that contains the genome. It may be observed directly by means of the electron microscope or indirectly by measuring the loss of ability of antiviral antiserum to arrest initiation of virus multiplication. The reason for this is that as long as virus particles remain outside the cell, combination with antibody significantly decreases their ability to cause productive infection; but once the particles are within the cell, they are no longer accessible to antibody.

Uncoating signifies the physical separation of viral nucleic acid from viral protein or, in the case of minus-stranded RNA viruses, the disruption of virus particles with resultant liberation of nucleocapsids. Double-stranded RNA-containing viruses also are not uncoated completely, but only to "subviral particles" (see below). Uncoating is of taxonomic significance, since viruses are the only intracellular infectious agents or parasites for which this is an obligatory step of the multiplication cycle.

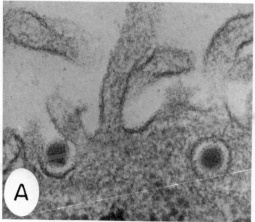
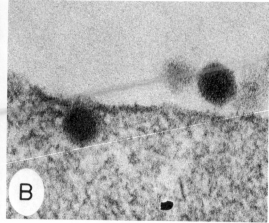

FIG. 5-3. The penetration and uncoating of adenovirus. Two alternative pathways are illustrated. **A.** Virus particles are engulfed by the cell membrane and enter the cytoplasm inside phagocytic vacuoles or vesicles which break down and release virus particles which have lost their pentons and fibers. Uncoating is completed near the nuclear pore complex and only viral DNA enters the nucleus. **B.** The virus particle passes directly across the cell membrane by a mechanism not yet understood. **C.** Diagrammatic representation of the uptake and uncoating of adenovirus particles. Once inside the cell, they migrate toward the nucleus, where the process of uncoating is completed. Free viral DNA appears only in the nucleus. (A, from Chardonnet and Dales: Virol 40:462, 1970; B, from Morgan, Rosenkranz, and Mednis: J Virol 4:777, 1969.)

Uncoating is best assessed by measuring physical and chemical changes in the adsorbed virus particles. Among these changes are progressive labilization of the capsid structure as judged by loss of its ability to shield the viral genome from hydrolysis by nucleases, development of susceptibility to reagents such as urea to which intact virus particles are resistant, loss of some antigenic determinants, and progressive loss of capsid protein.

There is no evidence for the involvement of enzymes in any of these changes; rather it seems that combination of virus particles with receptor molecules triggers conformational changes in the capsid that finally result in the liberation of the genome. The total time from adsorption to final uncoating ranges from several minutes to several hours.

The actual pathways of penetration and uncoating of the several different types of virus particles differ markedly—not surprising in view of the great diversity of virus structure.

Penetration and Uncoating of Virus Particles with Naked Icosahedral Nucleocapsids There is morphologic evidence for two quite distinct pathways of penetration. On the one hand, virus particles aligned with the outer surface of the cell seem to be engulfed by the cell membrane, which in essence flows over them. They are thus drawn into the cell within phagocytic vacuoles, which subsequently break down, liberating into the cytoplasm more or less intact virus particles, which are then uncoated either in the cytoplasm or in the nucleus, depending on the nature of the virus particle. Thus picornavirus particles, which multiply in the cytoplasm, are also uncoated in the cytoplasm, while adenovirus and papovavirus particles, both of which are assembled in the nucleus, are

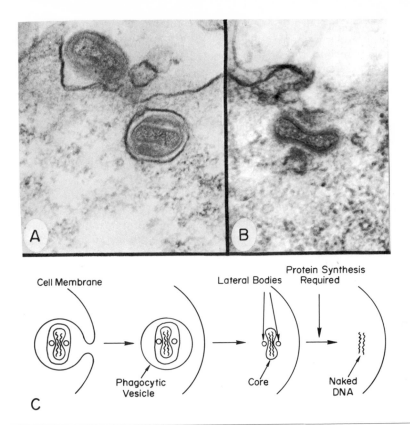

FIG. 5-4. The penetration and uncoating of vaccinia virus. Three stages are shown. **A.** One of the two virus particles is aligning itself with the cell membrane, which is preparing to flow around it. The other particle is already inside the cytoplasm, engulfed by a phagocytic vacuole. **B.** The phagocytic vacuole has broken down, as has the virus particle's outer protein coat. The core is now free, and the two lateral bodies have moved some distance away. The final stage of uncoating is the breakdown of the core, which results in the liberation of the DNA. This step is not achieved if protein synthesis is inhibited. This indicates that the synthesis of a special "uncoating protein" is required. Poxviruses are the only viruses which require the synthesis of a special protein for uncoating. The whole process is depicted diagrammatically in C. A and B, ×80,000. (A and B, from Dales: J Cell Biol 18:63, 1963; C, modified from Joklik: J Mol Biol 8:277, 1964.)

uncoated partly in the cytoplasm and partly in the nucleus, and exclusively in the nucleus, respectively. On the other hand, virus particles may be able to pass directly across the cell membrane without becoming engulfed in phagocytic vacuoles (Fig. 5-3).

Penetration and Uncoating Poxviruses Poxvirus particles have a highly specialized mechanism for penetration and uncoating. They enter cells by engulfment in phagocytic vacuoles and are broken down within them to cores that are liberated into the cytoplasm. The degradation of these cores, which results in the uncoating of the viral DNA, requires the synthesis of a special uncoating protein (Fig. 5-4).

Penetration and Uncoating of Enveloped Virus Particles Just as in the case of the naked icosahedral viruses, there seem to be two mechanisms by which enveloped viruses gain access to the interior of cells. The first involves fusion of the viral envelope with the outer cell membrane, thereby liberating the naked nucleocapsid into the cytoplasm, where the RNA is then either fully uncoated (the plus-stranded RNA-containing viruses) or transcribed (the minus-stranded RNA-containing viruses). There are several lines of evidence for this pathway, including direct morphologic evidence and the fact that cells infected with such viruses are killed by antiviral antibody plus complement (since fusion of the viral envelope with host cell membrane causes viral antigen to be incorporated into it). Paramyxoviruses enter cells by this mechanism. The second mechanism, for which there is also good morphologic evidence, is identical with that described above—namely, engulfment of the entire virus particle into phagocytic vacuoles and liberation of the nucleocapsid within the cytoplasm, rather than at the outer cell membrane. Rhabdoviruses and togaviruses are known to enter cells by this mechanism.

Eclipse

Adsorption, penetration, and uncoating result in loss of infectivity, which is referred to as "eclipse." The only residual infectivity is that due to the viral nucleic acid itself (or to the nucleocapsid, as the case may be), which, however, is never more than a small fraction of that of the virus particles themselves.

The first three stages of infection are usually inefficient processes. Virus particles often adsorb to portions of the cell surface at which penetration will not proceed, viral genomes may be damaged by ribonuclease which is frequently associated with outer (or plasma) cell membranes, and virus particles may fail to be released from the phagocytic vacuoles in which they have become engulfed. All these inefficiencies account in large part for the fact that the ratio of infectious to total animal virus particles is almost always far less than 1 (Chap. 1).

THE SYNTHETIC PHASE OF THE VIRUS MULTIPLICATION CYCLE

Once the viral genome is uncoated, the synthetic phase of the virus growth cycle commences. In essence, this encompasses, in a precisely regulated program, the replication of the viral genome, the synthesis of viral proteins, and the formation of progeny virus particles.

The location of viral genome replication is characteristic for each virus (Table 5-1). There is no correlation between this location and any other property, such as chemical nature or size of genome. Viral protein is always synthesized in the cytoplasm on polyribosomes composed of viral messenger RNA, host cell ribosomes, and host cell transfer RNA. In the case of RNA-

containing viruses most, if not all, their genetic information is expressed soon after uncoating. In the case of the double-stranded DNA-containing viruses, however, the multiplication cycle can be divided into clearly defined early and late periods, with the onset of viral DNA replication marking the beginning of the late period (see Fig. 5-1).

The Early Period

The early period of the synthetic phase is devoted primarily to the activation of reactions that are prerequisite for the initiation of viral genome replication. This activation proceeds as the result of the viruses exercising certain early functions. Among them are: (1) the inhibition of host DNA, RNA, and protein synthesis—which may involve the synthesis of virus-coded proteins that either inhibit or alter the specificities of the DNA-replicating, RNA-transcribing, and polypeptide-synthesizing systems—so that viral rather than host cell genetic information is processed; (2) the synthesis of proteins that form the matrix of inclusions, either within the nucleus or within the cytoplasm, within which viral nucleic acids replicate and viral morphogenesis proceeds; and (3) the synthesis of certain enzymes, primarily DNA and RNA polymerases.

The extent to which early functions are expressed varies greatly from virus to virus. Some viruses possess so little genetic information that only very few early functions are expressed; others possess so much that they may express from 30 to 50 early functions. Early functions are expressed through (early) virus-coded proteins that are transcribed from early viral messenger RNA species. Those viral genomes that are plus-stranded RNA themselves serve as messenger RNA. For all other viruses, plus-stranded messenger RNA must first be transcribed from infecting parental genomes by means of polymerases either associated with them or pre-existing in the host cell.

The Late Period

During the late period, the late viral functions are expressed. The late viral proteins are primarily the components of progeny virus particles and the enzymes and other nonstructural proteins that function during viral morphogen-

**TABLE 5-1   THE LOCATION OF VIRAL GENOME REPLICATION,
NUCLEOCAPSID FORMATION, AND VIRION MATURATION**

Virus	Genome Replication	Nucleocapsid Formation	Virion Maturation
Poxviruses	Cytoplasm	Cytoplasm	Cytoplasm
Herpesviruses	Nucleus	Nucleus	At nuclear and cytoplasmic membranes
Adenoviruses	Nucleus	Nucleus	Nucleus
Papovaviruses	Nucleus	Nucleus	Nucleus
Picornaviruses	Cytoplasm	Cytoplasm	Cytoplasm
Togaviruses	Cytoplasm	Cytoplasm	At membranes
Bunyaviruses	Cytoplasm	Cytoplasm	At membranes
Reoviruses	Cytoplasm	Cytoplasm	Cytoplasm
Orthomyxoviruses	Nucleus	Nucleus (?)	At membranes
Paramyxoviruses	Cytoplasm	Cytoplasm	At membranes
Rhabdoviruses	Cytoplasm	Cytoplasm	At membranes
RNA tumor viruses	Nucleus	Cytoplasm	At membranes
Arenaviruses	Cytoplasm	Cytoplasm	At membranes
Coronaviruses	Cytoplasm	Cytoplasm	At membranes

esis; they are encoded by late viral messenger RNA molecules that are transcribed from different sequences of the viral genome than early ones. Further, late messenger RNA molecules are transcribed from progeny genomes; and since there are always more progeny than parental genomes, far more late than early messenger RNA is always formed. The amount of late proteins that is synthesized therefore always greatly exceeds that of early ones. Activation of the portions of the viral genome that code for late functions may or may not be accompanied by deactivation of the portions that code for early functions. In either case, a mechanism exists that specifies that one set of genes (the early set) is transcribed from parental genomes, while a different set (the late set) is transcribed only from progeny genomes. The basis of this mechanism probably lies in the specificity of the enzyme(s) that transcribes DNA. Indeed, work with certain bacteriophages has indicated that the host-specified DNA-dependent RNA polymerase is modified early during infection to a form that can transcribe those sites on the phage genome that code for early functions, and that at the beginning of the late period its specificity is altered again so as to enable it to respond to those signals that specify late functions (Chap. 10). Similar mechanisms may also operate in the case of animal viruses.

During the late period, the newly formed virus genomes and capsid polypeptides are assembled into progeny virus particles, a process that is known as morphogenesis. This is a spontaneously occurring process, since most of the information for virus assembly resides in the amino acid sequences of the capsid polypeptides. Nucleic acid performs no essential function during morphogenesis, a fact demonstrated by the occurrence among the yield of most icosahedral viruses of empty virus particles—that is, virus particles that contain no nucleic acid—that are morphologically indistinguishable from mature virions.

The duration of the late period is generally limited by the ability of the host cell to supply energy for macromolecular synthesis. This is a critical factor, since infection with lytic viruses invariably interferes with the functioning of the host cell by multiple mechanisms that are discussed in Chapter 6. As a result, synthesis of viral nucleic acids and proteins slows down progressively, thereby limiting the amount of viral progeny.

THE RELEASE OF PROGENY VIRUS

The final step of the infection cycle is the release of progeny virus. There is no special mechanism for the release of unenveloped viruses and poxviruses; infected cells simply disintegrate more or less rapidly, liberating the viral progeny that has accumulated within them. The amount of cell-associated virus therefore exceeds the amount of released virus until the very last phase of the multiplication

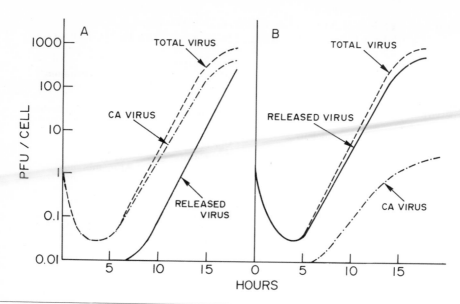

FIG. 5-5. The relationship between virus multiplication and release. **A.** This graph refers to viruses with icosahedral nucleocapsids. Such viruses are not released readily from cells. Viral progeny accumulates within cells, so that for much of the rise period the amount of cell-associated (CA) virus greatly exceeds the amount of released virus. Release of virus occurs only when cells break down at the end of the rise period. **B.** This graph refers to all enveloped viruses. Such viruses mature only in the process of being released, since it is only then that they acquire their envelope. The amount of liberated virus therefore always greatly exceeds the amount of cell-associated virus. The only cell-associated virus particles are those in the process of budding from the plasma membrane and those that bud into intracytoplasmic vacuoles.

cycle (Fig. 5-5). A special mechanism does, however, exist for the enveloped viruses, for which release is the final stage of morphogenesis. Here virus-coded envelope proteins are incorporated into certain areas of host cell membranes while nucleocapsids are being synthesized. The nucleocapsids then bud through these modified membrane patches and become enveloped by them (p. 111). Budding occurs both at the outer plasma cell membrane and at the membranes lining intracytoplasmic vacuoles, which then transport the virus to the exterior of the cell. These viruses do not exist in the mature infectious form within cells, and the amount of extracellular virus, therefore, greatly exceeds the amount of cell-associated virus at all stages of the multiplication cycle (Fig. 5-5).

The duration of the phases of the virus multiplication cycle varies greatly, depending on the nature of the virus and the nature of the host

TABLE 5-2   APPROXIMATE DURATION (HOURS) OF THE ECLIPSE PERIOD AND OF THE ENTIRE MULTIPLICATON CYCLE

Virus	Eclipse Period	Total Multiplication Cycle
Poxvirus: vaccinia virus	4	24
Herpesvirus: herpes simplex virus	3-5	12-30
Adenovirus	8-10	48
Papovavirus: polyoma virus	12-14	48
Poliovirus	1-2	6-8
Togavirus: Sindbis virus	2	10
Reovirus	4	15
Orthomyxovirus: influenza virus	3-5	18-36
Rhabdovirus: vesicular stomatitis virus	2	8-10

cell. Table 5-2 lists the minimum lengths of the eclipse periods and the complete multiplication cycles of some well-studied viruses.

The multiplication cycles of several important viruses

Of the many facets of virus multiplication, the two that are central are 1) the nature of the information encoded in viral genomes and 2) the manner in which this information is expressed. Not only does description of a virus in this manner define it in its most fundamental terms, but it also provides the framework of knowledge that is essential for a rational approach to antiviral chemotherapy. There follows a discussion of the strategy used by several important and intensively studied viruses to encode and express their information content.

THE MULTIPLICATION CYCLES OF DOUBLE-STRANDED DNA-CONTAINING VIRUSES

Knowledge concerning the multiplication of double-stranded DNA-containing viruses has expanded greatly in recent years owing to the advent of techniques for "mapping" their genomes by the use of bacterial restriction endonucleases. These enzymes, which are components of the restriction-modification systems in bacteria, cleave the DNA of animal viruses (and, indeed, any DNA that is not modified by its own specific methylase) at palindromic sequences that are highly specific for each enzyme. These sequences, which possess twofold rotational symmetry, may be as simple as

$$5' \text{ GG CC}$$
$$\text{CC GG } 5'$$

or as complex as

$$(A/T) \text{ G AA TT C } (T/A)$$
$$(T/A) \text{ C TT AA G } (A/T)$$

Clearly, the simpler the recognition site, the more often the DNA will be cut. The fragments that result may be ordered or "mapped" by digesting DNA under conditions when it is only partially cleaved, and analyzing the incompletely hydrolyzed pieces for neighboring fragments. Fig. 5-6 shows the 6 HindIII restriction endonuclease fragment profile of SV40 DNA, and Fig. 5-7 shows the order in which they are arranged, that is, the HindIII map. Several other restriction endonuclease maps of SV40 DNA are also shown. Given such maps—and more than 50 restriction endonucleases are now available—this type of analysis is clearly capable of defining precisely the position in the viral genome of even short DNA sequences. For example, one can now determine to which endonuclease fragment the messenger RNA that is transcribed at any time during the multi-

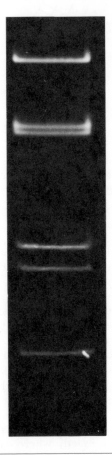

FIG. 5-6. Autoradiogram of a polyacrylamide gel in which there have been electrophoresed the six fragments that result when circular ^{32}P-labelled SV40 DNA is digested with restriction endonuclease Hind III. (Courtesy of Drs. J. K. Li and C. Huang.)

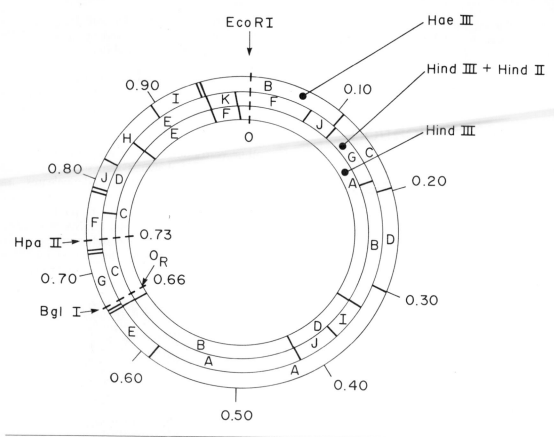

FIG. 5-7.   Restriction endonuclease cleavage maps of SV40 DNA, using Hind III, Hind III +
Hind II, and Hae III. EcoRI, Bgl I and Hpa II cleave only once. The origin of DNA replication is
at a large palindromic sequence, part of which forms the recognition sequence for Bgl I.

plication cycle will hybridize, which can provide information concerning when and how
frequently each portion of the virus genome is
transcribed during the multiplication cycle.
This technique is known as physical mapping,
since here the viral DNA is treated merely as a
sequence of nucleic acid bases. Physical maps
can be correlated with genetic maps if such are
available (see p. 92 and Chap. 7). In that case it
becomes possible to determine the order of the
various genes that comprise the viral genome
and the extent to which the genes are expressed. The ultimate aim of this type of analysis is to define exactly where each gene is located in the viral genome, when it is
transcribed during the virus multiplication cycle, how frequently it is transcribed and, finally, what controls or regulates the transcription program.

PAPOVAVIRUSES

Papovaviruses are the smallest double-stranded
DNA containing viruses. While practically
nothing is known of the multiplication cycles of
papilloma viruses (since they do not grow in
cultured cells), a great deal is known about how
polyoma virus and SV40 multiply. It should be
noted that several human viruses, namely BK
virus, JC virus and DAR virus (see Chap. 2), are
more or less closely related to SV40.

When these viruses infect cells of their natural hosts, viral DNA replicates, viral proteins
are made, progeny virus particles are formed
and the cells lyse; in other words, they initiate
productive infection. These host cells are
known as permissive cells. By contrast, cells of
other species are more or less nonpermissive—
that is, they do not support efficient multiplica-

tion; instead, the viral DNA becomes integrated into that of their host, thereby creating new, genetically different cells, namely virus-transformed cells. Such cells form tumors in animals. The transforming papovavirus-host cell interaction is discussed in Chap. 9; here the nature of their lytic multiplication cycles is considered.

The basic feature of the multiplication cycle of polyoma virus and SV40, like that of all double-stranded DNA containing viruses, is that it can be divided into well-defined early and late phases (see Fig. 5-8). During each phase, different portions of the viral genome are transcribed into messenger RNAs, which are then processed and translated into proteins by the cellular protein-synthesizing system, sometimes modified to translate viral messenger RNAs in preference to host messenger RNAs (see below).

The Early Period

The multiplication cycle of SV40 is best understood by reference to the genetic and transcription maps of its genome, a circular molecule of 5,226 nucleotide base pairs, which has been completely sequenced (Fig. 5-8). During the early phase, which lasts a surprisingly long time (14 to 18 hours), about 50% of the viral DNA, namely the region that extends from map position 0.65 (relative to the EcoR1 restriction endonuclease cleavage site) to map position 0.17, is transcribed into RNA. This region comprises a single gene, known as the A gene; however the transcripts of this region are processed not into one, but into three species of messenger RNA. The first is about 2,230 nucleotides long and is composed of two parts, about 330 and 1,900 nucleotides respectively, which map from map position 0.65 to 0.60 and from 0.54 to 0.17. RNA corresponding to the DNA between map positions 0.60 and 0.54 is not present in this messenger RNA; this region is "spliced" out of the original transcript by a mechanism that is not yet clear but that may well proceed by a scheme such as that depicted in Fig. 5-9. Splicing-out of regions of RNA transcripts is a recently discovered but common phenomenon even among transcripts of cellular genes; numerous cellular messenger RNAs are now known to comprise sequences from widely separated regions of DNA. A well-known example

is the messenger RNA of the β-globin gene of the mouse. This messenger RNA is derived from the transcript of a DNA sequence that is about 1,700 base pairs long, which is processed so that two sequences of 780 and 125 nucleotides respectively are spliced out. Thus the β-globin messenger RNA consists of three sequences of about 480, 205, and 155 nucleotides respectively, that are covalently joined but were originally separated. Splicing-out is a very widespread feature of the mechanism by which RNA transcripts of eukaryotic DNA are processed to messenger RNAs; other features of this mechanism are capping of their 5'-termini and polyadenylation of their 3'-termini.

The second species of early SV40 messenger is 2500 nucleotides long and is also composed of two parts, 630 and 1900 nucleotides long respectively, that map from map position 0.65 to 0.55 and from 0.54 to 0.17, respectively.

These two messenger RNAs code for the following polypeptides. First, the 2200 nucleotide-long messenger RNA codes for the T antigen, a protein with a molecular weight of about 95,000, that is located in the nucleus, binds to SV40 DNA near the origin of replication (see below) and may function in initiating or facilitating the initiation of its replication. The T antigen may also be responsible for inducing the synthesis of substantial amounts of cellular DNA (generally it replicates at least once) and of several enzymes involved in DNA synthesis during the initial stages of the multiplication cycle. These must be cellular enzymes since SV40 DNA is too small to code for them. Finally, the T antigen is also responsible for the maintenance of the transformed state when SV40 transforms cells (see Chap. 9).

The second species of early messenger RNA, that which is 2500 nucleotides long, codes for a protein that is known as the small t antigen, whose molecular weight is about 17,000 and which is essential for initiating the transformed cell state. Whereas the T antigen is coded by the sequences that extend from map position 0.65 to 0.60 and from 0.54 to 0.17, small t antigen is coded by the sequence from 0.65 to 0.55. T and t antigens therefore share amino acid sequences and they cross-react immunologically. The reason small t antigen is so short is that there are several termination codons near map position 0.55. These codons are spliced out in the messenger RNA species that codes for T antigen.

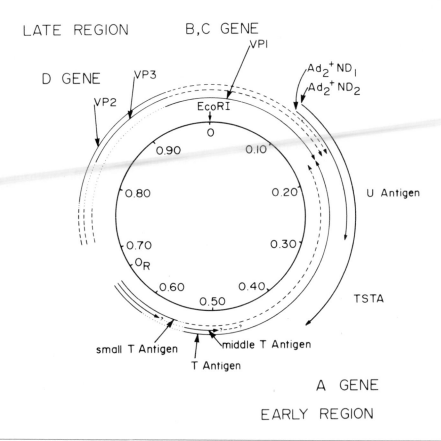

FIG. 5-8. The genetic and transcription map of SV40 DNA. It is located so that the restriction endonuclease EcoRI cleavage site is at 12 o'clock. O_R is the origin of replication of SV40 DNA. The coding sequences of messenger RNAs are indicated by solid lines, noncoding sequences by broken ones. Sequences spliced out during the processing of transcripts to messenger RNAs are indicated by dotted lines. The sequences present in the nondefective adenovirus 2-SV40 hybrids ND_1 and ND_2 are indicated by a thickened line, like SV40 DNA itself.

In polyoma virus-infected cells there is yet a third protein, middle t antigen, which shares antigenic determinants with the T antigen and is associated with the plasma cell membrane. Its molecular weight is about 55,000, and while it has the same amino terminal sequence as T and small t antigens, it also contains sequences that are unique to it. If these sequences are also coded by a portion of the A gene, which is likely, they are probably read from the sequences between map positions 0.54 and 0.17, but in a different reading frame from that used for T antigen.

Cells infected with SV40 also contain a protein, the U antigen (M.W. 28,000), which corresponds to the C-terminal portion of the T antigen. Its existence was first detected in cells infected with virus particles that are SV40-adenovirus hybrids (Ad2+ND₁, see Chap. 7) and that only contain SV40 DNA corresponding to map positions 0.28 to 0.11, so that the sequences between map positions 0.28 and 0.17 must therefore contain the U antigen code; but, as pointed out above, it is also present in cells infected with intact SV40 particles.

Finally, yet another protein is coded by the DNA sequences that comprise the A gene; it is the tumor-specific transplantation antigen, or TSTA, which is responsible for the induction of specific immunity against SV40-transformed tumor cells. It is located partly in the nucleus, but mostly in the plasma membrane. It also shares antigenic determinants with the T antigen, but the exact molecular relationship be-

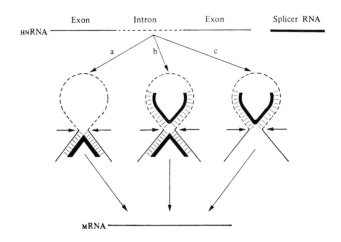

FIG. 5-9. Model for splicing out intervening sequences (introns) from RNA transcripts. Short RNA molecules (splicer molecules) are postulated to hybridize to exons (**a**) or to introns as well as to exons (**b**), or to introns **c**), in such a way as to bring the ends of introns into close juxtaposition; the sequences are then cut and ligated (spliced) (arrows) so as to produce messenger RNA. Recent evidence indicates that the sequences of splice junctions in different transcripts resemble each other, so that relatively small numbers of splicer RNAs could process many RNA transcript species. The splicer RNA function may be fulfilled by the "small nuclear RNAs (sn RNAs)", which are 90-220 nucleotides long and exist in the nuclei of eukaryotic cells. (From Murray and Holliday; Genet. Res., Camb. 34:173, 1979.)

tween these two proteins has not yet been defined. Its coding sequences are between map positions 0.39 and 0.17, because it is formed not only in cells transformed by wild type SV40 but also in cells infected with an SV40-DNA hybrid virus, Ad2+ND₂ (see Chap. 7), which contains only SV40 sequences between map positions 0.39 and 0.11.

The Late Period

During the late period the A gene continues to be transcribed, and in addition the remainder of the SV40 genome is also transcribed. This portion contains three genes, known as the B, C and D genes, which overlap as shown in Fig. 65-8; it is transcribed in the opposite direction, that is, from the opposite strand, as the A gene. Its transcripts are processed into three messenger RNAs that correspond to map positions 0.76 to 0.17, 0.83 to 0.17, and 0.95 to 0.17 respectively. In addition, all three messenger RNAs possess "leader" sequences that are about 120 nucleotides long and that are transcribed from the region between map positions 0.72 to 0.76. (The region that is spliced out in first messenger RNA species is very short, and

except for it, this messenger RNA species is a transcript of the entire late region of the SV40 genome.) These messenger RNAs are transcribed into three proteins. The map position 0.95-0.17 RNA species, which is the most abundant, is transcribed into VP1, a 46,000 dalton protein that is the major component of the SV40 capsid. The map position 0.83 to 0.17 messenger RNA species is translated only for part of its length—namely, the portion that corresponds to the region from map position 0.83 to about 0.97; this portion yields VP3, a 25,000 dalton protein that is a minor capsid component. Finally, the map position 0.76 to 0.17 RNA species also is translated only for part of its length—namely, the portion that corresponds to the region between map positions 0.76 and 0.97—and its translation product is VP2, a 35,000 dalton protein that is also a minor capsid component. The fascinating feature of this mode of gene expression is that the gene for VP3 lies entirely within the gene for VP2; their messenger RNAs are read in the same phase, and translation of both is terminated by the same termination codon which is a long way from the 3'-terminus of their messenger RNAs. Further, the initiation signal for VP1 lies about 120 nucleotides within the genes for VP2 and VP3, and VP1 is read in a different phase from VP2 and VP3.

The Replication of SV40 DNA

SV40 DNA, a circular molecule with 26 negative superhelical turns, exists in the capsid as a tightly coiled "chromatin" complex that also contains host cell histones H1, H3, H2A, H2B and H4. Upon removal of H1, this complex unfolds to yield 24 nucleosomes in the familiar "beads-on-a-string" conformation. The nucleosomes consist of about 140 base pairs each, coiled around the four remaining histones, and are connected to each other by DNA linkers some 30 base pairs long (Fig. 5-10).

The replication of SV40 DNA, in which host DNA polymerase α appears to play a key role, starts at map position 0.67, the origin of replication, where there is a remarkable palindromic sequence 25 base pairs long. Replication proceeds bidirectionally until the two forks meet

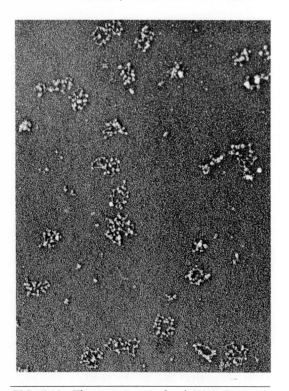

FIG. 5-10. Electronmicrograph of SV40 DNA nucleoprotein complex in the chromatinlike "beads-on-a-string" nucleosome conformation. ×50,000 (Courtesy of Drs W. Keller and U. Mueller: Science 201:406, 1978)

again, one DNA strand being synthesized in a continuous manner, the other in the form of Okazaki fragments that are then joined.

General Observations on the Strategy of Papovavirus Gene Expression

The preceding discussion of the multiplication cycle of SV40 (and indeed, almost as much is known about the multiplication cycle of polyoma virus) highlights the remarkable strategy according to which this virus expresses the information encoded in its genome. Its most astonishing feature is the fact that many of its coding sequences are used twice and even three times. Thus the early region of its DNA codes for no less than four or five proteins, the coding regions of most of which overlap; and the same is true of the late region, which codes for three proteins. One obvious advantage of this strategy is that it minimizes the amount of DNA that is necessary to encode all these proteins. It is also noteworthy that the functions of the early and late proteins are quite different. All early proteins have regulatory functions, and at least one, the T antigen, appears to act in a pleiotropic manner—that is, it seems to be multifunctional. By contrast, the late proteins are the structural components of the virus particle. Finally, it is remarkable that the DNA of this virus exists in the form of a complex that has many of the properties of cellular chromatin, and that it uses five cellular histones for this purpose.

ADENOVIRUSES

The size of adenovirus DNA (about 23×10^6 daltons) is 6 to 7 times that of SV40 and polyoma virus DNA; its transcription program is correspondingly more complex. However, the techniques that are now available are so powerful that spectacular progress has been made in analyzing it.

Some 17 virus-coded polypeptides can be detected in infected cells: they probably account for about 50% of the adenovirus coding potential. Six of these proteins—including a 72,000 dalton DNA-binding protein that appears to control the extent of early transcription—are transcribed during the early phase of the multi-

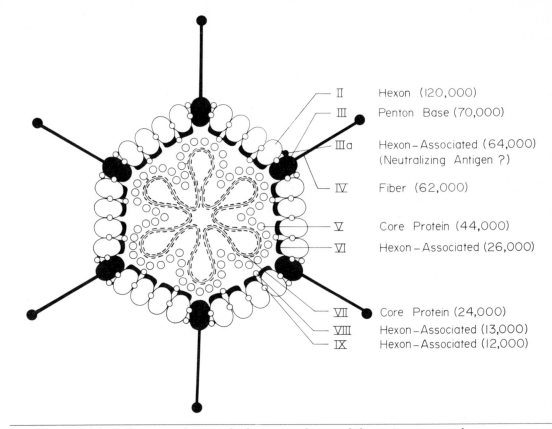

FIG. 5-11. Model of adenovirus showing the location and sizes of the various structural proteins. (Adapted from Brown DT et al: J Virol 16:366, 1975.)

plication cycle, the remainder during the late phase. Nine of the latter group are virion components (Fig. 5-11), and one is a nonstructural protein that is associated with polyribosomes.

The location of the genes for many of these proteins (Fig. 5-12) has been determined by a variety of techniques, among which are the following three. The first is straightforward ge-

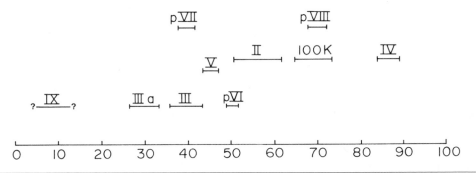

FIG. 5-12. Location of the genes for the structural proteins of adenovirus. All are late proteins. The 100K protein is a late nonstructural protein of unknown function. The prefix p denotes that the protein is synthesized in the form of a slightly larger precursor. The gene coding polypeptide IX is only about 500 nucleotide base pairs long; its precise location is not known with certainty. Note overlapping coding sequences.

netic recombination analysis: the relative order and location of genes can be specified by measuring the frequencies with which pairs of mutants in the various genes recombine with one another to yield wild type virus (see Chap. 7). If temperature-sensitive mutants are used, recombinants can easily be selected at the nonpermissive temperature (see Chap. 7). The other two gene-location techniques take advantage of the fact that completely ordered restriction endonuclease cleavage maps are now available for the DNAs of adenovirus strains of several serotypes. In one method, genes are mapped by crossing pairs of mutants (such as temperature-sensitive mutants; see Chap. 7) in different genes of virus strains that are related (for example, that belong to different serotypes), and characterizing the DNA of wild-type recombinants by restriction endonuclease analysis. The principle of the technique is illustrated in Fig. 5-13.

The other method is the most direct. Here messenger RNAs from infected cells are hybridized to separated restriction endonuclease fragments. They are then dissociated from the DNA, translated in cell-free protein synthesizing systems, and the proteins that are formed are identified by reference to authentic samples of adenovirus-coded proteins. Since the position on the viral genome of each restriction endonuclease cleavage fragment is known, the

location of the gene coding for the protein that is translated is also known. The location of several genes on the adenovirus genome is shown in Fig. 5-12.

The Adenovirus Transcription Program

Two transcription programs control the expression of adenovirus genes, one during the early phase, the other during the late phase. When messenger RNA isolated from polyribosomes *before* the beginning of DNA replication is hybridized to restriction endonuclease fragments, it is found to be transcribed from four rather short, widely separated regions of the genomes, two on each strand (Fig. 5-14). By contrast, messenger RNA species isolated *after* the beginning of DNA replication hybridize to almost the entire genome, and most are transcribed from one strand. As may be expected, not all regions of the DNA are transcribed into messenger RNA with equal frequency. Figure 5-15 shows the relative rates of accumulation of cytoplasmic messenger RNAs corresponding to various regions of the adenovirus genome.

So far we have considered only the messenger RNA species. Just as in the case of SV40 described above, these messenger RNAs are derived from primary transcripts that are

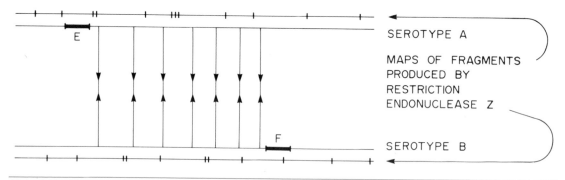

FIG. 5-13. Mapping of genes by restriction endonuclease analysis of recombinant DNA. Two temperature-sensitive mutants (see Chap. 7) are taken, one in gene E of serotype A, and one in gene F of serotype B. (Note that there are 31 serotypes of human adenovirus; see Table 2-4.) Wild-type recombinants (that is, recombinants able to grow at the normal temperature) are then isolated and their restriction endonuclease cleavage pattern profiles compared with those of the two mutants. Restriction endonuclease Z is used here as an example. Note that all recombination events must occur in the sequence between genes E and F. Clearly the DNAs of recombinants will have profiles partially characteristic of one parent and partially of the other. Fixing the distance between the two genes is thus a matter of defining the limits of the DNA segment between them; the more recombinants that are examined, and the more restriction endonucleases that are used, the more accurately can this distance be determined.

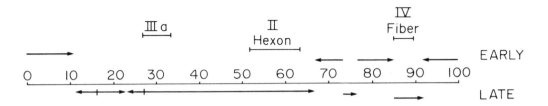

FIG. 5-14. Transcription map of adenovirus DNA. About one-third is transcribed during the early period and may be transcribed during the late period; two-thirds are transcribed only in the late period. Some sequences are transcribed from one strand, others from the other. The locations of the IIIa, II (hexon) and IV (fiber) genes are indicated as reference markers.

processed to messenger RNAs. The four regions of the genome that yield early messenger RNAs form independent transcription units. This is documented most directly by measuring their relative resistance to UV-irradiation. Cells infected with adenovirus are irradiated with increasing doses of ultraviolet light, and the amount of each early messenger RNA that accumulates is measured. It is found that the sensitivity to UV irradiation of the transcription of the four early regions is in each case proportional to its physical size. The situation is quite different for the late regions (or at least that part of them that is transcribed from the plus-strand). Here the sensitivity of accumulation of each messenger RNA species is not proportional to its size, but rather to the distance of its template from the left end of the whole transcription region (see Fig. 5-14). This and other evidence indicates that the entire late region is transcribed (in the nucleus) into one huge transcript, some 25,000 base pairs long, which is then processed, also in the nucleus, to at least 12 species of messenger RNA. These 12 species fall into five groups, each consisting of several species that possess different 5'-termini but the same 3'-terminus (Fig. 5-16). This fascinating processing pattern raises a host of questions concerning the mechanisms that specify the cleavage sites and that cause transcripts to be processed more often into some messenger RNAs than into others (since, for example, hexon messenger RNA is formed much more frequently than any other; see Fig. 5-15).

Just like late SV40 virus messenger RNAs, adenovirus messenger RNAs possess multipartite leader sequences; in fact, leader sequences were first discovered in adenovirus messenger RNAs. The basic discovery was that a single messenger RNA species is capable of hybridizing to several widely separated regions of the

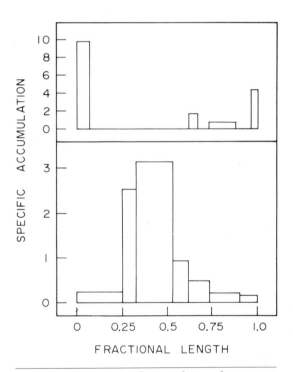

FIG. 5-15. Frequency of accumulation of messenger RNAs transcribed from various regions of the adenovirus genome during the early and late periods of the multiplication cycle. Cytoplasmic RNA, labeled with [³H]-uridine, was hybridized to the BamHI restriction endonuclease fragments (whose order in the adenovirus genome is known), and the relative amount of RNA hybridized to each fragment was determined. Note that we are dealing here with cytoplasmic RNA, presumably messenger RNA, not transcripts. Upper panel, early RNA (5.5 to 7.5 h after infection); lower panel, late RNA (24 to 26 h after infection). (From Smiley and Mak: J Virol 28:227, 1978.)

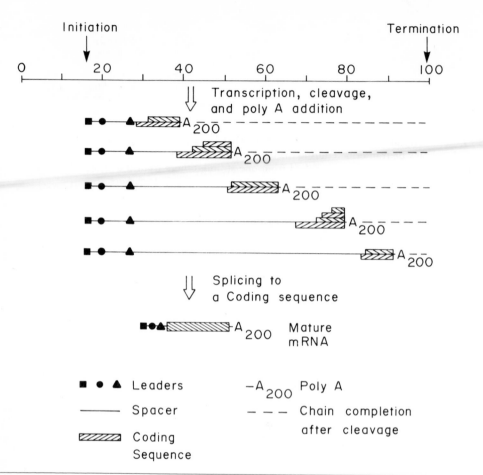

FIG. 5-16. Diagram depicting the proposed sequence of events in the processing of the transcript of the major late adenovirus transcription unit into messenger RNA molecules. The transcription unit extends from map position 16 to 99 and is transcribed from left to right. Each transcript is processed to one of 13 species of messenger RNA, probably as soon as the RNA polymerase has passed the right end of each coding sequence; but transcription always appears to go to completion (that is, to map position 99), irrespective of where each processed messenger RNA is destined to code. Five families of messenger RNAs are made, each with identical 3'-termini. Processing involves (a) splicing out of the two sequences between the leader sequences; (b) splicing out of the spacer sequence between the right-most leader and the 5'-terminus of the coding sequence; (c) capping of the 5'-terminus of the left-most leader sequence; (d) cleavage of the bond between the 3'-terminus of the coding sequence and the downstream portion of the transcript in the process of being completed; and (e) polyadenylation of the 3'-terminus of the messenger RNA. (From Nevins and Darnell: Cell 15:1477, 1978.)

adenovirus genome. One of the techniques for showing this was the so-called R-loop technique in which adenovirus DNA and RNA transcripts are mixed and partially melted; the DNA-DNA strands dissociate and more stable RNA:DNA hybrids are then formed instead. When these hybrids are examined with the electron microscope, the regions where RNA and DNA have hybridized are seen as loops (R-loops), one arm of which appears single-stranded (a DNA strand) and the other double-stranded (the RNA-DNA hybrid). This technique is extremely powerful for defining the precise locations of transcribed regions (since not only the length of R-loops but also their distance from the ends of DNA molecules can be measured very precisely; Fig. 5-17).

Using this technique it has been shown that

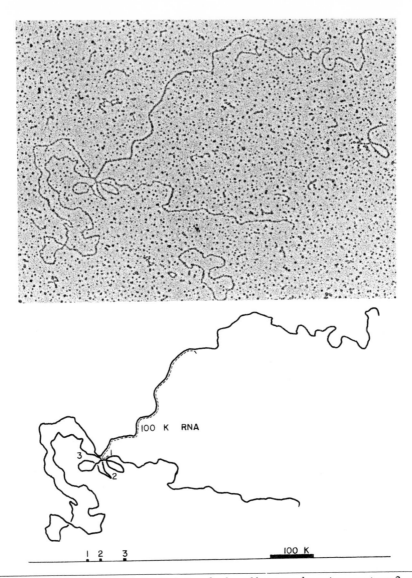

Fig. 5-17.   R loops.   An electron micrograph of a heteroduplex of human adenovirus serotype 2 (Ad2) RNA and single-stranded Ad2 DNA. The RNA was extracted from HeLa cells at late times after infection. The DNA is constraind into a series of deletion loops corresponding to the intervening sequences removed from the spliced RNA. The first, second, and third leader segments and the main coding body of the messenger RNA are readily visible. (Courtesy of Drs. L. T. Chow and T. R. Broker. Cell 15: 497-510, 1978)

most late adenovirus messenger RNA molecules begin with a leader sequence that is derived from three map positions, 16.6, 19.6 and 26.6, the total length of the sequence being about 200 nucleotides (Fig. 5-18). Presumably the original transcripts encompass not only the leader sequences and the coding sequences, but also those that lie between, and the latter are then spliced out.

Adenovirus DNA Replication

Adenovirus DNA, which is about 35,000 base pairs long, possesses terminally repeated inverted sequences 100-140 nucleotides long; further, the 5'-terminus of each strand is linked covalently to a 55,000 dalton protein molecule. Adenovirus DNA replication commences about 10 hours after infection and is probably effected by host cell DNA polymerase γ. Replication begins at each end (Fig. 5-19) and proceeds in the 5' to 3' direction by displacement and complementary strand synthesis, the products being two double-stranded and two single-stranded molecules. The latter then probably cyclize by virtue of their terminally repeated inverted sequences, which would again permit complementary strand synthesis by 5' to 3' displacement, the products being two more double-stranded DNA molecules. This replication mechanism, parts of which are known to occur while others are as yet hypothetical, is unique because it does not involve the formation of Okazaki fragments and because the transcription of the two parental strands does not proceed simultaneously, but sequentially. Conceivably the two 5'-terminally linked protein molecules play some critical role in adenovirus DNA replication.

There is good evidence that adenovirus DNA is integrated into host DNA during the lytic multiplication cycle just described. The significance of this phenomenon is not yet clear. It would seem that such integration is not an *essential* step for productive infection.

HERPESVIRUSES

The amount of genetic information stored in the genome of herpesviruses is some 30 to 50 times larger than that contained in polyomaviruses. The anatomy of herpes simplex virus DNA is remarkable; it consists of two large sequences, L and S, with molecular weights of about 68 and 10 million respectively, that are each flanked by repeated inverted sequences with molecular weights of about 6.5 and 3.5 million, some of which are terminally repeated (Fig. 5-20). In addition, the L and S segments may each be inverted relative to each other and to the repeated sequences. The result is a molecule with a molecular weight of about 100 million (about 150,000 base pairs) which exists in four different configurations. However, in the genome of pseudorabies virus, a porcine herpesvirus, only the S segment is flanked by inverted repeated sequences and only the S segment is inverted relative to the L segment, so that there are only two sequence isomers and the genome of herpesvirus saimiri, a simian herpesvirus, exists in only one configuration

which consists of one large sequence flanked by multiple reiterations of a simple (G+C)-rich sequence. The significance of the extraordinary sequence anatomy of herpes simplex virus DNA is not known, nor is it clear how all four isomers arise within the infected cell. Presumably intra- or intermolecular recombinational or replicational processes produce the observed flip-flops.

The early phase of the herpes simplex virus multiplication cycle lasts about three hours. Parental DNA moves to the nucleus and is there transcribed by cellular DNA-dependent RNA polymerase II into a series of transcripts that hybridize to about 50 percent of the genome. These transcripts are processed in the nucleus to a series of messenger RNA molecules that are translated in the cytoplasm into about 10 polypeptides that are termed immediate early, or alpha, polypeptides. These messenger RNAs represent no more than a small fraction of the transcripts from which they were processed; they hybridize to only about 10 percent of the genome, and most of them are derived from the terminally repeated segments that flank the L and S regions. Clearly extensive processing, such as was described above for polyoma and adenovirus messenger RNAs, also occurs for herpesvirus messenger RNAs.

If herpesvirus is caused to infect cells in the presence of inhibitors of protein synthesis, only immediate early messenger RNAs are formed. Otherwise the transcription program proceeds with the synthesis of a new class of messenger RNA, the early messenger RNAs, which are translated into early or beta proteins. The nature of the protein that is necessary for the inception of early messenger RNA transcription is not known; it is conceivable that it somehow alters the specificity of the RNA polymerase, so that it begins to recognize new transcription initiation signals. Early messenger RNAs hybridize to the entire genome. Once viral DNA replication has started, the transcription program changes again to that characteristic of the late phase, when the late, or gamma, polypeptides are synthesized. Late messenger RNA species also hybridize to the entire genome, but their relative amounts are different from those during the early phase.

Little is known of the function of the alpha, beta, and gamma class polypeptides. Several early enzymes are among the alpha and beta polypeptides: there is a deoxypyrimidine ki-

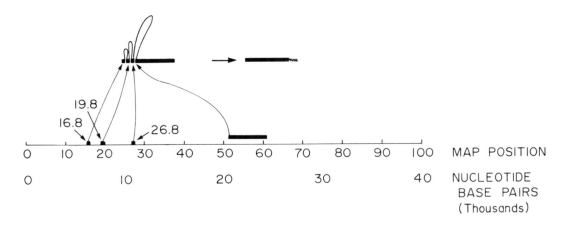

FIG. 5-18. Structure of the adenovirus hexon messenger RNA. It consists of 5 portions: (a) a leader sequence of about 50 nucleotides transcribed at map position 16.8; (b) a leader sequence of about 80 nucleotides transcribed at map position 19.8; (c) a leader sequence of about 110 nucleotides transcribed at map position 26.8; (d) a coding sequence of about 3800 nucleotides transcribed from a genome sequence beginning at map position 51.7; and (e) a poly(A) sequence of about 200 nucleotides. The spacer sequences between map position 16.8 and 19.8, 19.8 and 26.8, and 26.8 and 51.7, which are present in the primary transcript, are spliced out, possibly by a mechanism similar to that depicted in Fig. 5-9.

nase, a deoxyribonuclease, a ribonucleotide reductase, a deoxycytidylic acid deaminase and the DNA polymerase that replicates herpesvirus DNA. Another interesting early herpesvirus protein is an Fc receptor which may have a role in maintaining the latent nature of herpesvirus infections. Thus, the coating of infected cells with IgG molecules or immune complexes may protect cells from the destructive effect of potentially cytotoxic antibodies and lymphocytes; alternatively, anti-HSV IgG antibodies bound to herpesvirus-induced antigens on infected cells via their Fab regions may also bind to infected cells via their Fc regions, thereby rendering them unavailable for binding either to complement or to Fc receptors on effector cells capable of mediating antibody-dependent cell-mediated cytotoxicity. The other side of the coin is of course that cells infected with herpesvirus could be lysed nonspecifically by Fc receptor-positive killer cells; this has actually been observed.

The pattern of herpesvirus-induced polypeptides is best determined by labeling infected cells for brief periods of time (15–30 min) with a radioactively–labeled amino acid, separating their SDS-complexes by electrophoresis in polyacrylamide gels (see Chap. 2), and analyzing the gels by means of autoradiography. When this type of analysis is carried out at regular intervals during the multiplication cycle, profiles such as those depicted in Figure 5-21 are obtained. The synthesis of over 50 different species of proteins can be detected; they probably account for about one-half of the genetic information content of herpesvirus DNA. It is seen that different protein species are labeled and therefore synthesized at different stages of the multiplication cycle. Some proteins start being synthesized early, others later; some are synthesized for long periods of time, others for brief ones. There is obviously a complex program that controls when the various herpesvirus-coded proteins are synthesized. The existence of this very complex program of herpesvirus-coded protein synthesis at stages of the multiplication cycle when both early and late class cytoplasmic messenger RNAs hybridize to the entire herpesvirus DNA (see above) indicates that there must be a great deal of regulation at the translational level.

POXVIRUSES

Poxviruses are unique among animal viruses in that they require a newly-synthesized protein for uncoating; in the presence of inhibitors of protein synthesis, vaccinia virus cores do not

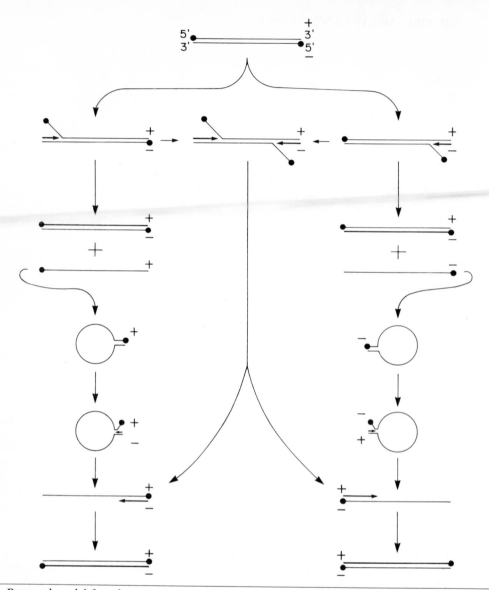

5-19. Proposed model for adenovirus DNA replication. Replication commences at either end and leads to displacement of either the plus or the minus strand. These are then postulated to cyclize via their terminally repeated inverted sequences (see Chap. 2) and to be again transcribed, this time yielding double-stranded molecules. Occasionally replication starts simultaneously at both ends, which would yield replication intermediates of the type indicated. •, the protein (M.W. 55,000) that is linked covalently to the 5'-termini of both DNA strands. (Adapted from Lechner and Kelly: Cell 12:1007, 1977.)

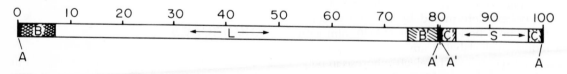

FIG. 5-20. The sequence arrangements in herpes simplex virus DNA. The distances are drawn to scale.

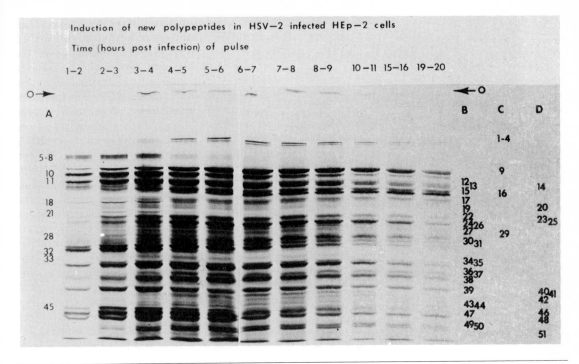

FIG. 5-21. Autoradiogram of a polyacrylamide gel in which the proteins synthesized at various stages of the herpesvirus infection cycle (1-2 hours, 2-3 hours, 3-4 hours, and so on) had been electrophoresed. The direction of electrophoresis was from O (origin) downward. A, B, C and D refer to kinetic groups to which individual proteins can be assigned. (Courtesy of Dr. Richard J. Courtney.)

liberate the DNA that they contain (see Chap. 2). This protein is coded by the virus itself. The reason why the viral genome can express itself prior to being uncoated is that vaccinia virus cores contain an RNA polymerase that transcribes almost one-half of the viral genome into very large transcripts which are processed to monocistronic capped and polyadenylated messenger RNAs prior to and concomitant with their extrusion into the cytoplasm; and one of the proteins that is translated from this messenger RNA population is the "uncoating" protein that causes cores to break down and liberate their DNA. (It may be noted that vaccinia virus cores may be an excellent system in which to study processing and splicing of messenger RNAs.)

Uncoated vaccinia DNA initiates the formation of inclusions or "factories" in the cytoplasm within which it is transcribed, replicates and is encapsidated into progeny virions. The inclusions, which are easily visible with the light microscope, are composed of fibrillar material and may be located anywhere in the cytoplasm. Their number per cell is proportioned to the multiplicity of infection, which suggests that each infecting virion initiates its own factory (Fig. 5-22).

During the early phase of the multiplication cycle about one half of the vaccinia virus genome is transcribed into messenger RNA. Among the early proteins are several enzymes, such as a thymidine kinase and a DNA polymerase, as well as several structural proteins.

Vaccinia virus DNA replication commences at about $1\frac{1}{2}$ hours after infection and is complete by about 5 hours (Fig. 5-23). It heralds the commencement of the late phase of the multiplication cycle, which is characterized by a series of new patterns of viral gene expression. As for the early proteins, synthesis of at least some of the early structural proteins continues, but synthesis of the early enzymes ceases. Cessation of the synthesis of early enzymes is due to the so-called switch-off phenomenon (Fig. 5-24), which is of interest since it provides one of the few well-documented examples of regulation of gene expression at the translational level. In essence, cessation of the synthesis of early enzymes is not due to inhibition of the transcrip-

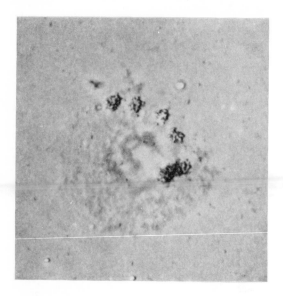

FIG. 5-22. Vaccinia virus factories in the cytoplasm of HeLa cell. Cells growing on a cover slip were infected at a multiplicity of 6 PFU per cell. At 6 hours after infection tritiated thymidine was added and at 7 hours the cells were fixed. Autoradiographic stripping film was then applied, and the slide stored for 2 weeks. On developing, the picture shown here was obtained. There are no grains (indicative of thymidine incorporation and therefore DNA replication) over the nucleus, but there are in the cytoplasm 5 labeled areas or factories (one is actually composed of two coalesced areas); this is where viral DNA is being synthesized. This cell had been stained with antibody to vaccinia virus coupled to fluorescein before autoradiography, and it was thereby demonstrated that the only areas in the cell that contained appreciable amounts of vaccinia virus antigens were the factories. Both viral DNA replication and viral morphogenesis therefore proceed within the factories. (From Cairns: Virol 11:603, 1960.)

zymes to be translated. If viral DNA replication and protein synthesis are inhibited, this inability does not develop. It is thought, therefore, that one of the first late proteins to be synthesized specifically prevents the translation of the messenger RNA molecules that code for early enzymes. This mechanism of controlling protein synthesis is obviously highly selective, since many other virus messenger RNA molecules—for example, those that code for structural proteins—continue to be translated. It is of potential significance for antiviral chemotherapy, since it implies the existence of a chemical difference between those messenger RNA molecules that continue to be translated and those that are switched off. It may prove possible to exploit this difference (see Chap. 8).

Another class of vaccinia proteins whose synthesis is switched off comprises certain late proteins that are formed for a period of only several hours, starting at about the time when viral DNA synthesis reaches its maximum. Together with the early and late proteins whose synthesis is never switched off, there are thus at least four different vaccinia virus protein translation patterns, and there may be many more (Fig. 5-25).

Somewhat more than one-half of vaccinia virus DNA codes for late proteins. By far the quantitatively most important of these are the structural viral proteins, most of which, like DNA, are usually formed in great excess. The remainder are synthesized in small amounts only, and little is known concerning their nature. Presumably they function during morphogenesis. The vaccinia virus-coded proteins that are synthesized at several periods throughout the multiplication cycle, as revealed by the technique of pulse-labeling followed by SDS-polyacrylamide gel electrophoresis and autoradiography, are shown in Fig. 5-26. It is seen that different proteins are synthesized at different stages of the multiplication cycle. It should

tion of the cognate genes, to instability of the corresponding messenger RNA species, or to instability of the enzymes themselves. Rather it is due to a suddenly developing inability of the messenger RNA species that code for early en-

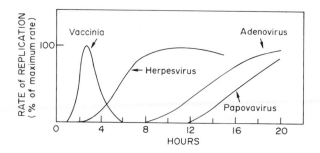

FIG. 5-23. The replication of vaccinia virus, herpesvirus, adenovirus, and papovavirus DNA. Vaccinia virus DNA is atypical in replicating during only a brief period of time early in the multiplication cycle. Progeny DNA molecules always form a pool from which individual molecules are selected at random for incorporation into progeny virus particles.

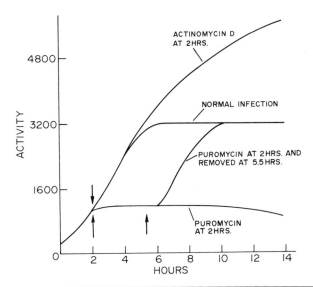

FIG. 5-24.   The switch-off phenomenon. This graph depicts the synthesis of "early" enzymes, such as thymidine kinase and DNA polymerase, during vaccinia virus infection. Under normal conditions these enzymes begin to be formed soon after infection, and their synthesis is "switched-off" at about 4 hours. If actinomycin D, which inhibits messenger RNA formation, is added at 2 hours, "switch-off" does not occur. This demonstrates, first, that the messenger RNAs from which these enzymes are translated are very stable, and, second, that switch-off itself requires the synthesis of some other messenger RNA. If protein synthesis is inhibited with puromycin at 2 hours, enzyme synthesis immediately ceases; if puromycin is removed at $5\frac{1}{2}$ hours, enzyme synthesis resumes and is again switched off after a time interval equivalent to that between the addition of puromycin and the onset of normal switch-off. This indicates that switch-off is due to the accumulation of a certain amount of some specific protein. (Modified from McAuslan: Virol 21:383, 1963.)

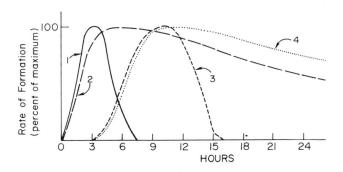

FIG. 5-25.   The synthesis of four classes of vaccinia-specified proteins. First, there are some early proteins whose synthesis is switched off when DNA replication commences. Early enzymes and two structural vaccinia virus proteins belong to this class. Second, there are those early proteins whose synthesis continues throughout the multiplication cycle. One of the major components of immature virus particles belongs to this class. Third, there are late proteins that are synthesized for some time following the onset of DNA replication and that are then switched off; finally, there are late proteins that are synthesized throughout the entire late period of the multiplication cycle. Most structural vaccinia virus proteins belong to this class. (Modified from Holowczak and Joklik: Virol, 33:726, 1967.)

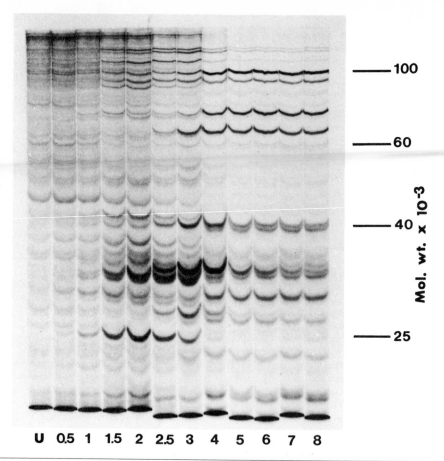

FIG. 5-26. Autoradiogram of a 9 percent polyacrylamide slab gel in which the proteins synthesized at various stages of the vaccinia virus multiplication cycle had been electrophoresed. At the times indicated (hours after infection) the cells were labeled for 15 min with [^{14}C]-protein hydrolysate. At the right is a molecular weight scale as determined by electrophoresing proteins of known size under identical conditions. (Courtesy of T. H. Pennington.)

be noted that very few host-coded proteins can be discerned even among the earliest profiles; like herpesvirus, vaccinia virus shuts down host-cell protein synthesis very efficiently (see Chap. 6).

Progeny genomes of DNA-containing viruses generally replicate faster than they are incorporated into virus particles. They therefore accumulate to form pools from which individual genomes are withdrawn at random for encapsidation; whereas some DNA molecules are withdrawn very soon after they are formed, others may remain naked for long periods of time. This is true particularly in the case of the vaccinia virus multiplication cycle, where DNA replication ceases at about 5 hours after infection, whereas virion morphogenesis contin-

ues for about 25 hours. The time necessary for a complete virus particle to be assembled around a vaccinia DNA molecule is about 1 hour.

The morphogenesis of vaccinia virus proceeds in the cytoplasm, where mature virus particles accumulate until they are liberated as the cells disintegrate (Fig. 5-27). The other three DNA-containing viruses are all assembled in the nucleus, where they often form paracrystalline arrays consisting of both complete and empty virus particles (Fig. 5-28). Excess protein also tends to form paracrystalline masses, particularly in the case of adenovirus (Fig. 5-29), and tubules that have the same diameter as virus particles are often found in cells infected with papovaviruses.

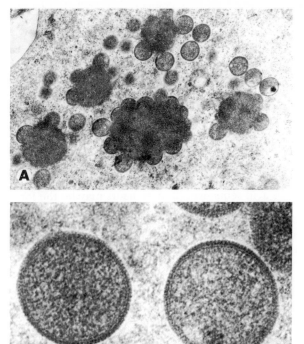

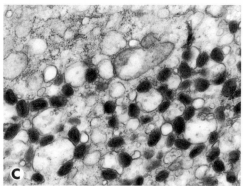

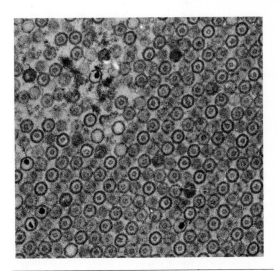

FIG. 5-27. Vaccinia virus particles in the cytoplasm of infected cells. **A.** Immature virus particles developing from intracytoplasmic inclusions 1 hour after reversal of vaccinia virus morphogenesis arrest by rifampicin (see Chap. 8). $\times 9,000$. **B.** Characteristic structure of immature vaccinia virus particles. $\times 48,000$. **C.** Mature vaccinia virus particles in the cytoplasm of infected cells. $\times 12,000$. (A and B, courtesy of Dr. T. H. Pennington; C, from Dales and Siminovitch: J Biochem Biophys Cytol 10:475, 1961.)

THE MULTIPLICATION CYCLE OF DOUBLE-STRANDED RNA CONTAINING VIRUSES

When reovirus infects cells, its outer capsid shell is partially degraded to form an entity known as a subviral particle (SVP) (Fig. 5-30). This process triggers the activation of an RNA polymerase (known as the transcriptase) within the SVP which transcribes all 10 genome segments into messenger RNA molecules. These messenger RNA molecules, which are exactly the same length as their double-stranded templates, are translated into protein after being capped (but not spliced or polyadenylated). The parental double-stranded RNA never escapes from the SVP, which persist in infected cells throughout the multiplication cycle.

After several hours the single-stranded messenger RNA molecules begin to associate with viral proteins to form complexes within which they are transcribed once, and once only, into minus-strands with which they remain associated, thereby giving rise to progeny double-stranded RNA molecules. These immature progeny virus particles then transcribe more

messenger RNA molecules which in time are translated into more proteins which gradually associate with the immature virus particles, causing them to mature, via several stages, into

FIG. 5-28. Intranuclear crystal composed of herpesvirus capsids and cores. $\times 35,000$. (From Nii, Morgan, and Rose: J Virol 2:517, 1968.)

complete progeny virus particles. It should be noted that the stages of the infection cycle when the plus and minus strands of progeny double-stranded RNA are synthesized are separated by an interval of several hours. Double-stranded RNA is thus synthesized by a mechanism that is quite different from that by which double-stranded DNA is formed.

THE MULTIPLICATION CYCLES OF SINGLE-STRANDED RNA CONTAINING VIRUSES

The principles involved in the multiplication of single-stranded RNA containing viruses differ in several respects from those described for the double-stranded DNA containing viruses. Most importantly, their multiplication cycles cannot be divided into the clearly defined early and late phases of those of the double-stranded DNA containing viruses. There follows a brief description of the multiplication cycles of several families of single-stranded RNA containing viruses.

THE MULTIPLICATION CYCLES OF PLUS-STRANDED RNA VIRUSES

Picornaviruses

We will consider the multiplication cycle of poliovirus, since it has been investigated in the greatest detail. The strategy of the poliovirus multiplication cycle is summarized schematically in Figure 5-31.

After uncoating, the parental poliovirus genome itself functions as messenger RNA and is translated into a single large protein with a molecular weight of between 200,000 and 300,000. This protein, the polyprotein, is then cleaved, in a precisely defined sequence, into at least 12 smaller proteins, among which are the enzyme or enzymes necessary to replicate poliovirus RNA, and the capsid proteins (Fig. 5-32). Little is known concerning the nature of the intermediate cleavage products or whether any of them exercise a specific function during the poliovirus multiplication cycle. They clearly possess the potential for doing so; and if they did, this would be another example of the multiple use of viral genetic information. It is of

interest in this respect that post-translational cleavage of viral proteins occurs frequently; there are numerous examples of vaccinia virus, adenovirus, myxovirus, togavirus, retrovirus and other viral capsid proteins being synthesized in the form of precursors larger than themselves. The reason in most cases is not expansion of the pool of genetic information, but rather one of the following: 1) viral capsid proteins may be too insoluble to be transported from the site where they are synthesized to the site of morphogenesis unless their solubility is increased by the addition of removable amino acid sequences; 2) precursors may be the forms in which capsid proteins must exist prior to assembly to enable them to interact correctly with other capsid proteins; and 3) cleavage of precursors to capsid proteins may provide at least part of the energy for morphogenesis.

The nature and source of the enzymes that cleave the precursors is not known with certainty. Clearly they must be very specific. Evidence is accumulating that they are virus-coded: thus protease activity is associated with a 40,000 dalton nonstructural protein coded by poliovirus, with the C protein of togaviruses (see below), and with the p15 protein of RNA tumor viruses (see below).

Poliovirus RNA replication starts as soon as sufficient polymerase has been formed. It occurs in two stages: first the parental plus-strand is transcribed into a minus-strand, and then the minus-strand serves as the template for repeated transcription into progeny plus-strands. A limited number of progeny plus-strands are then again transcribed into minus-strands, but this is not a common occurence; the total number of minus-strands in the infected cell probably does not exceed 1000, while up to a million plus-strands may be formed. RNA strands of both plus and minus polarity are covalently linked at their 5'-termini to a protein with a molecular weight of about 5000—via a tyrosine residue—which may serve as a primer for transcription, since it is also present on growing, less-than-full-size molecules. This protein is also linked to the RNA that is present inside virus particles; but it is not present on poliovirus RNA molecules that serve as messenger RNA. Thus progeny RNA molecules that are encapsidated have never served as messenger RNAs, and poliovirus RNA molecules that serve as messengers do not become encapsidated.

Whereas translation of the poliovirus genome

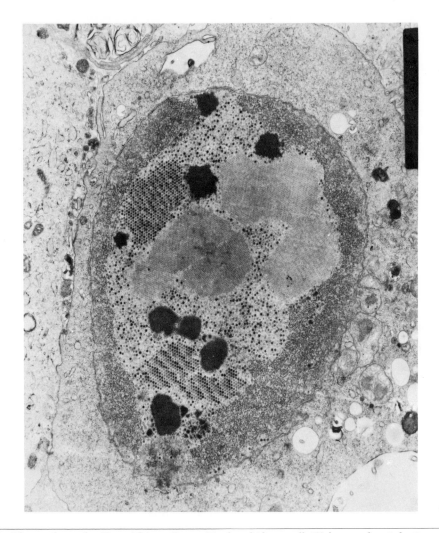

FIG. 5-29. The nucleus of a Vero African Green Monkey kidney cell, 70 hours after infection with adenovirus type 2. Paracrystalline arrays of virus particles, crystals of vertex or core proteins, and intranuclear inclusions (the densely stained masses) are all visible. ×10,000. (From Henry et al: Virol 44:215, 1971.)

into a single precursor protein has many advantages, there are also potential disadvantages. In particular, it implies that all portions of the viral genome are expressed with equal frequency. This is very wasteful, since many more capsid protein molecules are required than, for example, polymerase molecules. In practice, the various processed virus-specified proteins are not formed in equimolar amounts in infected cells: especially in the later stages of the multiplication cycle most of the proteins that are formed are indeed capsid components. The mechanism by which such translational control is achieved is probably as follows. It is known

that the capsid proteins are encoded in the 5'-terminal region of the poliovirus RNA molecule. This means that in the event of premature termination of translation, which is not unlikely for a long messenger RNA such as poliovirus RNA, the formation of capsid proteins would be favored over those like the polymerase that are encoded by sequences in its 3'-terminal half. Conceivably some virus-coded protein may act as a terminator of translation and thereby increase the likelihood of premature termination of translation.

As is the case for other icosahedral viruses, poliovirus progeny often accumulates in the

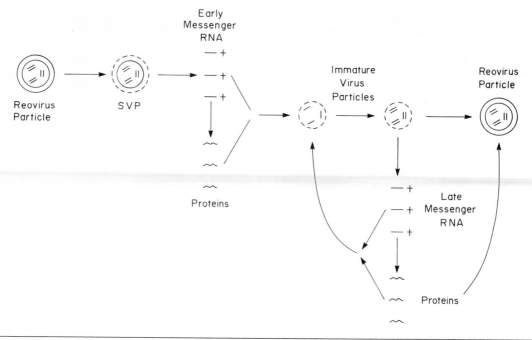

FIG. 5-30.   The reovirus multiplication cycle. For details, see text. Note that here early and late messenger RNA and protein molecules are identical. Since many immature virus particles are formed from each parental SVP, most viral messenger RNA molecules are transcribed from progeny templates.

form of large, intracytoplasmic, paracrystalline arrays (Fig. 5-33).

Togaviruses

The strategy of the togavirus multiplication cycle is quite different (Fig. 5-34). The togavirus genome is about 14,000 nucleotides long (M.W. 4.2–4.5 × 10⁶) and possesses a sedimentation coefficient of 42S. When this genome is uncoated, about two-thirds of it, starting from the 5'-terminus, is translated into one large protein which is cleaved into four proteins with molecular weights of about 86,000, 72,000, 70,000 and 60,000, two of which form the togavirus RNA polymerase. This enzyme transcribes the parental plus-strands into minus-strands which then act as templates for plus-strand synthesis. The crucial point here is that two types of plus-strands are transcribed: (a) full length 42S progeny plus-strands which, like progeny plus-strands of poliovirus RNA, can function either as messenger RNA, act as templates for additional minus-strand synthesis or be encapsidated into progeny virions, and (b) strands which are only about one-third as long and possess a sedimentation coefficient of 26S (M.W. 1.6–1.8 ×10⁶, about 5500 nucleotides long). These RNA molecules represent the 3'-terminal one third of the 42S RNA molecules, and they code for the three (or four, in certain virus strains) structural proteins of togavirus particles which are translated from it in the form of a 130,000 dalton precursor. This precursor is cleaved into the capsid or C protein (M.W. 33,000) which together with the RNA makes up the nucleocapsid, and two proteins (E1 and E2, for envelope) which, in their glycosylated state, form the spikes on the surface of togavirus particles. Cleavage of the 130,000 dalton precursor protein proceeds as follows (Fig. 5-35). While it is still being translated, the C protein is cleaved off and the new amino terminal end, which belongs to one of the envelope proteins, is inserted into internal cell membranes. This is followed by a second cleavage, and again the new amino terminal end, which belongs to the other envelope protein, is inserted into membrane. At the same

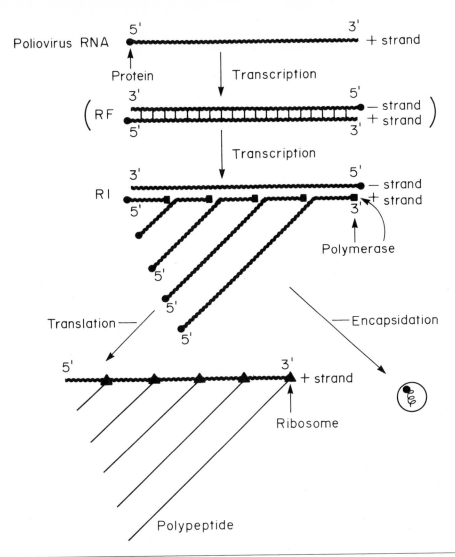

FIG. 5-31. Replication and functioning of poliovirus RNA. Parental viral RNA strands are translated into the polyprotein, and the polymerase derived from the polyprotein then transcribes them into strands of minus polarity, to yield double-stranded replicative forms (RF) which probably exist only very briefly. Progeny plus-strands are then transcribed repeatedly from these minus-strand templates by a peeling-off type of mechanism. The structure consisting of minus-stranded template and several plus-stranded transcripts in various stages of completion is known as the replicative intermediate (RI). Early during the infection cycle the number of RFs and therefore of RIs increases, so that the rate of formation of progeny plus-strands first increases and then becomes constant. Progeny plus-strands are either translated or encapsidated.

time the two envelope glycoproteins become glycosylated. Recent evidence suggests that the proteolytic enzyme that effects these cleavages may be the C protein.

In essence, then, the togavirus genome expresses itself via two messenger RNAs instead of via only one, as do the picornaviruses. The special 26S capsid protein messenger RNA is transcribed about three times as frequently as the 42S RNA which acts as the messenger RNA only for the nonstructural protein precursor, only two-thirds of it actually being translated.

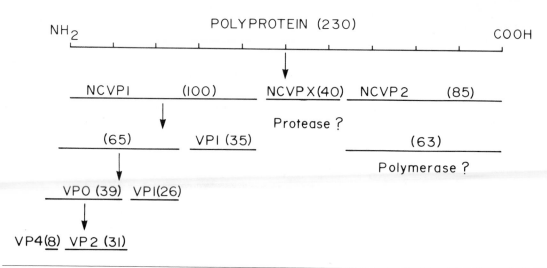

FIG. 5-32. The cleavage pattern of the poliovirus polyprotein. The 4 capsid proteins, VP1-4, are coded by the aminoterminal half. Recent evidence suggests that the polymerase may be a protein with a molecular weight of 63,000, and that a virus-coded protein with a molecular weight of about 40,000 may possess protease activity. Together with the four capsid proteins, these two enzymes would account for most of the coding capacity of the poliovirus RNA molecule. The order of these various proteins in the polyprotein has been determined by the pactamycin mapping technique (see Chap. 7).

These relative transcription frequencies are controlled by at least two proteins: one is one of the nonstructural proteins described above, which promotes transcription of 26S RNA; the other is the C protein, which represses it.

The morphogenesis of togaviruses involves the formation of cores or nucleocapsids which consist of 42S plus-stranded RNA as well as C protein, and budding through patches of cell membrane into which the glycosylated envelope proteins have been inserted. This process is discussed in more detail below.

Little is known concerning the molecular events that occur during the multiplication cycles of coronaviruses and caliciviruses. Like togavirus-infected cells, cells infected with caliciviruses contain not only genome-length, but also a species of subgenome-length viral RNA.

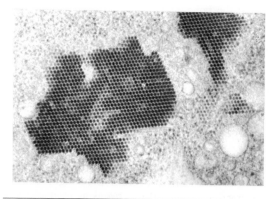

FIG. 5-33. A large crystal of poliovirus in the cytoplasm of a HeLa cell infected for 7 hours. ×50,000. (From Dales et al: Virol 26:379, 1965.)

THE MULTIPLICATION CYCLES OF MINUS-STRANDED RNA VIRUSES

Rhabdoviruses

The most intensively investigated minus-stranded RNA virus is the rhabdovirus vesicular stomatitis virus (VSV). VSV enters the cell via invagination into phagocytic vacuoles, which results in its envelope being removed and its nucleocapsid being liberated into the cytoplasm. The envelope consists of the lipid bilayer, the M or matrix protein (M.W. 29,000) and the G or spike protein, which is a 70,000 dalton glycoprotein; the nucleocapsid consists of the nucleocapsid or N protein (M.W. 50,000),

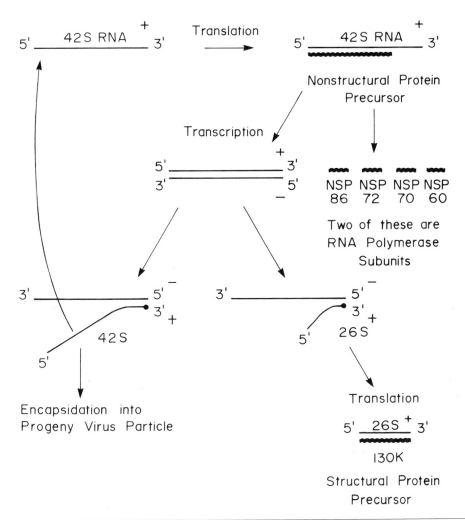

FIG. 5-34.   The strategy of the togavirus multiplication cycle. For details see text. For fate of the 130K structural protein precursor, see Fig. 5-35.

and small amounts of the phosphorylated NS protein (M.W. 40,000) and the very large L protein (M.W. 170,000). The RNA of VSV is a linear single strand of minus polarity that comprises five genes, the order of which, from the 3' end (and therefore from the 5' end of its plus-stranded transcript) is N-NS-M-G-L. VSV RNA does not code for any nonstructural proteins; these five genes, which code for proteins with an aggregate molecular weight of about 360,000, is all that VSV RNA (M.W. 4×10^6) can encode.

Removal of the envelope activates an RNA polymerase in the nucleocapsid, which tran-

scribes the minus-stranded RNA within it into plus-stranded RNA that can be translated. This polymerase has not yet been isolated; however, it seems that the RNA-N protein complex serves as its template (rather than the RNA alone), and that the NS and L proteins are responsible for the enzyme activity.

The polymerase transcribes the minus-strand in nucleocapsids by two distinct mechanisms. The first consists of the faithful transcription of the minus-strands into plus-strands, which in turn serve as templates for the synthesis of numerous progeny minus-strands. This is the process of minus-strand replication. The second

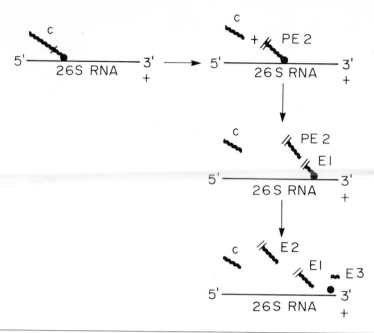

FIG. 5-35. The processing of the togavirus 130K structural protein precursor. The precursor only accumulates to a significant extent within cells if its processing is inhibited by protease inhibitors. Under normal conditions the C or nucleocapsid protein is cleaved off the aminoterminal end of the growing precursor chain, and at the same time the newly generated aminoterminus of the E2 (spike protein) precursor (PE2) is inserted into membrane. PE2 is then cleaved off the growing precursor chain and the aminoterminus of E1 (another spike protein) is inserted into membrane in similar manner. Finally a small protein E3 is liberated into the medium as membrane-associated PE2 is cleaved to yield membrane-associated E2.

mechanism involves the transcription of minus-strands into plus-stranded RNA that can serve as the messenger RNA through which the VSV genome expresses itself. For this purpose a different type of transcript is synthesized. In its simplest terms, one may envisage this type of transcription process as follows: the first gene to be transcribed is the N gene. Upon reaching some signal at the end of this gene, about 350 A residues are added to form a poly(A) sequence. Faithful transcription then resumes with transcription of the NS gene, which is also followed by the addition of a poly(A) sequence; and the same pattern is then followed for the remaining M, G and L genes. Very soon after transcription, or perhaps even before it is completed, this huge transcript is then cleaved at the 3'-termini of the poly(A) sequences to yield the five species of messenger RNA which are then capped at their 5'-termini, and being already polyadenylated at their 3'-termini, are ready to be translated into the five capsid proteins. Interestingly enough, all five messenger RNA species begin with the sequence 5'-AACAG. . ., which may provide at least part of the signal for polyadenylation when it is encountered by the polymerase. A further interesting feature of the transcript is that the coding sequence for the N gene is preceded by a "leader" sequence of about 70 nucleotides which is cleaved off before it becomes a messenger RNA. The function of this leader sequence is unknown.

Transcription of parental minus-strands into plus-stranded RNA molecules is referred to as "primary" transcription; since the nucleocapsid contains the polymerase it proceeds without requiring the synthesis of new virus-specified proteins. Transcription of plus-strands into progeny minus-strands, and of progeny minus-strands into additional genome length plus-strands or messenger RNA precursor, is referred to as "secondary" transcription and does require the synthesis of newly synthesized viral proteins (L and NS proteins at the very least). Nothing is known of the mechanism that regulates the relative frequency of transcription of minus strands into genome length plus strands on the one hand or messenger RNA precursors on the other.

The five VSV proteins are synthesized in about the same molecular ratios as those in which they exist in virus particles; for example, about 20 times as many N as L protein molecules are synthesized. These ratios appear to be controlled at the level of transcription rather than translation. This presents a paradox, for the manner in which VSV RNA is transcribed implies that equal numbers of each of the five species of messenger RNA molecules should be formed. It seems, however, that there is a polarity effect; that is, as the polymerase transcribes the RNA it tends to "fall off," so that the further a gene is situated from the origin of transcription, the less of it is transcribed. Indeed, the gene order 3'-N-NS-M-G-L-5' in the VSV template parallels the frequency, in a decreasing sense, of their transcription into messenger RNA.

Like other enveloped viruses, VSV has a special mechanism for being liberated from cells, namely budding. Enveloped viruses code for two types of structural proteins, nucleocapsid proteins and envelope proteins. The nucleocapsid proteins (the N, NS and L proteins in the case of VSV) are assembled into nucleocapsids either in the nucleus (in the case of orthomyxoviruses) or in the cytoplasm (all other enveloped RNA viruses), where they can often be seen in large numbers. The envelope proteins (the M and G proteins in the case of VSV) follow a different pathway. The G protein is synthesized on ribosomes that either are attached to membranes, or become attached to membranes very soon after they begin to be translated. Like other proteins that must pass through membranes, the G protein probably possesses a very hydrophobic amino acid "leader" sequence of some 15 to 20 amino acids at its aminoterminal end which is cleaved off once it has penetrated the membrane; in fact, it seems that before the protein is 50 amino acids long its amino terminus has already passed through membrane. As it continues to be synthesized, the protein is then also glycosylated. This process proceeds in several stages, the first of which is the transfer to two of its asparagine residues of a mannose-rich precursor oligosaccharide from dolichol pyrophosphate, which is then processed via a series of reactions to the final prosthetic group (Fig. 5-36). As discussed in Chap. 2, the precise nature of the final oligosaccharide is determined in part by the nature and the relative amounts of the various cellular glycosyl transferases and in part, particularly beyond the mannose fork, by the nature of the protein that is being glycosylated. While glycosylation is proceeding, the G protein is transported through the cellular membrane system to patches on the plasma membrane and membranes lining intracytoplasmic vacuoles, where it usually completely replaces host-specified membrane proteins. Nucleocapsids then migrate to these modified areas which they recognize with great specificity, align themselves with them, and then bud through them, becoming coated by them in the process (Figs. 5-37 and 5-38). The M protein is incorporated into virus particles immediately prior to budding. While budding is very efficient in some strains of cells, it is very inefficient in others. This sometimes leads to the accumulation in the cytoplasm of very large numbers of nucleocapsids (Fig. 5-39).

The budding process itself does not harm the host cell significantly. Many cells persistently infected with enveloped viruses (see Chap. 6) remain normal in appearance and continue to multiply for many generations while viruses bud from their surfaces. This is not to say that many enveloped viruses are not cytopathic; it merely says that budding per se is not a factor in cytopathogenicity.

Paramyxoviruses

Paramyxoviruses differ from most other enveloped viruses in that they enter cells by fusing with cell membranes, with resultant liberation of their nucleocapsids into the cytoplasm. It is likely that the ability of the envelopes of these viruses to fuse with cell membranes is responsible for their pronounced ability to cause cell fusion (see Chap. 6).

Paramyxovirus RNA (M.W. 5×10^6) codes for six polypeptides, all of which have counterparts among the rhabdovirus structural proteins; there is one major, NP, and two minor, P and L, nucleocapsid proteins, and three enve-

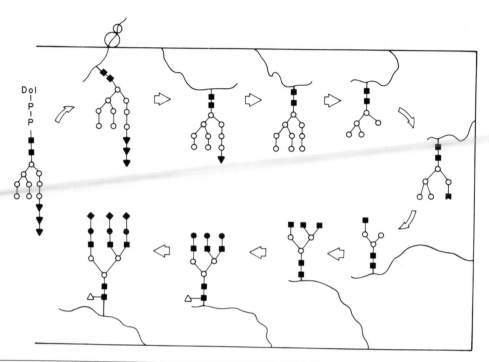

FIG. 5-36. Proposed sequence for the synthesis of the oligosaccharide subunit of the G spike protein of VSV. The first step is the transfer of a mannose-rich precursor from dolichol pyrophosphate to an asparagine residue of the G protein; this precursor is then processed to yield the final oligosaccharide unit. Dol, dolichol; ■, N-acetylglucosamine; ○, Mannose; ▲, glucose; ●, galactose; ◆, sialic acid (N-acetylneuraminic acid-galactose); △, fucose. (From Kornfeld, Li and Tarabus: J Biol Chem 253:7771, 1978.)

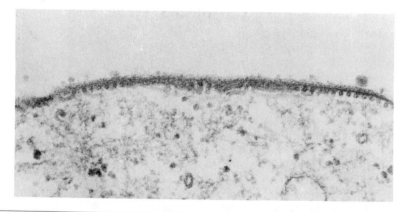

FIG. 5-37. Modification of the plasma membrane of a monkey kidney cell infected with SV5. A layer of dense material resembling the spikes present on viral envelopes is present on the outer surface of the membrane. Nucleocapsid strands, many seen in cross section, are aligned immediately beneath modified patches of the cell membrane. In due course they will bud through these membrane patches, becoming enveloped by them in the process. ×67,000. (From Compans et al: Virol 30:411, 1966.)

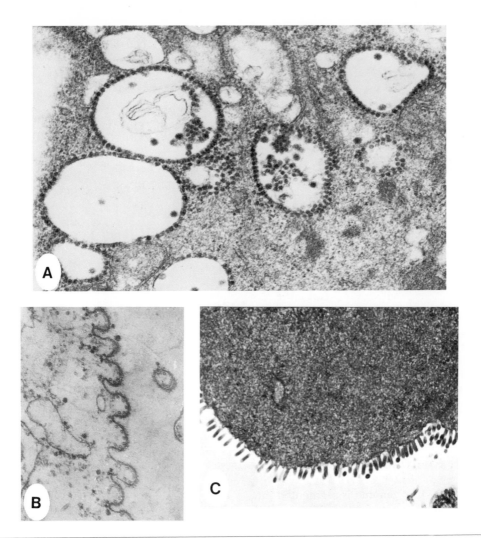

FIG. 5-38. Budding of enveloped viruses. **A.** A chick embryo cell infected with the togavirus Semliki Forest virus (SFV). Numerous nucleocapsids lining cytoplasmic vacuoles prior to budding into them can be seen. ×22,000. **B.** A row of SV5 particles budding from the plasma membrane of a monkey kidney cell, showing many cross sections of nucleocapsids. ×50,000. **C.** VSV budding from the plasma membrane of a mouse L cell. In L cells the majority of VSV particles bud from the plasma membrane; in other cells, such as chick embryo fibroblasts and pig kidney cells, VSV buds mostly into cytoplasmic vacuoles. ×21,500. (A, from Grimley, Berezesky, and Friedman: J Virol 2:1326, 1968; B, from Compans et al: Virol 30:411, 1966; C, from Zee, Hackett, and Talens: J Gen Virol 7:95, 1970.)

lope proteins, a "matrix" protein M and two glycoproteins, F and HN. Together, these proteins require the total coding capacity of the virus genome. Their order, from the 3'-terminus of the minus strand, is NP, F, M, P, HN and L.

The strategy of the paramyxovirus multiplication cycle resembles that of the rhabdovirus multiplication cycle.

A remarkable feature of the two paramyxovirus glycoproteins, F and HN, which form the two types of spikes on the surface of paramyxovirus particles—the former being responsible for the fusion and hemolytic, the latter for the hemagglutinin and neuraminidase activities (see Chap. 2)—is that they must be cleaved once, and once only, in order for virus particles to be infectious. The resulting fragments are

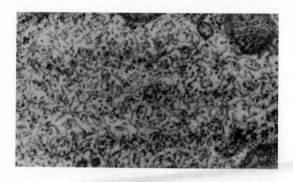

FIG. 5-39. Accumulation of SV5 nucleocapsids in the cytoplasmic matrix of BHK-21 cells. Such accumulation does not occur in monkey kidney cells from which nucleocapsids bud as rapidly as they are synthesized. ×34,000. (From Compans et al: Virol 30:411, 1966.)

not lost, but remain covalently bound to each other via -SS- bonds. The cleavage is accomplished by a protease that is present in some cells and not in others (which therefore produce noninfectious virus particles); cleavage can also be accomplished by certain proteases in vitro (which results in the activation of noninfectious virus particles).

Orthomyxoviruses

The influenza virus genome is segmented into eight components (Fig. 5-40), each of which codes for a single protein. These eight proteins are: P1, P2 and P3, three large proteins that are minor nucleocapsid components and are thought to be active in RNA transcription and replication; the NP protein, which is the major nucleocapsid component; two glycoproteins, HA and NA, that form the two types of glycoprotein spikes; the M (matrix) protein, which is the envelope protein; and the NS protein, which is a nonstructural protein of uncertain function.

Like most other enveloped viruses, influenza virus enters the cell via phagocytic vesicles; however, the liberated nucleocapsids do not remain in the cytoplasm, but migrate to the nucleus where they replicate and are transcribed. The overall strategy of the influenza virus multiplication cycle resembles that of rhabdoviruses and paramyxoviruses, with pri-

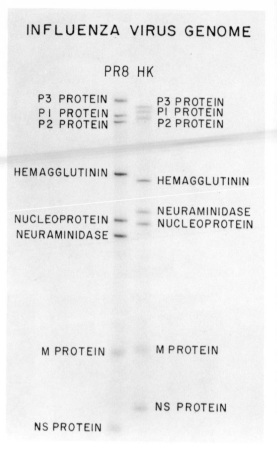

FIG. 5-40. Electropherogram in a polyacrylamide gel of the eight minus-stranded segments of influenza A (strains PR8 and HK) RNA, together with the proteins which they encode. Proteins P1, P2, and P3 function in RNA replication and transcription; the nucleoprotein is the nucleocapsid protein; the M protein is associated with the internal side of the envelope membrane; the hemagglutinin and neuraminidase are the two spike glycoproteins; and the NS protein is a nonstructural protein with unknown function. (From Ritchey, Palese and Schulman: J Virol 20:307, 1976.)

mary transcription catalyzed by parental nucleocapsids leading to the formation of plus-stranded messenger RNA molecules and plus-stranded templates for the transcription of progeny minus-strands, followed by secondary transcription of both plus- and minus strands catalyzed by newly synthesized polymerase. A remarkable feature of the influenza virus multiplication cycle is that it is dependent on transcription of host DNA; influenza virus will not multiply in enucleated or UV-irradiated cells, and it is inhibited by actinomycin D (a classical inhibitor of DNA-dependent RNA polymerase) and α-amanitin (an inhibitor of cellular DNA-dependent RNA polymerase II). The reason influenza virus multiplication is dependent on cellular DNA transcription appears to be the fact that the transcription of influenza RNA must be primed by newly synthesized capped cellular RNAs.

Influenza virus minus-stranded RNA is transcribed into two types of transcripts. The first are messenger RNAs. These are all incomplete transcripts; transcription stops before the polymerase reaches the 5'-end of the minus-strand templates. Interestingly enough, the untranscribed nucleotide sequence seems to be highly conserved (that is, very similar) among all eight RNA segments. All messenger RNA species are polyadenylated at their 3'-termini and capped at their 5'-termini, the caps perhaps being transferred from the primer host cell RNAs described above. The second type of transcripts are faithful plus-stranded copies of the minus-strands; they, in turn, serve as the templates for the synthesis of progeny minus-strands.

Just as in the case of rhabdoviruses and paramyxoviruses, the various influenza virus-coded proteins are synthesized in roughly the same relative amounts as those in which they exist in virus particles. The regulation that operates here is at the level of transcription, not of translation; but nothing is known of its nature in molecular terms.

Finally, a word about the function of the influenza virus neuraminidase. It seems that the primary function of this enzyme is to remove neuraminic acid from the oligosaccharide moieties of the HA protein. If this function is prevented (either by the addition of inhibitors, or by the use of appropriate mutants), the neuraminic acid-containing hemagglutinin serves as the receptor for other virus particles, with resultant extensive aggregation. Further, the HA protein is inserted into the influenza virus envelope in a biologically inactive precursor form; just like the HA protein of paramyxoviruses, it must be cleaved into two components by a protease, otherwise the virus particle is noninfectious. This cleavage is prevented unless neuraminic acid is removed from the oligosaccharides of the HA protein.

FURTHER READING

Books and Reviews

GENERAL

Butterworth BE: Proteolytic processing in animal virus proteins. Curr Top Microbiol Immunol 77:1, 1977

Casjens S, King J: Virus assembly. Ann Rev Biochem 44:555, 1975

Dales S: Early events in cell-animal virus interactions. Bacteriol Rev 37:103, 1973

Hershko A, Fry M: Post-translational cleavage of polypeptide chains: role in assembly. Ann Rev Biochem 44:775, 1975

Nathans D, Smith HO: Restriction endonucleases in the analysis and restructuring of DNA molecules. Ann Rev Biochem 44:273, 1975

PAPOVAVIRUSES

Acheson NH: Transcription during productive infection with polyoma virus and simian virus 40. Cell 8:1, 1976

Fareed GC, Davoli D: Molecular biology of papovaviruses. Ann Rev Biochem 46:471, 1977

Sambrook J: The molecular biology of the papovaviruses. In Nayak DP (ed): The Molecular Biology of Animal Viruses. New York, Marcel Dekker, 1977

ADENOVIRUSES

Doerfler W: Integration of viral DNA into the host genome. Curr Top Microbiol Immunol 71:1, 1975

Flint J: The topography and transcription of the adenovirus genome. Cell 10:153, 1977

Kelly TJ Jr, Nathans D: The genome of simian virus 40. Adv Virus Research 21:86, 1977

Levine AJ: SV40 and adenovirus early functions involved in DNA replication and transformation. Biochem Biophys Acta 458:213, 1976

Lewis AM Jr: Defective and nondefective Ad2-SV40 hybrids. Prog Med Virol 23:96, 1977

Sharp PA, Flint SJ: Adenovirus transcription. Curr Top Microbiol Immunol 74:137, 1976

Winnacker E: Adenovirus DNA: Structure and function of a novel replicon. Cell 14:761, 1978

Wold WSM, Green M, Buttner W: Adenoviruses. In Nayak DP (ed): The Molecular Biology of Animal Viruses. Vol 2:673. New York, Marcel Dekker, 1978

HERPESVIRUSES

Clements JB, Hay J: RNA and protein synthesis in herpesvirus-infected cells. J Gen Virol 35:1, 1977

Honess RW, Watson DH: Unity and diversity in the herpesviruses. J Gen Virol 37:15, 1977

Kaplan AS (ed): The Herpesviruses. New York and London, Academic Press, 1973

Roizman B: The herpesviruses. In Nayak DP (ed):

The Molecular Biology of Animal Viruses. New York, Marcel Dekker, 1977

Roizman B: The structure and isomerization of herpes simplex virus genomes. Cell 16:481, 1979

Roizman B, Furlong D: The replication of herpesviruses. In Fraenkel-Conrat H, Wagner RR (eds): Comprehensive Virology, Vol 3, p 229. New York and London, Plenum, 1974

POXVIRUSES
Moss B: Poxviruses. In Nayak DP (ed): The Molecular Biology of Animal Viruses, Vol 2:849. New York, Marcel Dekker, 1978

PARVOVIRUSES
Rose JA: Parvovirus reproduction. In Fraenkel-Conrat H, Wagner RR (eds): Comprehensive Virology. Vol 3, p 1. New York and London, Plenum, 1974

Salzman LA: The Parvoviruses. In Nayak DP (ed): The Molecular Biology of Animal Viruses, Vol 2:539. New York and Basel, Marcel Dekker, 1978

Siegl G: The parvoviruses. Virology Monographs 15. Vienna and New York, Springer-Verlag, 1976

Ward DC, Tattersall P (eds): Replication of mammalian parvoviruses. New York, Cold Spring Harbor Lab, 1978

PICORNAVIRUSES
Levintow L: Reproduction of picornaviruses. In Fraenkel-Conrat H, Wagner RR (eds): Comprehensive Virology. Vol 2, p 109. New York and London, Plenum, 1975

Rekosh DMK: The molecular biology of picornaviruses. In Nayak DP (ed): The Molecular Biology of Animal Viruses, Vol 1:63. New York, Marcel Dekker, 1977

Rueckert RR: On the structure and morphogenesis of picornaviruses. In Fraenkel-Conrat H, Wagner RR (eds): Comprehensive Virology, Vol 6, p 131. New York and London, Plenum, 1976

TOGAVIRUSES
Kaariainen L, Soderlund H: Structure and replication of alphaviruses. Curr Top Microbiol Immunol 82:15, 1978

Pfefferkorn ER, Shapiro D: Reproduction of togaviruses. In Fraenkel-Conrat H, Wagner RR (eds): Comprehensive Virology, Vol 2, p 171. New York and London, Plenum, 1974

Strauss JH, Strauss EG: Togaviruses. In Nayak DP (ed): The Molecular Biology of Animal Viruses, Vol 1:111. New York, Marcel Dekker, 1977

BUNYAVIRUSES
Obijeski JF, Murphy FA: Bunyaviridae: Recent biochemical developments. J Gen Virol 37:1, 1977

REOVIRUSES
Flewett TH, Woode GN: The rotaviruses. Arch Virol 57:1, 1978

Joklik, WK: Reproduction of reoviridae. In Fraenkel-Conrat H, Wagner RR (eds): Comprehensive Virology, Vol 2, p 231. New York and London, Plenum, 1974

Ramig RF, Fields BN: Reoviruses. In Nayak DP (ed): The Molecular Biology of Animal Viruses, Vol 1:383, New York, Marcel Dekker, 1977

Silverstein SC, Christman JK, Acs G: The reovirus replicative cycle. Ann Rev Biochem 45:375, 1976

ORTHOMYXOVIRUSES
Barry RD, Almond JW, McGeoch DJ, Inglis SC, Mahy BWJ: Structure and function of the influenza virus genome. In Mahy BWJ, Barry RD (eds): Negative Strand Viruses and the Host Cell, p 1. London, Academic Press, 1978

Compans RW, Choppin PW: Reproduction of myxoviruses. In Fraenkel-Conrat H, Wagner RR (eds): Comprehensive Virology, Vol 4, p 179. New York and London, Plenum, 1975

Nayak DP: The biology of myxoviruses. In Nayak DP (ed): The Molecular Biology of Animal Viruses, Vol 1:281. New York, Marcel Dekker, 1977

Palese P: The genes of influenza virus. Cell 10:1, 1977

Scholtissek C: The genome of influenza virus. Curr Top Microbiol Immunol 80:139, 1978

Skehel JJ, Hay AJ: Influenza virus transcription. J Gen Virol 39:1, 1978

PARAMYXOVIRUSES
Choppin PW, Compans RW: Reproduction of paramyxoviruses. In Fraenkel-Conrat H, Wagner RR (eds): Comprehensive Virology, Vol 4, p 95. New York and London, Plenum, 1975

Ishida N, Homma M: Sendai virus. Adv Virus Res 23:349, 1978

Kingsbury DW: Paramyxoviruses. In Nayak DP (ed): The Molecular Biology of Animal Viruses, p 349. New York, Marcel Dekker, 1977

Mahy BWJ, Barry RD (eds): Negative strand viruses and the host cell. London, Academic Press, 1978

Morgan EM, Rapp F: Measles virus and its associated diseases. Bact Rev 41:636, 1977

RHABDOVIRUSES
Banerjee AK, Abraham G, Collonno RJ: Vesicular stomatitis virus: mode of transcription. J Gen Virol 34:1, 1977

Bishop DHL, Smith MS: Rhabdoviruses: In Nayak DP (ed): The Molecular Biology of Animal Viruses, Vol 1:167. New York, Marcel Dekker, 1977

Bishop DHL (ed): The Rhabdoviruses. CRC Press, 1979

Emerson SU: Vesicular stomatitis virus: structure and

function of virion components. Curr Top Microbiol Immunol 73:1, 1976

Schneider LG, Diringer H: Structure and molecular biology of rabies virus. Curr Top Microbiol Immunol 75:153, 1976

Wagner RR: Reproduction of rhabdoviruses. In Fraenkel-Conrat H, Wagner RR (eds): Comprehensive Virology, Vol 4, p 1. New York and London, Plenum, 1975

ARENAVIRUSES

Lehmann-Grube F (ed): Lymphocytic choriomeningitis virus and other arenaviruses. Berlin, Heidelberg and London, Springer-Verlag, 1973

Selected Papers

PAPOVAVIRUSES

Acheson NH: Transcription during productive infection with polyoma virus and simian virus 40. Cell 8:1, 1976

Müller U, Zentgraf H, Eicken I, Keller W: Higher order structure of simian virus 40 chromatin. Science 201:406, 1978

Reddy VB, Dhar R, Weissman SM: Nucleotide sequence of the genes for the simian virus 40 proteins VP2 and VP3. J Biol Chem 253:621, 1978

Smart JE, Ito Y: Three species of polyoma virus tumor antigens share common peptides probably near the amino termini of the proteins. Cell 15:1427, 1978

Volckart G, Van de Voorde A, Fiers W: Nucleotide sequence of the simian virus 40 small-t gene. Proc Natl Acad Sci USA 75:2160, 1978

ADENOVIRUSES

Brown DT, Burlingham BT: Penetration of host cell membranes by adenovirus 2. Virol 12:386, 1973

Burger H, Doerfler W: Intracellular forms of adenovirus DNA. III. Integration of the DNA of adenovirus 2 into host DNA in productively infected cells. J Virol 13:975, 1974

Carter TH, Blanton RA: Autoregulation of adenovirus type 5 early gene expression. II. Effect of temperature-sensitive early mutation on virus RNA accumulation. J Virol 28:450, 1978

Chardonnet Y, Dales S: Early events in the interaction of adenoviruses with HeLa cells. Virol 40:462, 478, 1970

Chow LT, Broker TR: The spliced structures of adenovirus 2 fiber message and other late mRNAs. Cell 15:497, 1978

Chow LT, Gelinas RE, Broker TR, Roberts RJ: An amazing sequence arrangement at the 5' ends of adenovirus 2 messenger RNAs. Cell 12:1, 1977

Chow LT, Roberts JM, Lewis JB, Broker TR: Map of cytoplasmic RNA transcripts from lytic adenovirus

type 2, determined by electron microscopy of RNA:DNA hybrids. Cell 11:819, 1977

Darnell JE Jr: Implications of RNA:RNA splicing in evolution of eukaryotic cells. Science 202:1257, 1978

Evans RM, Fraser N, Ziff E, Weber J, Wilson M, Darnell JE Jr: The initiation sites for RNA transcription in ad2 DNA. Cell 12:733, 1977

Goldberg S, Weber J, Darnell JE Jr.: The definition of a large viral transcription unit late in ad2 infection of HeLa cells: mapping by effects of ultraviolet irradiation. Cell 10:618, 1977

Grodzicker T, Anderson C, Sambrook J, Mathews MB: The physical locations of structural genes in adenovirus DNA. Virol 80:111, 1977

Ishibashi M, Maizel JV: The polypeptides of adenovirus. VI. Early and late glycopolypeptides. Virol 58:345, 1974

Lassam NJ, Bayley ST, Graham FL: Synthesis of DNA, late polypeptides and infectious virus by host-range mutants of adenovirus 5 in nonpermissive cells. Virol 87:463, 1978

Lechner RL, Kelley TJ: The structure of replicating adenovirus 2 DNA molecules. Cell 12:1007, 1977

Nevins JR, Darnell JE Jr: Steps in the processing of ad2 mRNA: poly(A)$^+$ nuclear sequences are conserved and poly(A) addition precedes splicing. Cell 15:1477, 1978

Philipson I, Lonberg-Holm K, Pettersson N: Virus-receptor interactions in an adenovirus system. J Virol 2:1064, 1968

van der Vliet PC, Levine AJ: DNA-binding proteins specific for cells infected by adenovirus. Nature [New Biol] 246:170, 1973

Werner G, zur Hausen H: Deletions and insertions in adenovirus type 12 DNA after viral replication in Vero cells. Virol 86:66, 1978

Ziff E, Fraser N: Adenovirus type 2 late mRNAs: structural evidence for 3'-co-terminal species. J Virol 25:897, 1978

HERPESVIRUSES

Ben-Porat T, Kervina M, Kaplan AS: Early functions of the genome of herpesvirus: V. Serological analysis of "immediate-early" proteins. Virol 65:335, 1975

Clements JB, Watson RJ, Wilkie NM: Temporal regulation of herpes simplex virus type 1 transcription: location of transcripts of the viral genome. Cell 12:275, 1977

Fenwick M, Roizman B: Regulation of herpesvirus macromolecular synthesis. VI. Synthesis and modification of viral polypeptides in enucleated cells. J Virol 22:720, 1977

Honess RW, Roizman B: Regulation of herpesvirus macromolecular synthesis: 1. Cascade regulation of the synthesis of three groups of viral proteins. J Virol 14:8, 1974

Honess RW, Roizman B: Regulation of herpesvirus macromolecular synthesis: sequential transition of polypeptide synthesis requires functional viral polypeptides. Proc Natl Acad Sci USA 72:1276, 1975

Knipe DM, Ruyechan WT, Roizman B: Molecular genetics of herpes simplex virus. III. Fine mapping of a genetic locus determining resistance to phosphonoacetate by two methods of marker transfer. J Virol 29:698, 1979

Kozak M, Roizman B: Regulation of herpesvirus macromolecular synthesis: nuclear retention of nontranslated viral RNA sequences. Proc Natl Acad Sci USA 71:4322, 1974

Nii S, Morgan C, Rose HM, Hsu KC: Electron microscopy of herpes simplex virus. IV. Studies with ferritin-conjugated antibodies. J Virol 2:1172, 1968

Preston VG, Davison AJ, Marsden HS, Timbury MC, Subak-Sharpe JH, Wilkie, NM: Recombinants between herpes simplex virus types 1 and 2: analyses of genome structures and expression of immediate early polypeptides. J Virol 28:499, 1978

Tralka TS, Costa J, Rabson A: Electron microscopic study of herpesvirus saimiri. Virol 80:158, 1977

POXVIRUSES

Cooper JA, Moss B: Transcription of vaccinia virus mRNA coupled to translation in vitro. Virol 88:149, 1978

Dales S, Milovanovitch V, Pogo BGT, Weintraub SB, Huima T, Wilton S, McFadden G: Biogenesis of vaccinia: isolation of conditional lethal mutants and electron microscopic characterization of their phenotypically expressed defects. Virol 84:403, 1978

Hruby DE, Guarino LA, Kates JR: Vaccinia virus replication. I. Requirement for the host-cell nucleus. J Virol 29:705, 1979

Katz E, Moss B: Vaccinia virus structural polypeptide derived from a high molecular weight precursor: formation and integration into virus particles. J Virol 6:717, 1970

Lake JR, Silver M, Dales S: Biogenesis of vaccinia: Complementation and recombination analysis of one group of conditional-lethal mutants defective in envelope cell-assembly. Virol 96:9, 1979

McAuslan BR: The induction and repression of thymidine kinase in the poxvirus-infected HeLa cell. Virol 21:383, 1963

Oda K, Joklik WK: Hybridization and sedimentation studies on "early" and "late" vaccinia messenger RNA. J Mol Biol 27:395, 1967

Pelham HRB: Use of coupled transcription and translation to study mRNA production by vaccinia cores. Nature 269:533, 1977

Pennington TH: Vaccinia virus polypeptide synthesis: sequential appearance and stability of pre- and post-replicative polypeptides. J Gen Virol 25:433, 1974

Pogo BGT, Dales S: Biogenesis of vaccinia: Separation of early stages from maturation by means of hydroxyurea. Virol 43:144, 1971

PICORNAVIRUSES

Butterworth BE, Rueckert RR: Kinetics of synthesis and cleavage of encephalomyocarditis virus-specific proteins. Virol 50:535, 1972

Flanegan JB, Baltimore D: Poliovirus polyuridylic acid polymerase and RNA replicase have the same viral polypeptide. J Virol 29:352, 1979

Korant B, Chow N, Lively M, Powers J: Virus specified protease in poliovirus infected HeLa cells. Proc Natl Acad Sci USA 76:2992, 1979

McGregor S, Rueckert RR: Picornaviral capsid assembly: similarity of rhinovirus and enterovirus precursor subunits. J Virol 21:548, 1977

Medrano L, Green H: Picornavirus receptors and picornavirus multiplication in human-mouse hybrid cell lines. Virol 54:515, 1973

Paucha AE, Colter JS: Evidence for control of translation of the viral genome during replication of mengovirus and poliovirus. Virol 67:300, 1975

Rekosh D: Gene order in poliovirus capsid proteins. J Virol 9:479, 1972

Summers DF, Maizel JV, Darnell JE: Evidence for virus-specific noncapsid proteins in poliovirus-infected HeLa cells. Proc Natl Acad Sci USA 56:505, 1965

TOGAVIRUSES

Birdwell CR, Strauss EG, Strauss JH: Replication of Sindbis virus. III. An electron microscopic study of virus maturation using the surface replica technique. Virol 56:429, 1973

Birdwell CR, Strauss JH: Distribution of the receptor sites for Sindbis virus on the surface of chicken and BHK cells. J Virol 14:672, 1974

Bonatti S, Cancedda R, Blobel G: Membrane biogenesis. In vitro cleavage, core glycosylation and integration into microsomal membranes of Sindbis virus glycoproteins. J Cell Biol 80:219, 1979

Brzeski H, Clegg JCS, Atkins GJ, Kennedy SIT: Regulation of the synthesis of Sindbis virus-specified RNA: role of the virion core protein. J Gen Virol 38:461, 1978

Brzeski H, Kennedy SIT: Synthesis of Sindbis virus nonstructural polypeptides in chick embryo fibroblasts. J Virol 22:420, 1977

Helenius A, Morein B, Fries E, Simons K, Robinson P, Schirrmacher V, Terhorst C, Strominger JL: Human (HLA-A and HLA-B) and murine (H-2K and H-2D) histocompatibility antigens are cell surface receptors for Semliki Forest virus. Proc Natl Acad Sci USA 75:3846, 1978

Schlesinger MJ, Schlesinger S: Large-molecular-weight precursors of Sindbis virus proteins. J Virol 11:1013, 1973

Simmons DT, Strauss JH: Translation of Sindbis virus 26S RNA and 49S RNA in lysates of rabbit reticulocytes. J Mol Biol 86:397, 1974

BUNYAVIRUSES

Gentsch JR, Bishop DHL: The small viral RNA segments of bunyaviruses codes for viral nucleocapsid protein. J Virol 28:417, 1978

REOVIRUSES

Acs G, Klett H, Schonberg M, Christman J, Levin DH, Silverstein SC: Mechanism of reovirus double-stranded ribonucleic acid synthesis in vivo and in vitro. J Virol 8:684, 1971

Bellamy AR, Joklik WK: Studies on reovirus RNA. J Mol Biol 29:19, 27, 1967

Chang C, Zweerink HJ: Fate of parental reovirus in infected cell. Virol 46:544, 1971

Huismans H, Joklik WK: Reovirus-coded polypeptides in infected cells: isolation of two native monomeric polypeptides with affinity for single-stranded and double-stranded RNA, respectively. Virol 70:411, 1976

McCrae MA, Joklik WK: The nature of the polypeptide encoded by each of the ten double-stranded RNA segments of reovirus type 3. Virol 89:578, 1978

Schonberg M, Silverstein SC, Levin DH, Acs G: Asynchronous synthesis of the complementary strands of the reovirus genome. Proc Natl Acad Sci USA 68:505, 1971

Silverstein SC, Schonberg M, Levin DH, Acs G: The reovirus replicative cycle: conservation of parental RNA and protein. Proc Natl Acad Sci USA 67:275, 1970

Watanabe Y, Millward S, Graham AF: Regulation of transcription of the reovirus genome. J Mol Biol 36:107, 1968

Zweerink HJ: Multiple forms of SS → DS RNA polymerase activity in reovirus-infected cells. Nature 247:313, 1974

Zweerink HJ, McDowell MJ, Joklik WK: Essential and nonessential noncapsid reovirus proteins. Virol 45:716, 1971

ORTHOMYXOVIRUSES

Almond JW: A single gene determines the host range of influenza virus. Nature 270:617, 1977

Bouloy M, Plotch SJ, Krug RM: Globin mRNAs are primers for the transcription of influenza viral RNA in vitro. Proc Natl Acad Sci USA 75:4886, 1978

Caliguiri LA, Holmes KV: Host-dependent restriction of influenza virus maturation. Virol 92:15, 1979

Inglis SC, Mahy BWJ: Polypeptides specified by the influenza virus genome. III. Control of synthesis in infected cells. Virol 95:154, 1979

Klenk H, Rott R, Orlich M: Further studies on the activation of influenza virus by proteolytic cleavage of the hemagglutinin. J Gen Virol 36:151, 1977

Lamb RA, Choppin PW: Synthesis of influenza virus proteins in infected cells: translation of viral polypeptides, including three P polypeptides, from RNA produced by primary transcription. Virol 74:504, 1976

Mowshowitz SL: P1 is required for initiation of cRNA synthesis in WSN influenza virus. Virol 91:493, 1978

Palese P, Ritchey MB, Schulman JL: P1 and P3 proteins of influenza virus are required for complementary RNA synthesis. J Virol 21:1187, 1977

Schulman JL, Palese P: Virulence factors of influenza A viruses: WSN virus neuraminidase required for plaque production in MDBK cells. J Virol 24:170, 1977

PARAMYXOVIRUSES

Collins PL, Hightower LE, Ball LA: Transcription and translation of Newcastle disease virus mRNA's in vitro. J Virol 28:324, 1978

Gimenez HB, Pringle CR: Seven complementation groups of respiratory syncytial virus temperature-sensitive mutants. J Virol 27:459, 1978

Glazier K, Raghow R, Kingsbury DW: Regulation of Sendai virus transcription: evidence for a single promoter in vivo. J Virol 21:863, 1977

Graves MC, Silver SM, Choppin PW: Measles virus polypeptide synthesis in infected cells. Virol 86:254, 1978

Lamb RA, Choppin PW: Determination by peptide mapping of the unique polypeptides in Sendai virions and infected cells. Virol 84:469, 1978

Mountcastle WE, Compans RW, Lackland H, Choppin PW: Proteolytic cleavage of subunits of the nucleocapsid of the paramyxovirus Simian virus 5. J Virol 14:1253, 1974

RHABDOVIRUSES

Ball LA, White CN: Order of transcription of genes of vesicular stomatitis virus. Proc Natl Acad Sci USA 73:442, 1976

Birdwell CR, Strauss JH: Maturation of vesicular stomatitis virus: Electron microscopy of surface replicas of infected cells. Virol 59:587, 1974

Clinton GM, Little SP, Hagen FS, Huang AS: The matrix (M) protein of vesicular stomatitis virus regulates transcription. Cell 15:1455, 1978

Erving RA, Toneguzzo F, Rhee SH, Hoffmann T, Ghosh HP: Synthesis and assembly of membrane glycoproteins: presence of leader peptide in nonglycosylated precursor of membrane glycoprotein of vesicular stomatitis virus. Proc Natl Acad Sci USA 76:570, 1979

Freeman GJ, Rose JK, Clinton GM, Huang AS: RNA synthesis of vesicular stomatitis virus. VII. Complete separation of the mRNA's of vesicular stomatitis virus by duplex formation. J Virol 21:1094, 1977

Hale AH, Witte ON, Baltimore D, Eisen HN: Vesicular stomatitis virus glycoprotein is necessary for H-2-restricted lysis of infected cells by cytotoxic T lymphocytes. Proc Natl Acad Sci USA 75:970, 1978

Herman RC, Adler S, Lazzarini RA, Collonno RJ, Banerjee AK, Westphal H: Intervening polyadenylate sequences in RNA transcripts of vesicular stomatitis virus. Cell 15:587, 1978

Knipe DM, Baltimore D, Lodish HF: Separate pathways of maturation of the major structural proteins of vesicular stomatitis virus. J Virol 21:1128, 1977

Li E, Tabas I, Kornfeld S: The synthesis of complex-type oligosaccharides. I, II, and III. J Biol Chem 253:7762, 7771 and 7779, 1978

Pennica D, Lynch KR, Cohen PS, Ennis HL: Decay of vesicular stomatitis virus mRNAs in vivo. Virol 94:484, 1979

Rose JK: Nucleotide sequences of ribosome recognition sites in messenger RNAs of vesicular stomatitis virus. Proc Natl Acad Sci USA 74:3672, 1977

Rothman JE, Lodish HF: Synchronized transmembrane insertion and glycosylation of a nascent membrane protein. Nature 269:775, 1977

Schloemer RH, Wagner RR: Mosquito cells infected with vesicular stomatitis virus yield unsialylyated virions of low infectivity. J Virol 15:1029, 1975

Scholtissek C, Rott R, Ham G, Kaluza G: Inhibition of the multiplication of vesicular stomatitis and Newcastle disease virus by 2-deoxy-glucose. J Virol 13:1186, 1974

Toneguzzo F, Ghosh HP: In vitro synthesis of vesicular stomatitis virus membrane glycoprotein and insertion into membranes. Proc Natl Acad Sci USA 75:715, 1978

Villarreal LP, Breindl M, Holland JJ: Determination of molar ratios of vesicular stomatitis virus-induced RNA species in BHK_{21} cells. Biochem 15:1663, 1976

Zee YC, Hackett AJ, Talens L: Vesicular stomatitis virus maturation sites in six different host cells. J Gen Virol 7:95, 1970

CELL RECEPTORS

Fan DP, Sefton BM: The entry into host cells of Sindbis virus, vesicular stomatitis virus and Sendai virus. Cell 15:985, 1978

Lomberg-Holm K, Crowell RL, Philipson L: Unrelated animal viruses share receptors. Nature 259:679, 1976

CHAPTER 6

The Effect of Virus Infection on the Host Cell

The effect of virus infection on host cells is far more difficult to study in molecular terms than the process of virus multiplication. Study of the multiplication process requires merely the ability to recognize and measure virus-specified macromolecules; study of the effect on host cells requires a detailed knowledge of the functioning of the normal host cell. In fact, studies that have focused attention on the effect of virus infection on host cells have widened our knowledge of the functioning of uninfected cells, and the acquisition of such knowledge has been one of the important spin-offs of the study of virus-cell interactions.

Whatever the reason that lytic viruses destroy their host cells, several causes can be ruled out. One is that virus synthesis creates an excessive demand for protein and nucleic acid precursors, so that competition causes a shortage of building blocks that prevents synthesis of host cell macromolecules. This is unlikely, since the amount of viral material that is synthesized rarely exceeds 10 percent of total host cell material. Considerations of this nature also rule out the necessity for breakdown of host cell material in order to provide precursors for the synthesis of viral macromolecules.

Since both nucleic acid and protein synthesis are absolutely dependent on the supply of energy, the largest virus yields would be expected in cells that are damaged least for the longest periods of time. Many highly lytic viruses are, however, very successful. This is primarily because they take over the host cell's synthetic apparatus very rapidly and multiply extensively in the brief period of time for which the apparatus can function.

Cytopathic effects

The most easily detected effects of infection with lytic viruses are the cytopathic effects, which can be observed both macroscopically and microscopically. Plaque formation is due to the cytopathic effect of viruses; viruses kill the cells in which they multiply, and plaques are the areas of killed cells. The light microscope, as well as the electron microscope, often reveals changes in a variety of cell organelles soon after infection. Frequently the nucleus is affected first, with pyknosis, changes in nucleolar structure, and margination of the chromatin. Changes in the cell membrane usually follow; cells gradually lose their ability to adhere to supporting surfaces and therefore round up, and sometimes develop a strong tendency to fuse with one another (see below). This is then often followed by the appearance, either in the nucleus or in the cytoplasm, of distinct spreading foci which are generally composed of fibrillar material; these are the classic inclusion bodies that have long been described by cytologists, and they represent the sites of virus-directed biosynthesis and morphogenesis. Finally, at about the time when structural viral protein synthesis proceeds at its maximum pace, necrotic and grossly degradative changes become noticeable. They may be attributed to at least three causes. First, by this time interference with host cell macromolecular biosynthesis is generally complete (see below). It is known that all host cell macromolecules turn over to a greater or lesser extent—that is, they are continually broken down and resynthesized by a mechanism that operates no matter whether the cells grow or not. Inhibition of resynthesis in the presence of continuing breakdown could clearly lead to structural and functional failure. Second, plasma membrane function declines, probably because host protein and lipid synthesis ceases and because patches of host membrane protein are replaced by virus-coded ones. Certainly permeability increases soon after infection, and before long loss of plasma membrane function results in failure to maintain the proper intracellular ionic environment and in diminished transport of essential nutrients into the cell and of waste products out of it. Third, the membranes lining lysosomes also begin to fail, and as a result the degradative hydrolytic enzymes that they contain begin to leak out into the cytoplasm, thereby exacerbating the effects caused by the other mechanisms.

The net result of cell necrosis is the release of those viruses that do not bud from the cell membrane. In general, the smaller the virus, the more readily it is released. Large viruses, such as poxviruses, are often retained in the ghosts of infected cells for considerable periods of time. In the body, the situation may be different, for damaged cells may become phagocytosed, thereby providing an additional mechanism for the dissemination of viral progeny.

Whereas the reasons for the necrotic and degradative changes that occur during the late phases of the virus multiplication cycle are reasonably well understood, it is not as obvious why some viruses cause cytopathic effects soon after infection. For some viruses it is likely that early cytopathic effects are caused by early viral proteins—that is, proteins that are synthesized after infection; for others, it seems that such effects are caused by components of the invading virus particles. It is interesting in this regard that the adenovirus fiber is known to be cytotoxic, and that as few as ten heat-inactivated vaccinia virus particles (which cannot possibly express any genetic function) can kill cells. On the other hand, it is also known that empty reovirus particles—that is, particles that contain no RNA—cause no cytopathic effects, even at very high multiplicities.

Inhibition of host macromolecular biosynthesis

The induction of cytopathic effects early after infection may be intimately related to the more or less rapidly developing inhibition of host protein, DNA and RNA synthesis, which is a fundamentally important factor in the lytic virus-cell interaction.

Host protein synthesis is inhibited first. This is often missed if only the overall rate of protein synthesis is measured, since viral protein synthesis generally takes over as host protein synthesis declines. Inhibition of host protein synthesis then quickly causes inhibition of host DNA replication, which is known to depend on the activity of several short-lived proteins. Host RNA synthesis also soon ceases, probably not as a result of any direct effect on DNA-dependent RNA polymerases, but because of changes in the physical state of the DNA [margination of

chromatin, changes in nucleolar structure (see above)] which prevent it from acting as a template for transcription. As a result, not only is progressively less and less host cell messenger RNA synthesized, but the supply of new ribosomes is also quickly interrupted; in fact, the only ribosomes that are usually available for protein synthesis during the virus multiplication cycle are those that are present in the cytoplasm at the time of infection.

The reasons why infection with viruses inhibits host protein synthesis have been studied extensively, but are not well understood. Sometimes the reason may be that less and less host cell messenger RNA reaches the cytoplasm; however, it is clear that in many virus infections the rate of host protein synthesis decreases far more rapidly than would be expected on the basis of the decay rate of messenger RNA in uninfected cells. For example, infection with high multiplicities of vaccinia virus inhibits host protein synthesis by well over 90 percent within two hours (Fig. 6-1). This is not a general toxic effect on the host cell, since viral protein is already being synthesized vigorously at this time; nor is inhibition of host protein synthesis usually due to destruction of host cell messenger RNA, since the continued existence of this RNA in undiminished amounts can often be demonstrated by its ability to be translated within in vitro protein synthesizing systems prepared from uninfected cells. Rather, shortly after infection a mechanism must develop that actively prevents from being translated the cellular messenger RNA which is present in the cytoplasm at the time of infection.

There is probably not one, but several such mechanisms, depending on the nature of the virus. Viral messenger RNAs are, in general, more efficient initiators of protein synthesis than host cell messenger RNAs; in vitro protein-synthesizing systems presented with mixtures of host and viral messenger RNAs usually translate the latter almost exclusively. Competition with viral messenger RNAs is therefore undoubtedly a major reason why translation of cellular messenger RNAs gradually ceases; simple competition, however, cannot account for the very rapid inhibition such as that illustrated in Fig. 6-1. In the case of some viruses, a component of parental virus particles appears to inhibit protein synthesis; in the case of others,

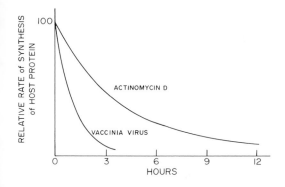

FIG. 6-1. Effect of infection with vaccinia virus, as well as of actinomycin D, on cellular protein synthesis in mouse L fibroblasts. Actinomycin D inhibits RNA transcription; the decrease in the rate of host protein synthesis in its presence therefore reflects the stability of cellular messenger RNA, whose average half-life is about three hours. Infection with vaccinia virus inhibits host protein synthesis much more rapidly; presumably the infection either causes host messenger RNA to be inactivated or it actively prevents host messenger RNA from being translated.

such as rhabdoviruses and poxviruses, the inhibitor seems to be newly synthesized RNA or a protein translated from it. Infection with picornaviruses, such as poliovirus, causes inactivation of the cap-dependent recognition mechanism of messenger RNAs during protein synthesis initiation; and it has also been suggested that infection may cause an influx of sodium ions into the cell, which would inhibit initiation of translation of host cell messenger RNAs and enhance that of viral messenger RNAs. Finally, some viruses, such as some herpesviruses, do indeed cause host cell messenger RNA to be degraded.

While the hypothesis that inhibition of host DNA and RNA synthesis is secondary to inhibition of host protein synthesis is probably correct for most viral infections, several exceptions have been reported. Thus VSV and Sindbis virus rapidly shut off host RNA and DNA synthesis respectively by mechanisms that do not appear to involve protein synthesis inhibition; a component of parental poxvirus particles has been reported to inhibit host DNA synthesis; and the isolated G glycoprotein of VSV inhibits cell DNA, but not protein synthesis.

Changes in the regulation of gene expression

Virus infection may also affect the regulation of host genome expression. Thus the activity of certain enzymes on the pathway of nucleic acid biosynthesis often increases after infection; for example, infection with the papovaviruses polyoma and SV40 causes increases in the activity of at least six enzymes. One of these enzymes is a deoxypyrimidine kinase that phosphorylates both deoxyuridine and deoxycytidine. Cells possess two forms of this enzyme, one of which is synthesized in resting, the other in growing cells. When resting cells are infected with SV40 or adenovirus, it is the latter form of the enzyme whose formation is induced. Further, infection with almost all viruses leads to the synthesis of a new protein, interferon, which will be discussed in Chapter 8. These lines of evidence suggest that infection may upset the mechanisms that usually prevent the expression of a large portion of cellular DNA, and that there may be a period of time during the early part of the infection cycle when more of the cell's genome expresses itself in infected than in uninfected cells.

Appearance of new antigenic determinants on the cell surface

Sooner or later following infection, the outer cell membrane is modified. This manifests itself in a variety of ways: the cells' morphology changes, they become more agglutinable by the lectin concanavalin A (Chap. 9), permeability increases, and new antigenic determinants appear on the cell surface. When the infecting virus is an enveloped virus, these new determinants are likely to be viral envelope proteins that have become incorporated into the cell membrane, but new antigenic determinants also appear on the surfaces of cells that are infected with nonenveloped viruses. The presence of these determinants serves to alert the immune mechanism that the cells are infected and should therefore be eliminated (Fig. 6-2). They are discussed further in Chapter 9 in relation to tumor and transplantation antigens.

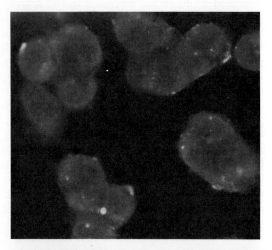

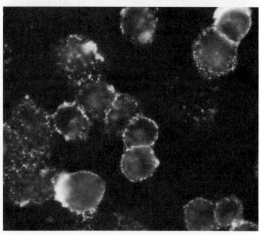

FIG. 6-2. Virus-specified surface antigen on HeLa cells infected with vaccinia virus. Top. Uninfected cells. Bottom. Cells infected for 8 hours. The cells were unfixed and were stained with rabbit antiserum to rabbit cells infected with vaccinia virus. (Courtesy of Dr. Yoshiaki Ueda.)

Cell fusion

Several viruses, particularly paramyxoviruses such as Sendai virus and herpesviruses, cause cells to fuse with one another, which results in the formation of giant syncytia, masses of cytoplasm bounded by one membrane that may contain hundreds and even thousands of nuclei. Fusion seems to be caused by changes induced in cell membranes as a result of interaction with certain glycoproteins in viral envelopes (glycoprotein B2 in herpesvirus envelopes and the F spike glycoprotein of paramyxoviruses); it is therefore caused not only by active, but also by inactivated virus particles. Further, it can be induced not only among identical, but also among different cells. The products of fusion are either heterokaryons—cells containing several nuclei of different types (Fig. 6-3)—or hybrid cells that contain the fused nuclei of the parents. Hybrid cells are frequently viable and have become widely used for studies in somatic cell genetics. They are produced by fusing mix-

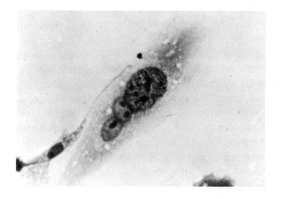

FIG. 6-3. Cell fusion induced by UV-inactivated Sendai virus. Three chick embryo fibroblasts have fused to yield the heterokaryon shown here, which contains three nuclei. The two small ones are normal chick nuclei; the large one is from a chick cell transformed with Rous sarcoma virus. It has been labeled with ^{3}H-thymidine; the grains which it seems to contain are actually silver grains, caused by the disintegration of ^{3}H atoms in a thin layer of photographic emulsion which overlies this cell. Detection of radioactive label by this technique is known as autoradiography. ×400. (From Svoboda and Dourmashkin: J Gen Virol 4:523, 1969.)

tures of the two cell lines to be hybridized with inactivated Sendai virus (or, more commonly nowadays, by treatment with polyethylene glycol) and are then cloned. They are particularly useful for determining the chromosomal location of specific genes; as hybrid cells multiply, they often lose chromosomes, and one can readily correlate the loss of a specific chromosome with the loss of a particular gene function. In recent years, many human genes have been assigned to chromosomes in this manner, using human-mouse hybrid cells that lose human chromosomes. Further applications of hybrid cells are: (1) hybrids of mutant cells permit genetic complementation analysis, (2) hybridization of virus-transformed cells with normal cells permits the rescue of tumor virus genomes integrated into the host cell genome (see Chap. 9), (3) hybrid cells can be used to determine whether events such as the induction of the synthesis of specific proteins are under positive or negative control, and (4) the genetic factors that control susceptibility to virus infection and the expression of tumorigenicity can be studied with the aid of somatic cell hybrids. The technique has also become very important in immunology for the production of "hybridomas" that produce individual species of antibodies (monoclonal antibodies) (see Chap. 4).

Abortive infection

Most of the changes described so far relate to the lytic multiplication cycle under conditions of productive infection—that is, when virus multiplies to high titer in permissive cells. However, viruses can also infect cells that are not fully permissive and even cells that are nonpermissive. In such cells, viruses cannot multiply because some essential step of the multiplication cycle cannot proceed. Examples of this type are the abortive infection of HeLa cells by influenza virus, of dog kidney cells by herpes simplex virus, of pig kidney cells by certain mutants of rabbitpox virus, of monkey cells by human adenovirus, and many others. The infection of permissive cells in the presence of antiviral agents (see Chap. 8) is also abortive. In almost all such cases the viral genome begins to express itself, the alterations in the host cell that were described above occur, and the cells die.

Persistent infections

As a rule, the infection of permissive cells with lytic viruses leads to productive infection and cell death. Occasionally, however, cell cultures are observed which, while multiplying more or less normally, nevertheless release significant amounts of virus. Such cultures are said to be persistently infected. Stable relationships between cell growth and virus multiplication, which occur not only in vitro, but also in vivo, are of two kinds.

The first involves infection by viruses that cause a minimum of the type of cell damage discussed above. A good example is provided by the paramyxovirus SV5. This virus interacts with many cells by means of a lytic, cytocidal interaction. However, when it infects monkey kidney cells, it causes practically no cell damage and permits the infected cells to grow freely while multiplying itself. In such cultures, all cells multiply, all cells are infected, and all cells produce virus. Infection of new cells plays no role in this situation; consequently this type of infection cannot be cured by the addition of neutralizing antibody.

The second situation is quite different. Here the virus-cell interaction is lytic—that is, the cells that are infected always die—but the extent of virus multiplication is limited, so that the yield is small. In addition, various factors reduce the probability of reinfection, so that the proportion of infected cells in the total cell population is kept small and constant. The factors that induce and maintain this type of persistent infection are as follows. First, this type of infection occurs most readily in cells that are almost, but not quite, nonpermissive and that, therefore, produce only small amounts of virus. Second, it occurs in situations where factors in the medium, such as antibody or interferon, prevent the majority of released progeny virus from infecting new cells. Third, many of the virus strains that have been isolated from persistently infected cells and animals are less virulent, less cytopathic, and, interestingly enough, temperature-sensitive with respect to their ability to multiply. This has now been found for foot-and-mouth disease virus, Coxsackie virus, Sindbis virus and WEE, influenza virus, Newcastle disease virus, Sendai virus, mumps and measles virus, VSV and herpesvirus, and others. Attempts to isolate viruses from the tissues of patients with suspected chronic and persistent virus infections should take account of the possibility that they may grow much better at 31C-33C than at 37C. Finally, virus yields from persistently infected cells sometimes contain, in addition to infectious virus, deletion mutants, that is, virus particles that lack a portion of their genome (see Chap. 8). These virus particles, which are unable to multiply on their own, are nevertheless capable of interfering with the multiplication of infectious virus and are therefore known as defective interfering or DI particles. Their presence has the effect of dampening the effect of virus infection, that is, it reduces both cell damage and the amount of infectious virus that is produced. Their role in initiating and maintaining persistent infections is readily demonstrated as follows: if DI particles of VSV, the so-called T particles (see Chap. 7), are inoculated into animals together with normal, infectious VSV, an otherwise fatal disease is converted to a slowly progressing persistent infection. The formation of DI particles has now been implicated in chronic infections with Sendai virus, NDV and measles, lymphocytic choriomeningitis virus, Sindbis virus and WEE.

In summary, the replacement of virulent by less virulent virus strains and the inhibition of virus multiplication by DI particles are major factors in the establishment and maintenance of persistent virus infections. Persistent infections of this class are important in the intact organism, where the presence of antiviral antibody and interferon in amounts too low to eliminate virus may provide conditions favoring low-level persistent infections. It should be noted that this kind of persistent infection can be cured by the addition of large amounts of neutralizing antibody.

We have so far considered only persistently infected cell populations that release significant amounts of virus. However, there are also persistently infected cells that never release virus, or do so only rarely. One type of such cells are cells transformed by RNA tumor viruses; they will be discussed in Chap. 9. Another type are cells that are almost nonpermissive—that is, virus multiplication within them can start, but is arrested at some stage or other. Examples of this type are as follows. Most herpesviruses have a pronounced tendency to infect cells in which their development is normally blocked, but from which they can be released in infectious form if conditions are right. The released virus particles can then initiate productive infection in some other type of cell. This is the phenomenon of latency, the best known example of which is the latent herpes simplex virus type I infection of the neurons of sensory ganglia, from which the virus emerges to cause infections of the skin such as fever blisters (see Chap. 13). The form in which herpesvirus persists in ganglion cells is not known. This type of persistent infection, whether by herpes simplex virus or varicella-zoster virus or cytomegalovirus, is generally not progressive; that is, the virus multiplies for some time, then becomes latent, and then reinitiates productive infection. Other persistent infections of this type tend to cause progressive disease. Examples of this type are subacute sclerosing panencephalitis (SSPE), which is an infection of the central nervous system by a virus closely related to measles virus, in which large amounts of viral nucleocapsids are present in infected cells (see Chap. 25); and progressive multifocal leucoencephalopathy (PML), which is a papovavirus infection (see Chap. 15).

Modification of cellular permissiveness

Occasionally, virus infection can alter to an amazing degree the permissiveness of cells for completely unrelated viruses. The effects are always highly specific. Thus, adeno-associated virus (AAV) can multiply only in cells infected with adenovirus; SV40 enables human adenovirus to grow in monkey cells; and poxviruses such as vaccinia and fibroma virus enable VSV to multiply in rabbit cells. The nature of the helper functions is not known, although it is clear that in each case the helper virus must be able to express at least some early function. However, it may well be that in all these situations the helper virus does not provide a gene product that is used in a specific manner by the other virus, but rather that the helper modifies the cell in some way essential for the ability of the second virus to multiply.

FURTHER READING

Books and Reviews

Blattner RJ, Williamson AP, Heys FM: Role of viruses in the etiology of congenital malformations. Prog Med Virol 15:1, 1973

Davis FM, Adelberg EA: Use of somatic cell hybrids for analysis of the differentiated state. Bacteriol Rev 37:197, 1973

Docherty JJ, Chopan M: The latent herpes simplex virus. Bacteriol Rev 38:337, 1974

Fucillo DA, Sever JL: Viral teratology. Bacteriol Rev 37:19, 1973

Handmaker SD: Hybridization of eukaryotic cells. Ann Rev Microbiol 27:189, 1973

Harris H: Cell Fusion. Cambridge, Mass, Harvard Univ Press, 1975

Huang AS: Viral pathogenesis and molecular biology. Bact Rev 41:811, 1977

Huang AS, Baltimore D: Defective interfering animal viruses. In Fraenkel-Conrat and Wagner RR (eds): Comprehensive Virology. Vol 10:73, New York, Plenum, 1977

Nichols WW: Virus-induced chromosome abnormalities. Ann Rev Microbiol 24:479, 1970

Poste G: Virus-induced polykaryocytosis and the mechanism of cell fusion. Adv Virus Res 16:303, 1970

Rifkin DB, Quigley JP: Virus-induced modification of cellular membranes related to viral structure. Ann Rev Microbiol 28:325, 1974

Smith H: Mechanisms of virus pathogenicity. Bacteriol Rev 36:291, 1972

Stevens JG, Todaro GJ, Fox CF (eds): Persistent viruses. New York, San Francisco, London, Academic Press, 1978

Selected Papers

CYTOPATHIC EFFECTS

Bablanian R, Baxt B, Sonnabend JA, Esteban M: Studies on the mechanisms of vaccinia virus cytopathic effects. II. Early cell rounding is associated with vi-

rus polypeptide synthesis. J Gen Virol 39:403, 1978

Bablanian R, Estaban M, Baxt B, Sonnabend JA: Studies on the mechanisms of vaccinia virus cytopathic effects. I. Inhibition of protein synthesis in infected cells is associated with virus-induced RNA synthesis. J Gen Virol 39:391, 1978

Baxt B, Bablanian R: Mechanisms of vesicular stomatitis virus-induced cytopathic effects. I. Early morphologic changes induced by infectious and defective-interfering particles. Virol 72:370, 1976

Baxt B, Bablanian R: Mechanisms of vesicular stomatitis virus-induced cytopathic effects. II. Inhibition of macromolecular synthesis induced by infectious and defective-interfering particles. Virol 72:383, 1976

Ebina T, Satake M, Ishida N: Involvement of microtubules in cytopathic effects of animal viruses: early proteins of adenovirus and herpesvirus inhibit formation of microtubular paracrystals in HeLa-S3 cells. J Gen Virol 38:535, 1978

Hecht TT, Summers DF: Effect of vesicular stomatitis virus infection on the histocompatibility antigen of L cells. J Virol 10:578, 1972

Homma M: Specific chromosome aberrations in cells persistently infected with hemadsorption type 2 virus. Virol 34:60, 1968

Kjellén L, Ankerst J: Cytotoxicity of adenovirus-antibody aggregates: sensitivity to different cell strains, and inhibition by hexon antiserum and by complement. J Virol 12:25, 1973

Koschel K, Aus HM, ter Meulen V: Lysosomal enzyme activity in poliovirus-infected HeLa cells and vesicular stomatitis virus-infected L cells: biochemical and histochemical comparative analysis with computer-aided techniques. J Gen Virol 35:359, 1974

McDougall JK: Adenovirus-induced chromosome aberrations in human cells. J Gen Virol 12:43, 1971

Marcus PI, Sekellick MJ: Cell killing by viruses. 1. Comparison of cell-killing, plaque-forming, and defective-interfering particles of vesicular stomatitis virus. Virol 57:321, 1974

Marvaldi JL, Lucas-Lenard J, Sekellick MJ, Marcus PI: Cell killing by viruses. IV. Cell killing and protein synthesis inhibition by vesicular stomatitis virus require the same gene functions. Virol 79:267, 1977

Nachtigal M, Duff R, Rapp F: Chromosome aberrations in Syrian hamster embryo cells transformed after exposure to ultraviolet-irradiated herpes simplex virus 1 or 2. J Natl Cancer Inst 54:97, 1975

Taylor MW, Cordell B, Souhrada M, Prather S: Viruses as an aid to cancer therapy: regression of solid and ascites tumors in rodents after treatment with bovine enteroviruses. Proc Natl Acad Sci USA 68:836, 1971

CELL FUSION

Deisseroth A, Burk R, Picciano D, et al: Synthesis in somatic cell hybrids: globin gene expression in hybrids between mouse erythroleukemia and human marrow cells or fibroblasts. Proc Natl Acad Sci USA 72:1102, 1975

Ege T, Krondahl U, Ringertz NR: Introduction of nuclei and micronuclei into cells and enucleated cytoplasms by Sendai virus induced fusion. Exp Cell Res 88:428, 1974

Manservigi R, Spear PG, Buchan A: Cell fusion induced by herpes simplex virus is promoted and suppressed by different viral glycoproteins. Proc Natl Acad Sci USA 74:3913, 1977

Sekiguchi T, Sekiguchi F, Tomii S: Complementation in hybrid cells derived from mutagen-induced mouse clones deficient in hypoxanthine-guanine phosphoribosyl-transferase activity. Exp Cell Res 88:410, 1974

Tedesco TA, Diamond R, Orkwiszewski KG, Roedecker HJ, Croce CM: Assignment of the human gene for hexose-1-phosphate uridylyltransferase to chromosome 3. Proc Natl Acad Sci USA 71:3483, 1974

Wiener F, Klein G, Harris H: The analysis of malignancy by cell fusion. VI. Hybrids between different tumor cells. J Cell Sci 16:189, 1974

INHIBITION OF HOST MACROMOLECULAR
BIOSYNTHESIS

Abreu SL, Lucas-Lenard J: Cellular protein synthesis shutoff by mengovirus translation of nonviral and viral mRNA's in extracts from uninfected and infected Ehrlich ascites tumor cells. J Virol 18:182, 1976

Atkins GJ: The effect of infection with Sindbis virus and its temperature-sensitive mutants on cellular protein and DNA synthesis. Virol 71:593, 1976

Baglioni C, Simili M, Shafritz DA: Initiation activity of EMC virus RNA, binding to initiation factor eIF-4B and shut-off of host cell protein synthesis. Nature 275:240, 1978

Bienz K, Egger D, Rasser Y, Loeffler H: Differential inhibition of host cell RNA synthesis in several picornavirus-infected cell lines. Intervirol 10:209, 1978

Carrasco L, Smith AE: Sodium ions and the shut-off of host cell protein synthesis by picornaviruses. Nature 264:807, 1976

Celma ML, Ehrenfeld E: Effect of poliovirus double-stranded RNA on viral and host-cell protein synthesis. Proc Natl Acad Sci USA 71:2440, 1974

Drillien R, Spehner D, Kirn A: Host range restriction of vaccinia virus in Chinese hamster ovary cells: Relationship to shut-off of protein synthesis. J Virol 28:843, 1978

Ehrenfeld E, Lund H: Untranslated vesicular stomatitis virus messenger RNA after poliovirus infection. Virol 80:297, 1977

Farmilo AJ, Stanners CP: Mutant of vesicular stomatitis virus which allows deoxyribonucleic acid synthesis and division in cells synthesizing viral ribonucleic acid. J Virol 10:605, 1972

Fenwick ML, Walker MJ: Suppression of the synthesis of cellular macromolecules by herpes simplex virus. J Gen Virol 41:37, 1978

Golini F, Thach SS, Birge CH, Safer B, Merrick WC, Thach RE: Competition between cellular and viral mRNAs in vitro is regulated by a messenger discriminatory initiation factor. Proc Natl Acad Sci USA 73:3040, 1976

Helentjaris T, Ehrenfeld E: Inhibition of host cell protein synthesis by UV-inactivated poliovirus. J Virol 21:259, 1977

Helentjaris T, Ehrenfeld E: Control of protein synthesis in extracts from poliovirus-infected cells. I. mRNA discrimination by crude initiation factors. J Virol 26:510, 1978

Jen G, Birge CH, Thach RE: Comparison of initiation rates of encephalomyocarditis virus and host protein synthesis in infected cells. J Virol 27:640, 1978

Koizumi S, Simizu B, Hashimoto K, Oya A, Yamada M: Inhibition of DNA synthesis in BHK cells infected with Western equine encephalitis virus. 1. Induction of an inhibitory factor of cellular DNA polymerase activity. Virol 94:314, 1979

Levine AJ, Ginsberg HS: Role of adenovirus structural proteins in the cessation of host-cell biosynthetic functions. J Virol 2:430, 1968

McCormick W, Penman S: Inhibition of RNA synthesis in HeLa cells and L cells by mengovirus. Virol 31:135, 1967

McSharry JJ, Choppin PW: Biological properties of the VSV glycoprotein. I. Effects of the isolated glycoprotein or host macromolecular synthesis. Virol 84:172, 1978

Marvaldi J, Sekellick MJ, Marcus PI, Lucas-Lenard J: Inhibition of mouse L cell protein synthesis by ultraviolet-irradiated vesicular stomatitis virus requires viral transcription. Virol 84:127, 1978

Moss B: Inhibition of HeLa cell protein synthesis by the vaccinia virion. J Virol 2:1028, 1968

Nishioka Y, Silverstein S: Alterations in the protein synthetic apparatus of Friend erythroleukemia cells infected with vesicular stomatitis virus or herpes simplex virus. J Virol 25:422, 1978

Nuss DL, Koch G: Translation of individual host mRNA's in MPC-11 cells is differentially suppressed after infection by vesicular stomatitis virus. J Virol 19:572, 1976

Pogo BGT, Dales S: Biogenesis of poxviruses: further evidence for inhibition of host and virus DNA synthesis by a component of the invading inoculum particle. Virol 58:377, 1974

Rose JK, Trachsel H, Leong K, Baltimore D: Inhibition of translation by poliovirus: inactivation of a specific initiation factor. Proc Natl Acad Sci USA 75:2732, 1978

Rosemond - Hornbeak H, Moss B: Inhibition of host protein synthesis by vaccinia virus: fate of cell mRNA and synthesis of small poly(A)-rich polyribonucleotides in the presence of actinomycin D. J Virol 16:34, 1975

Steiner-Pryor A, Cooper PD: Temperature-sensitive poliovirus mutants defective in repression of host protein synthesis are also defective in structural protein. J Gen Virol 21:215, 1973

Weck PK, Wagner RR: Inhibition of RNA synthesis in mouse myeloma cells infected with vesicular stomatitis virus. J Virol 25:770, 1978

PERSISTENT INFECTIONS AND LATENCY

Dimmock NJ, Kennedy SIT: Prevention of death in Semliki forest virus-infected mice by administration of defective-interfering Semliki forest virus. J Gen Virol 39:231, 1978

Galloway DA, Fenoglio C, Shevchuk M, McDougall JK: Detection of herpes simplex RNA in human sensory ganglia. Virol 95:?, 1979

Gerdes JC, Marsden HS, Cook ML, Stevens JG: Acute infection of differentiated neuroblastoma cells by latency-positive and latency-negative simplex virus ts mutants. Virol 94:430, 1979

Holland JJ, Villarreal LP: Persistent noncytocidal vesicular stomatitis virus infections mediated defective T particles that suppress virion

transcriptase. Proc Natl Acad Sci USA 71:2956, 1974

Kimura Y, Ito Y, Shimokata K, et al: Temperature-sensitive virus derived from BHK cells persistently infected with HVJ (Sendai virus). J Virol 15:55, 1975

Knight P, Duff R, Rapp F: Latency of human measles virus in hamster cells. J Virol 10:995, 1972

Preble OT, Youngner JS: Temperature-sensitive defects of mutants isolated from L cells persistently infected with Newcastle disease virus. J Virol 12:472, 1973

Price RW, Schmitz J: Route of infection, systemic host resistance, and integrity of ganglionic axons influence acute and latent herpes simplex virus infection of the superior cervical ganglion. Inf Imm 23:373, 1979

Thacore HR, Youngner JS: Persistence of vesicular stomatitis virus in interferon-treated cell cultures. Virol 63:345, 1975

Youngner JS, Preble OT, Jones EV: Persistent infection of L cells with vesicular stomatitis virus: evolution of virus populations. J Virol 28:6, 1978

CHAPTER 7
The Genetics of Animal Viruses

The genetic approach to studying the virus-cell interaction in general and the virus multiplication cycle in particular has long been recognized as being extremely powerful in the bacteriophage field (see Chap. 10); during the last decade it has become very useful in animal virology as well.

The principal techniques of viral genetics are the isolation of mutants in the various genes, their characterization, and the study of how they interact. In essence, these techniques permit evaluation of the role of individual viral genes and of the proteins which they encode, in the complex series of interactions that is initiated when viruses infect cells.

Types of virus mutants

Viruses are encapsidated segments of genetic material; and like other genetic systems, viral genomes are not invariate, but are subject to change by mutation. Spontaneous mutations occur constantly in the course of virus multiplication; and while many are lethal, others are not. Virus populations may thus be regarded as genetically heterogeneous, since they are likely to contain mutants, at an average rate of 10^{-6}, at each locus. Mutant virus strains can also be generated in the laboratory as a result of mutagenesis; and it is from mutagenized virus populations that mutant isolation is usually undertaken. Among the procedures for mutagenizing RNA-containing viruses are treatment of virus particles with nitrous acid, hydroxylamine, N-methyl-N'-nitro-N-nitrosoguanidine (NTG), or ethane methane sulfonic acid, and propagation in the presence of 5-fluorouracil, 5-azacytidine, or proflavine (for the double-stranded RNA-containing reovirus). For DNA-containing viruses, treatment of virus particles with nitrous acid, hydroxylamine, or ultraviolet irradiation, and growth in the presence of 5-bromo-deoxyuridine, NTG, or proflavine are used most commonly.

There are two principal types of virus mutants: point mutants, in which there is a change in a single nucleotide base, and deletion mutants, in which a sequence of nucleic acid has

been deleted. Deletion mutants will be considered below when defective virus particles are discussed. The most important point mutants are the conditional lethal mutants.

CONDITIONAL LETHAL MUTANTS

Conditional lethal mutants can multiply under some conditions, but not under others when wild-type virus can do so. There are two classes of these mutants. The first comprises mutants that are temperature-sensitive (ts) with respect to ability to multiply. Wild-type animal viruses can generally multiply over a temperature range that extends from a lower limit of about 20C to 24C to an upper limit of about 39.5C for mammalian viruses and 40C to 41C for avian ones. In ts mutants there is a nucleic acid base substitution that causes an amino acid replacement in some virus-coded protein, as a result of which it cannot assume or maintain the structural conformation necessary for activity at elevated or nonpermissive (restrictive) temperature, though it can still do so at lower or permissive temperatures. The typical ts mutation thus causes the formation of an enzyme or a structural protein that cannot function in a temperature range (typically from about 36C to 41C) where the corresponding protein of the wild-type strain can function.

The second class of conditional lethal mutants are the nonsense mutants, which have been extremely useful in the bacterium/bacteriophage fields. In these mutants, the codon for some amino acid is changed to a termination codon (UAG, UAA or UGA); as a result, polypeptides are formed that are shorter than those specified by wild-type virus and therefore cannot function. However, mutant bacterial strains exist which contain mutated tRNA molecules that recognize these termination signals as some amino acid (generally different from the original one), which is therefore inserted into the amino acid sequence, thereby again permitting full-length polypeptides to be formed. If the amino acid change is such that the altered protein can still function, the effect of the original nonsense mutation is therefore suppressed. The mutated tRNA molecules are known as suppressor tRNAs and the mutant bacterial strains as suppressor strains. Since their expression is influenced by the genetic constitution of the host cells, these mutations are sometimes referred to as host-dependent mutations. Incidentally, the mutations that give rise to UAG, UAA and UGA are known as *amber, ochre* and *opal* respectively.

Since the appropriate suppressors are not nearly as prevalent in eukaryotic as in prokaryotic cells, nonsense mutants of animal viruses have so far not been used. The reason conditional lethal mutants are so important in studies seeking to define the reactions essential for virus multiplication is that these mutants permit study of the virus multiplication cycle with one, and only one, reaction unable to proceed. Use of such mutants, therefore, permits both the assessment of the role of any particular known reaction in the course of virus multiplication and also the detection of hitherto unknown functions.

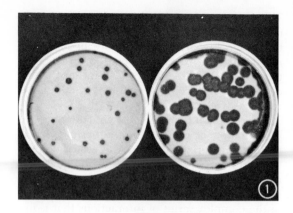

FIG. 7-1. Comparative sizes of two Mengovirus plaque size variants, S-Mengo (left) and L-Mengo (right). The plaques are 48 hours old. The L mutant is more virulent in animals and more cytopathic in cell cultures. (From Amako and Dales: Virol 32:184, 1967.)

MUTANTS WITH OTHER COMMONLY OBSERVED PHENOTYPES

In addition to temperature-sensitive and nonsense mutants, several other mutant phenotypes are commonly observed, primarily because they are easy to isolate. Among them are the following.

Plaque-Size Mutants

Many virus strains give rise to spontaneous mutants that form smaller plaques than wild-type virus because their adsorption is inhibited by sulfated polysaccharides present in agar (Fig. 7-1). Large plaque mutants are also known, and in this case the ability of wild-type virus to adsorb is inhibited by the polysaccharides, whereas that of the mutant is not. In either case, the site of the mutation is in a capsid polypeptide that functions in adsorption.

Drug-resistant Mutants

Drugs capable of inhibiting the multiplication of certain viruses are known (Chap. 8), and mutants exist that are resistant to these drugs. Examples are poliovirus mutants resistant to guanidine, herpesvirus mutants resistant to phosphonoacetic acid, and vaccinia virus mutants resistant to rifampicin and IBT. Poliovirus mutants dependent on guanidine and vaccinia virus mutants dependent on IBT also exist.

Enzyme-deficient Mutants

Viruses code for several enzymes essential for virus multiplication; and mutations that result in loss of this ability are obviously lethal. Some viruses also code for enzymes that are not essential, and mutants lacking the ability to code for them are viable. For example, poxviruses and herpesviruses code for enzymes that phosphorylate thymidine (thymidine kinases). Virus mutants that are deficient in the ability to induce the synthesis of these enzymes are known; they multiply well, demonstrating that the survival advantage conferred by the ability to code for them is small.

Hot Mutants

These are mutants that can grow at temperatures higher than needed by wild-type virus. For example, whereas 41C is near the upper limit of the temperature growth range of wild-type poliovirus, mutant strains exist that grow as well at 41C as at 37C, or even better. Not surprisingly, such strains are very virulent, since they can multiply rapidly in patients with higher fever, when the multiplication of wild-type virus is at least partially inhibited.

Interactions among viruses

Under conditions of multiple infection, cells may become infected with two or more virus particles with different genomes. If they are sufficiently closely related—that is, if they belong to the same virus family—they may be able to interact. There are several types of such interactions.

RECOMBINATION

The detection of recombination between two virus strains depends on the availability of techniques that permit the differentiation of the phenotypes of the two parents from those of the recombinants. If the two parents are single-step mutants of some wild-type strain, each differing from it in some recognizable manner, some of the recombinants will have the wild-type genotype and be easily detectable. The detection of other recombinants, such as the reciprocal recombinants or recombinants between viruses that differ in several loci, is usually more difficult and requires the use of selective conditions.

Viruses differ greatly in the ease with which they undergo recombination, the principal relevant factor being the nature of their genomes. With the exception of poliovirus and foot-and-mouth-disease virus, viruses that possess a single molecule of single-stranded RNA do not recombine. Viruses that contain a single molecule of double-stranded DNA recombine efficiently, most probably by a mechanism analogous to that by which the genomes of bacteria and higher organisms recombine. The most efficiently recombining animal viruses are those whose genomes consist of several nucleic acid segments. In such cases recombination proceeds not by classical recombination involving breakage and reformation of covalent bonds, but by simple reassortment of segments into new sets (Fig. 7-2). Both single-stranded and double-stranded RNA segments participate in this type of recombination, as shown by the fact that pairs of reovirus and influenza virus mutants generate wild-type virus with high frequency. Recombination of this type may account for the marked antigenic variability of influenza virus, which is illustrated by the fact that influenza A_0, A_1, and A_2 virus strains prevalent during the periods from 1933 to 1947, 1947 to 1957, and since 1957 possess quite different hemagglutinins and neuraminidases (H0N1, H1N1, and H2N2 as well as H3N2, respectively; Chap. 2). It is known that influenza virus strains pathogenic for man can recombine with strains pathogenic for other animal species, and it has been suggested that new human pathogens have arisen in this manner. In fact, the Hong Kong influenza virus strain of 1968 may have derived part of its genome from an equine or duck influenza virus.

MULTIPLICITY REACTIVATION

Viruses that contain double-stranded nucleic acid frequently exhibit multiplicity reactivation after being subjected to ultraviolet irradiation. This phenomenon is recognized by the fact

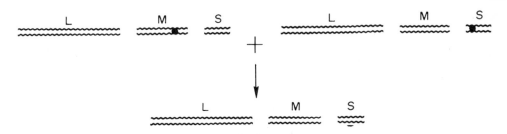

FIG. 7-2. Generation of wild-type genomes by reassortment of damaged genome segments. Two genomes, each consisting of three segments of double-stranded nucleic acid, are shown. One carries a mutation in an M segment, the other in an S segment. When they are introduced into the same cell, sets of undamaged segments are generated by reassortment. This type of mechanism can account for the generation of new genotypes among reoviruses, influenza viruses and bunyaviruses.

that the frequency of virus survivors increases sharply with multiplicities of infection above 1. It is due to cooperation between viral genomes that have been damaged by the irradiation and that can therefore no longer multiply on their own. The nature of the cooperation is probably recombination; that is, the damaged genomes recombine until an intact genome arises, which can then replicate and form progeny.

COMPLEMENTATION

Viral genomes can also interact indirectly by means of complementation. A typical example of this type of interaction is provided by infection of cells at the restrictive temperature with two virus mutants that bear temperature-sensitive mutations in different cistrons and neither of which can multiply alone. If complementation occurs, progeny comprising both mutants is produced. The explanation of this phenomenon is that each mutant produces functional gene products of all cistrons except that bearing the temperature-sensitive mutation, so that in cells infected with both mutants all gene products necessary for virus multiplication are formed, and both mutants can therefore multiply. Complementation plays a major role in permitting the survival of viruses with genomes that contain damaged or nonfunctional genes.

A special case of complementation is the phenomenon of poxvirus reactivation. As outlined in Chapter 5, the second stage of poxvirus un-

coating—that is, the uncoating of the viral core—requires the synthesis of an uncoating protein. This protein is not formed if the infecting virus has been subjected to protein-denaturing conditions, such as heat; heat-inactivated poxvirus particles are therefore not uncoated and cannot multiply. If, however, cells are simultaneously infected with denatured and undenatured virus particles, both are uncoated, because the uncoating protein elicited by the undenatured particles also uncoats the denatured particles which are therefore enabled to multiply. This phenomenon was initially interpreted as reactivation of the denatured virus by the active virus, hence its name.

PHENOTYPIC MIXING AND PHENOTYPIC MASKING

Another special case of complementation is the dual phenomenon of phenotypic mixing and phenotypic masking. When two closely related viruses—for example, poliovirus type 1 and poliovirus type 3—infect the same cell, the two resulting sets of progeny genomes may become encapsidated not only by their own capsids but also by hybrid capsids—that is, capsids composed of polypeptide chains characteristic of both genomes (phenotypic mixing)—or even by capsids entirely specified by the other genome (phenotypic masking or transcapsidation) (Fig. 7-3). This situation is most readily detected by antigenic analysis, for the former class of virus

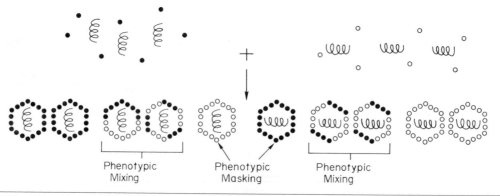

FIG. 7-3. Phenotypic mixing and phenotypic masking. Simultaneous infection with two related viruses is illustrated. Either genome can be encapsidated in capsids that are composed exclusively of homologous capsomers, or mixed capsomers (phenotypic mixing), or of exclusively heterologous capsomers (phenotypic masking). The method of detecting the latter two classes of particles is described in the text.

particles is neutralized by antiserum to either of the two parents, while virus particles of the latter class are neutralized by antiserum to one of the parents, and their progeny is neutralized by antiserum to the other.

A similar phenomenon occurs among enveloped viruses; but here, it involves not only viruses that are related, but also viruses that are completely unrelated. In particular, the nucleocapsid of the rhabdovirus vesicular stomatitis virus (VSV) possesses a remarkable ability to become encapsidated in envelopes that are only partially, or sometimes not at all, specified by it. For example, among the yield of cells simultaneously infected with both VSV and the paramyxovirus SV5 there are bullet-shaped particles that contain VSV nucleocapsids encased in envelopes that bear not only VSV-specified glycoprotein spikes but also both types of SV5-specified spikes. Another example of great current interest is provided by VSV nucleocapsids completely encased in RNA tumor virus envelopes. Since such particles are easily and rapidly quantitated, they have great potential for studies on RNA tumor virus host range (which is specified by the envelope) and for detecting the presence of RNA tumor virus-specified envelope proteins, particularly in connection with the search for human tumor viruses (Chap. 9). Finally, VSV nucleocapsids can even be encased in herpesvirus envelopes. Nucleocapsids of one virus enclosed in envelopes specified by another are known as pseudotypes. Viruses differ in their propensity to form pseudotypes; those that do so most readily are VSV and RNA tumor viruses.

Defective virus particles

There are several types of virus particles that cannot multiply on their own but can multiply in cells simultaneously infected with some infectious virus. They can be subdivided into two classes: those that interfere extensively with the multiplication of their helper virus, and those that do not.

DEFECTIVE INTERFERING (DI) VIRUS PARTICLES

When viruses are passaged repeatedly at high multiplicity, the progeny frequently includes, in addition to mature virus particles, defective virus particles that are capable of interfering with the multiplication of homologous virus. Such virus particles have the following properties: (1) They contain the normal structural capsid proteins, (2) they contain only a part of the virus genome—that is, they are deletion mutants, (3) they can reproduce only in cells infected with homologous virions, which act as helpers, (4) although unable to reproduce on their own, they can nevertheless express a variety of functions in the absence of helper, such as inhibition of host biosynthesis, synthesis of viral proteins, and transformation of cells, and (5) they specifically interfere with the multiplication of homologous virus.

The following are some examples of defective interfering virus particles that have been characterized in some detail.

Defective Interfering Influenza Virus Particles

Under conditions of repeated passaging at high multiplicity, the infectivity of successive yields of influenza virus gradually decreases a million-fold or even more, though the total number of virus particles that is produced remains roughly the same; in other words, noninfectious, defective virus particles gradually replace virions in the yields. This phenomenon was first described in 1952 by von Magnus and bears his name. Defective particles are not formed if influenza virus is passaged at low multiplicity, and they are readily eliminated from virus stocks by passaging at a multiplicity of less than 1, which shows that defective particles cannot multiply on their own. The ability of the defective particles to inhibit the multiplication of infectious virus is demonstrated by the fact that the addition of defective particles to influenza virus preparations free of them immediately reduces the yields of infectious virus in most types of cells.

The essential difference between infectious and defective influenza virus particles is that the latter have lost their largest RNA segments and acquired instead at least six new RNA segments that are smaller than the smallest segment of wild-type virus and that are in all probability derived from the larger segments. The loss of infectivity is a consequence of the deletions; the ability to interfere with the multiplication of wild-type virus is due to the presence of the new small RNA segments. The mechanism of the interference and the reason why these deletion mutants compete successfully with wild-type virus will be discussed below.

Defective Interfering Particles of Other Viruses

Defective interfering particles are produced not only by viruses with segmented genomes but also by viruses with genomes that consists of a single nucleic acid molecule. The defective particles then contain nucleic acid molecules that are shorter than those of infectious virus particles. For example, when VSV is passaged at high multiplicity, virions among the progeny are gradually replaced by particles that are about one-third as long and contain RNA molecules about one-third as long as VSV genomes (Fig. 7-4). These particles, the so-called T (truncated) particles, can multiply only in the presence of virions but interfere extensively with their multiplication.

Another well-studied example is that of the DI particles of the togaviruses Semliki forest virus and Sindbis virus. Both generate, during a few passages at high multiplicity, particles that contain RNA molecules that are no more than about one-fifth as long as those of wild-type virus. These particles arise via a well-defined series of stepwise deletions that is illustrated schematically in Fig. 7-5. Although this progressive sequence is followed in many cell types, the nature of the cell can nevertheless influence profoundly the nature of the DI particle that finally emerges as a stable population; in some cells no DI particles are generated, in others the deletion sequence is arrested early, and the relative stability of each intermediate often differs markedly in different cell types. Interestingly, all DI togavirus particles contain about the same amount of RNA, which means that the smaller the RNA molecule, the more copies of it are enveloped in each virus particle. Defective interfering particles occur also in high passage yields of poliovirus, pseudorabies virus, the papovaviruses polyoma and SV40, and probably most other viruses. All are deletion mutants that require the presence of wild-type viruses as helpers, but outgrow them quickly.

The two central puzzles concerning DI particles are, first, how do deletions arise? and sec-

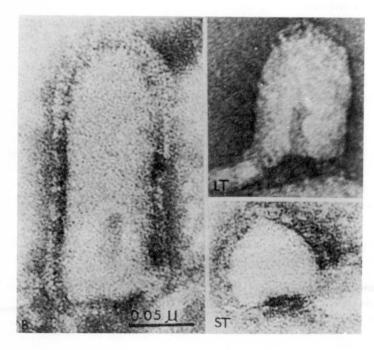

FIG. 7-4. Defective interfering particles of VSV. B, a bullet-shaped normal virus particle; LT, a *l*ong *t*runcated particle; ST, a *s*hort *t*runcated particle. LT particles are about half as long as normal virus particles; ST particles, which are round, about one-third as long. The defective particles contain RNA molecules that are proportionately shorter than the RNA molecules in normal virus particles. (Courtesy of Dr. C. Y. Kang.)

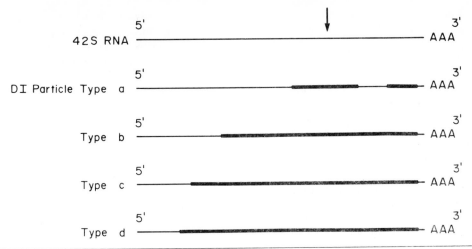

FIG. 7-5. Diagrammatic representation of the nucleotide sequence relationships of a series of DI Semliki forest virus (SFV) particles that arise sequentially in a strain of HeLa cells. ▬ denotes deleted sequences, drawn approximately to scale; the arrow indicates the junction between the nonstructural and structural genes (see Chap. 5). The deletions in all types of DI particles are seen to span sequences in both genes. Type *a* particles appear to be precursors of type *b* particles, and so on; that is, each DI particle persists for a few passages and then disappears, giving rise to the particle with the next larger deletion. Some cells (such as another strain of HeLa cells) do not generate SFV DI particles at all. Others, such as a strain of pig kidney cells, give rise only to DI particles types *a* and *b*, but can support the multiplication of types *c* and *d* when infected with them. (From Stark and Kennedy: Virol 89:285, 1978.)

ond, why do DI particles interfere so successfully with the multiplication of wild-type virus? The answer to neither question is known with certainty. Some DI RNA molecules appear to contain internal deletions. They could conceivably arise via intramolecular recombination. Other deletions are thought to result from transcription errors which occur when the RNA polymerase molecule begins to transcribe backwards the nascent RNA strand that it is in the process of transcribing. Interestingly, DI particles of VSV are not formed in cells simultaneously infected with fibroma virus, but the mechanism of this effect is not known.

As for why DI particles interfere, it is almost certain that the much shorter DI RNA molecules compete very successfully with wild-type virus RNA molecules for the limited number of RNA polymerase molecules; not only would the shorter molecules replicate much more rapidly, but they also appear to have a greater affinity for the RNA polymerase. It is interesting, in this regard, that once DI DNA molecules have arisen, they form their own replication complexes; and it is easy to imagine how such DI RNA complexes could attract most of the limited number of RNA polymerase molecules in the cell, thereby preventing the replication of wild-type RNA.

Finally, it is interesting to speculate that the generation and propagation of DI particles may be a mechanism for ensuring the survival of viruses by permitting them to establish persistent infections. If this is true, DI RNA should not only be smaller than wild-type RNA, possess higher affinity for RNA polymerase and replicate more rapidly, but it should also preferably not code for any viral polypeptide (so as to minimize cytopathic effects). The DI RNAs that have been best characterized fulfill all of these criteria.

DEFECTIVE ADENOVIRUS AND POLYOMAVIRUS PARTICLES

The DI particles described in the preceding section are those formed by RNA-containing viruses. DNA-containing viruses also form defective particles, but usually they do not interfere strongly.

Defective particles are commonly present among the yields of most DNA-containing viruses. Thus, when adenovirus type 2 is passaged in KB cells, about 10 percent of the yield consists of virus particles that contain somewhat less DNA than virions; and of this DNA, only about 3 percent of the DNA is viral DNA, the rest being host cell DNA that is covalently linked to it. This is no doubt a consequence of the fact that adenovirus DNA is integrated into that of the host cell during the course of multiplication (see Chap. 5). The small amount of viral DNA in these defective particles is not confined to a specific region of the viral genome, but all viral DNA sequences are represented more or less equally in it. Incomplete virus particles that contain less than the entire viral genome but no host DNA are formed during the multiplication of all DNA-containing viruses.

Another example of deletion mutants are the evolutionary mutants of SV40, which evolve in stepwise fashion when SV40 is passaged repeatedly at high mutiplicity; like other deletion mutants, these mutants can be propagated in the presence of wild-type helper virus. In its most extreme form the DNA of these mutants consists of three tandemly linked units each of which comprises a sequence of no more than 150 nucleotides of SV40 DNA around the origin of replication linked to a sequence of about 1500 nucleotides of host cell DNA.

Adenovirus-SV40 Hybrid Particles

Human adenovirus cannot multiply in monkey cells but will do so in the presence of the simian papovavirus SV40, which performs some helper function. The progeny of such mixed infections sometimes includes particles that contain, within an adenovirus capsid, a hybrid DNA molecule that contains both adenovirus and SV40 DNA sequences, but does not contain *all* the sequences of either. One well-known example of such particles is the E46+ strain of adenovirus serotype 7, which was generated several years ago during attempts to adapt human adenovirus to grow in monkey kidney cells for vaccine production. It turned out that the monkey kidney cells contained SV40, which recombined with adenovirus type 7 to produce genomes that contain covalently linked sequences of both adenovirus and SV40 DNA: these genomes lack the adenovirus DNA sequences between map positions 5 and 21 and contain instead two SV40 DNA sequences joined end-to-end—namely, those between map positions 0.50 to 0.71 and between 0.11 to 0.66. Thus these particles contain the entire early region of SV40, and two doses of the sequence between 0.50 and 0.66. These particles cannot, of course, multiply on their own, since they contain neither a complete adenovirus genome nor a complete SV40 genome; but they can perform that function which enables human adenoviruses to multiply in monkey cells. They are therefore called PARA (particles aiding the replication of adenovirus). Further, they can themselves multiply in the presence of the human adenovirus, which presumably supplies the function(s) coded by the piece of adenovirus DNA which is missing in the hybrid particle DNA. Thus, here are two types of virus particles, human adenovirus and the hybrid PARA particle, neither of which can multiply in monkey cells by itself, but which can both multiply if they infect monkey cells simultaneously.

Several other types of adenovirus-SV40 hybrids are also known. One example is provided by hybrids between adenovirus type 2 and SV40 that contain not partial but complete SV40 genomes covalently linked to incomplete adenovirus genomes. Such hybrids enable adenovirus type 2 to multiply in monkey cells and are themselves complemented by adenovirus type 2. In addition they yield infectious SV40, since they contain the entire SV40 genome.

TABLE 7-1   NONDEFECTIVE ADENOVIRUS TYPE 2-SV40 HYBRIDS

| Hybrid | Adenovirus DNA Deleted | | SV40 DNA Inserted | |
	Map Position	*M.W.*	*Map Position*	*M.W.*
Ad2+ ND$_1$	80.6 – 86.0	1.24×10^6	0.11 – 0.39	0.58×10^6
Ad2+ ND$_2$	79.9 – 86.0	1.40×10^6	0.11 – 0.43	1.02×10^6
Ad2+ ND$_3$	80.7 – 86.0	1.22×10^6	0.11 – 0.18	0.22×10^6
Ad2+ ND$_4$	81.5 – 86.0	1.04×10^6	0.11 – 0.54	1.38×10^6
Ad2+ ND$_5$	78.9 – 86.0	1.63×10^6	0.11 – 0.39	0.90×10^6

Even more interesting adenovirus-SV40 hybrids are the infectious adenovirus type 2-SV40 hybrids, five of which have been characterized in detail (Table 7-1). In these hybrids, a portion of the adenovirus genome is deleted from a region that is not essential for virus multiplication. This region comprises 4.5 to 7.1 percent of the adenovirus genome and ends at map position 86 in all hybrids. In its place is inserted a portion of SV40 DNA that codes for early functions; this portion varies from 7 to 43 percent of the SV40 genome and also starts at map position 0.11 in all five hybrids. The largest piece (43 percent) of SV40 DNA expresses all early SV40 functions, while the smallest piece (7 percent) can code for no more than 10,000 to 15,000 daltons of protein; yet this is sufficient to endow even this hybrid with the capacity to multiply in monkey cells. Interestingly, human adenovirus variants have recently been isolated that contain no SV40 sequences at all and yet can grow in monkey cells without a helper. These variants appear to be point mutants in a gene that maps between 59 and 80 on the adenovirus map and that codes for a protein that controls late adenovirus messenger RNA expression; for whereas wild-type human adenovirus cannot express late genetic information in monkey cells, these mutants can do so.

These five hybrids provide a set of virus particles with different but precisely defined amounts of SV40 genetic material. They have been very useful in studies of the arrangement of genetic information in the early region of the SV40 genome and of how the information is expressed (see Chap. 5). Thus cells infected with hybrid ND_3 do not synthesize any recognizable SV40 protein; cells infected with ND_1 synthesize the U antigen (see Chap. 5); cells infected with ND_4 and ND_5 synthesize the U antigen as well as the tumor-specific transplantation antigen (TSTA), which, according to this analysis, is therefore also a portion of the T antigen; and cells infected with ND_4 synthesize not only the U antigen and the TSTA, but also the T antigen.

Other Viruses

Other examples of viruses that require helpers are found among the RNA tumor viruses. In particular, many mammalian sarcomagenic viruses depend on leukemia viruses for the provision of their envelopes. These viruses are discussed in detail in Chapter 9.

Viruses that can multiply only in cells also harboring other viruses may be of great importance in causing human diseases of as yet undefined etiology. These viruses are very difficult to detect, but detailed studies of systems such as those described above may provide valuable clues in the search for additional ones.

Interference between viruses

It has been known for a long time that when two different viruses infect the same cell, they may interfere with each other and diminish each other's yield. Although the precise nature of the inhibition is usually not known, it appears that there are two primary causes of interference. First, sometimes the first virus inhibits the ability of the second virus to adsorb, by either blocking its receptors (as is the case for certain pairs of enteroviruses) or by destroying its receptors (as with certain pairs of myxoviruses). Second, one virus may prevent the messenger RNA of the second virus from being translated. Thus, just as poliovirus inhibits the translation of host-cell messenger RNA by inactivating the cap-dependent mRNA recognition mechanism (see Chap. 6), so does it inhibit the translation of VSV messenger RNA and interfere with the ability of VSV to multiply. Similarly, the translation of vaccinia virus messenger RNA is prevented in cells infected with adenoviruses. It is likely that the ability of VSV to interfere with the multiplication of pseudorabies virus, of Sindbis virus to interfere with the multiplication of VSV and NDV, and of rubella virus to interfere with the multiplication of NDV, is also due to interference with the ability of viral messenger RNA to be translated. This mechanism of interference would account for the marked specificity of such inhibitory effects; thus whereas cells infected with rubella virus become resistant to NDV, they remain susceptible to a variety of other viruses.

Although so far there have been meager results of studies that are directed specifically at determining why certain pairs of viruses interfere with each other's ability to multiply, it does seem that there is great potential for using one virus to interfere with some specific reaction essential for the multiplication of another virus. Such studies may become increasingly feasible as we learn more about the ability of viruses to interfere with the expression of genomes other than their own.

Mapping of the genomes of animal viruses

The basic aims of virologists are (1) to characterize the various functions involved in virus multiplication and (2) to identify the portions of the viral genome that encode these functions. The first of these aims depends primarily on the availability of temperature-sensitive mutants. As pointed out above, these mutants permit the virus multiplication cycle to be examined when one of its component reactions fails to function; these reactions can therefore be identified and characterized one by one. In practice, this approach to investigating the virus multiplication cycle involves the isolation of ts mutants and their characterization, which usually means identification and characterization of protein coded by the mutant gene.

Great strides have recently been made in the mapping of animal virus genomes—that is, in identifying those portions of their genomes that encode specific proteins. The following is a brief summary of the techniques that are most useful for mapping the genes for DNA- and RNA-containing viruses.

DNA-Containing Viruses

1. Recombination analysis. This has been described on p. 92.
2. Analyzing, by restriction endonuclease analysis, the DNA of recombinants of pairs of mutants (for example ts mutants) in different genes of virus strains that are related (for example, belonging to adenovirus serotypes). This was described in Fig. 5-13.

3. Since naked viral double-stranded DNA can be introduced into cells and can usually express itself (see Chap. 2), marker rescue analysis can be carried out. Cells are infected with the naked DNA of a ts mutant together with individual restriction endonuclease fragments of wild-type virus DNA. Wild-type recombinants will be formed and can be isolated if the fragment of wild-type DNA carries the gene, or portion of the gene, where the mutation was in the ts mutant DNA. Since the location on the genome of the restriction endonuclease fragment is known, the location of the mutated gene is known.
4. It is sometimes possible to isolate, by gel electrophoresis and density gradient centrifugation, virus-coded messenger RNAs from infected cells, to translate them in cell-free protein-synthesizing systems, and to identify the proteins for which they code. Since the origin of these messenger RNAs can be specified by determining with which restriction endonuclease fragment they can hybridize, the regions of the viral genome that code for specific viral proteins can thus be identified. Another version of this analysis is described on p. 92.

RNA-Containing Viruses

The gene order on viral genomes that consist of RNA can be identified as follows.

UV-Transcriptional Mapping

UV-irradiation damages RNA (see Chap. 3) and inactivates it as a template for transcription. Since transcription of viral RNA is initiated from a single site (at its 3'-terminus), the effect of irradiation is to interfere differentially with the transcription of its various genes; transcription of those furthest from the transcription initiation site is inhibited most, while transcription of those closest to it will be inhibited least. The relative amounts of each species of messenger RNA that is transcribed can be determined readily by measuring the amount of protein that is translated from it. It was in this manner that the gene orders of the VSV and paramyxovirus genomes were determined (see Chap. 5).

Pactamycin Translation Inhibition Mapping

This technique is capable of mapping the gene order of viral genomes that act as messenger RNAs (such as genomes of picornaviruses). The technique depends on the fact that these genomes are multigenic but monocistronic; that is, there is a single site (at their 5'-terminus) where translation is initiated, the RNA being translated into a large precursor which is then cleaved into the various functional proteins (see Chap. 5). Pactamycin inhibits initiation of translation, but not elongation. Thus, if pactamycin is added to infected cells and then at intervals of time thereafter a radioactively-labeled amino acid is added, its incorporation into the region of polyprotein closest to the 5'-end will be inhibited first, and its incorporation into the region of the polyprotein that is furthest from the 5'-end will be inhibited last. It has been found in this way that the 5'-terminal half of poliovirus RNA codes for the structural proteins, while the 3'-terminal half codes for the polymerase.

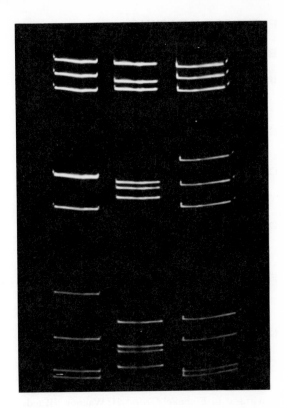

FIG. 7-6.   Separation by electrophoresis in polyacrylamide gels (Maizel Tris-Glycine system) of the ten segments of double-stranded RNA in reovirus particles of strains of (from left to right) serotypes 3, 2 and 1. Note that two segments in the reovirus type 3 genome overlap. Clearly the homologous segments differ slightly but significantly in electrophoretic mobility (that is, in size) and can therefore be identified. Similar patterns are obtained if the structural protein species of these strains are compared. (Courtesy of Dr. R. Gaillard.)

Mapping of Genes in Segmented Genomes

Identification of the genes encoded by the individual segments of the reovirus and influenza virus genomes can be accomplished in two ways. The first is the direct one: it is possible to translate the individual RNA species in cell-free protein-synthesizing systems and to identify the proteins for which they code. The second way is analogous in principle to that described above (see p. 142): one takes advantage of the fact that the individual RNA segments and proteins of different serotypes of reovirus and influenza virus can be identified by observing their migration rates when electrophoresed in polyacrylamide gels (Fig. 7-6). By crossing pairs of virus strains that belong to different serotypes and examining the RNA and protein patterns of the recombinants that are produced, one can determine which RNA segment codes for each individual protein. In the simplest case, one would look for recombinants in which all RNA segments except one are derived from one parent; then there would be one, and only one, protein that would be characteristic of the other parent. This technique also permits ts mutants to be assigned to RNA segments. To do this, ts mutants of one serotype are crossed with a wild-type strain of another serotype; progeny are then plaqued at the restrictive (normal) temperature, and the pattern of RNA segments in electrophoretograms is then examined for a number of plaque isolates. Since recombination in reovirus and influenza virus occurs by random assortment of RNA segments (see p. 135), one would expect that all RNA segments would be randomly derived from either one parent or the other, except one—namely, that which bears the ts lesions; that segment will always be derived from the wild-type parent.

FURTHER READING

Books and Reviews

Bratt MA, Hightower LE: Genetics and paragenetic phenomena of paramyxoviruses. In Fraenkel-Conrat H, Wagner RR (eds): Comprehensive Virology, Vol 9, p 457. New York, Plenum, 1977

Cooper PD: Genetics of picornaviruses. In Fraenkel-Conrat H, Wagner RR (eds): Comprehensive Virology, Vol 9, p 133. New York, Plenum, 1977

Cross RK, Fields BN: Genetics of reoviruses. In Fraenkel-Conrat H, Wagner RR (eds): Comprehensive Virology, Vol 9, p 291. New York, Plenum, 1977

Doerfler W: Animal virus-host genome interactions. In Fraenkel-Conrat H, Wagner RR (eds): Comprehensive Virology, Vol 10, p 279. New York, Plenum, 1977

Eckhart W: Genetics of polyoma virus and simian virus 40. In Fraenkel-Conrat H, Wagner RR (eds): Comprehensive Virology, Vol 9, p 1. New York, Plenum, 1977

Ginsberg HS, Young CSH: Genetics of adenoviruses. In Fraenkel-Conrat H, Wagner RR (eds): Comprehensive Virology, Vol 9, p 27. New York, Plenum, 1977

Hightower LE, Bratt MA: Genetics or orthomyxoviruses. In Fraenkel-Conrat H, Wagner RR (eds): Comprehensive Virology, Vol 9, p 535. New York, Plenum, 1977

Huang AS, Baltimore D: Defective interfering animal viruses. In Fraenkel-Conrat H, Wagner RR (eds): Comprehensive Virology, Vol 10, p 73. New York, Plenum, 1977

Lewis AM Jr: Defective and nondefective Ad2-SV40 hybrids. Prog Med Virol 23:96, 1977

Pfefferkorn ER: Genetics of togaviruses. In Fraenkel-Conrat H, Wagner RR (eds): Comprehensive Virology, Vol 9, p 209. New York and London, Plenum, 1977

Pringle CR: Genetics of rhabdoviruses. In Fraenkel-Conrat H, Wagner RR (eds): Comprehensive Virology, Vol 9, p 239. New York and London, Plenum, 1977

Subak-Sharpe JH, Timbury MC: Genetics of herpesviruses. In Fraenkel-Conrat H, Wagner RR (eds): Comprehensive Virology, Vol 9, p 89. New York, Plenum, 1977

Vogt PK: Genetics of RNA tumor viruses. In Fraenkel-Conrat H, Wagner RR (eds): Comprehensive Virology, Vol 9, p 341. New York, Plenum, 1977

Selected Papers

MUTANTS, VARIANTS AND RECOMBINANTS

Desselberger U, Nakajima K, Alfino P, Pedersen FS, Haseltine WA, Hannoun C, Palese P: Biochemical evidence that "new" influenza virus strains in nature may arise by recombination (reassortment). Proc Natl Acad Sci USA 75:3341, 1978

Dubbs DR, Kit S: Isolation and properties of vaccinia mutants deficient in thymidine kinase inducing activity. Virol 22:214, 1964

Gorman BM, Taylor J, Walker PJ, Young PR: The isolation of recombinants between related orbiviruses. J Gen Virol 41:333, 1978

Klessig DF, Hassel JA: Characterization of a variant of human adenovirus type 2 which multiplies efficiently in simian cells. J Virol 28:945, 1978

Sambrook JF, Padgett BL, Tomkins JKN: Conditional lethal mutants of rabbitpox virus. 1. Isolation of host-cell dependent and temperature-dependent mutants. Virol 28:592, 1966

Schuerch AR, Matsuhisa T, Joklik WK: Temperature-sensitive mutants of reovirus. VI. Mutant ts447 and ts556 particles that lack either one or two genome RNA segments. Intervirol 3:36, 1974

Tjia S, Fanning E, Schick J, Doerfler W: Incomplete particles of adenovirus type 2. III. Viral and Cellular DNA sequences in incomplete particles. Virol 76:365, 1977

Werner G, zur Hausen H: Deletions and insertions in adenovirus type 12 DNA after viral replication in vero cells. Virol 86:66, 1978

INTERACTIONS AMONG VIRUSES

Chen C, Crouch NA: Shope fibroma virus induced facilitation of vesicular stomatitis virus adsorption and replication in nonpermissive cells. Virol 85:43, 1978

Choppin PW, Compans RW: Phenotypic mixing of envelope proteins of the parainfluenza virus SV5 and vesicular stomatitis virus. J Virol 5:609, 1970

Huang AS, Palma EL, Hewlett N, Roizman B: Pseudotype formation between enveloped RNA and DNA viruses. Nature 252:743, 1974

Joklik WK, Woodroofe GM, Holmes IH, Fenner F: The reactivation of poxviruses. I. The demonstration of the phenomenon and techniques of assay. Virol 11:168, 1960

Joklik WK, Abel P, Holmes IH: Reactivation of poxviruses by a nongenetic mechanism. Nature 186:992, 1960

Lake JR, Priston RAJ, Slade WR: A genetic recombination map of foot-and-mouth disease virus. J Gen Virol 27:355, 1975

Laver WG, Webster RG: Studies on the origin of pandemic influenza: evidence implicating duck and equine influenza viruses as possible progenitors of the Hong Kong strain of human influenza. Virol 51:383, 1973

Salo RJ, Mayor HD: Adenovirus-associated polypeptides synthesized in cells coinfected with either adenovirus or herpesvirus. Virol 93:237, 1979

Trautman R, Sutmoller P: Detection and properties of a genomic masked viral particle consisting of foot-and-mouth disease virus nucleic acid in bovine enterovirus protein capsid. Virol 44:537, 1971

Zavadova Z, Zavada J: Unilateral phenotypic mixing of envelope antigens between togaviruses and vesicular stomatitis virus or avian RNA tumor virus. J Gen Virol 37:557, 1977

GENE MAPPING

Almond JW, McGeoch D, Barry RD: Method for assigning temperature-sensitive mutations of influenza viruses to individual segments of the genome. Virol 81:62, 1977

Ball A: Transcriptional mapping of vesicular stomatitis virus in vivo. J Virol 21:411, 1977

Frost E, Williams J: Mapping temperature-sensitive and host-range mutations of adenovirus type 5 by marker rescue. Virol 91:39, 1979

Khoury G, Fareed GC, Berry K, et al: Characterization of a rearrangement in viral DNA: mapping of the circular Simian virus 40-like DNA containing a triplication of a specific one-third of the viral genome. J Mol Biol 87:289, 1974

Morse LS, Pereira L, Roizman B, Schaffer PA: Anatomy of herpes simplex virus (HSV) DNA. X. Mapping of viral genes by analysis of polypeptides and functions specified by HSV-1 × HSV-2 recombinants. J Virol 26:389, 1978

Mustoe TA, Ramig RF, Sharpe AH, Fields, BN: A genetic map of reovirus. III. Assignment of the double-stranded RNA-positive mutant groups A, B, and G to genome segments. Virol 85:545, 1978

Oxford JS, McGeoch DJ, Schild DGC, Beare AS: Analysis of virion RNA segments and polypeptides of influenza A virus recombinants of defined virulents. Nature 273:778, 1978

Palese P, Ritchey MB, Schulman JL: Mapping of the influenza virus genome. II. Identification of the P1, P2, P3 genes. Virol 76:114, 1977

Ramig RF, Fields BN: Revertants of temperature-sensitive mutants of reovirus: evidence for frequent extragenic suppression. Virol 92:155, 1979

Ramig RF, Mustoe TA, Sharpe AH, Fields BN: A genetic map of reovirus. II. Assignment of the double-stranded RNA-negative mutant groups C, D, and E to genome segments. Virol 85:531, 1978

Ritchey MB, Palese P: Identification of the defective genes in three mutant groups of influenza virus. J Virol 21:1196, 1977

Sharpe AH, Ramig RF, Mustoe TA, Fields BN: A genetic map of reovirus. I. Correlation of genome RNAs between serotypes 1, 2 and 3. Virol 84:63, 1978

Weiner HL, Ramig RF, Mustoe TA, Fields BN: Identification of the gene coding for the hemagglutinin of reovirus. Virol 86:581, 1978

Wilson MC, Fraser NW, Darnell JE Jr: Mapping of RNA initiation sites by high doses of UV irradiation: evidence for three independent promoters within the left 11% of the Ad-2 genome. Virol 94:175, 1979

DEFECTIVE VIRUS PARTICLES

Ben-Porat T, Demarchi JM, Kaplan AS: Characterization of defective interfering viral particles present in a population of pseudorabies virions. Virol 61:29, 1974

Cole CN, Baltimore D: Defective interfering particles of poliovirus. II. Nature of the defect. III. Interference and enrichment. J Mol Biol 76:325, 345, 1973

Cole CN, Smoler D, Wimmer E, Baltimore D: Defective interfering particles of poliovirus. I. Isolation and physical properties. J Virol 7:478, 1971

Crumpton WM, Dimmock NJ, Minor PD, Avery RJ: The RNAs of defective-interfering influenza virus. Virol 90:370, 1978

Huang AS, Greenawalt JW, Wagner RR: Defective T particles of vesicular stomatitis virus. Virol 30:161, 173, 1966

Huang AS, Baltimore D: Defective viral particles and viral disease processes. Nature 226:325, 1970

Janda JM, Davis AR, Nayak DP, De BK: Diversity and generation of defective interfering influenza virus particles: Virol 95:48, 1979

Kang CY, Glimp T, Clewley JP, Bishop DHL: Studies on the generation of vesicular stomatitis virus (Indiana serotype) defective interfering particles. Virol 84:142, 1978

Kennedy SIT, Bruton CJ, Weiss B, Schlesinger S: Defective interfering passages of Sindbis virus: nature of the defective virion RNA. J Virol 19:1034, 1976

Kingsbury DW, Portner A: On the genesis of incomplete Sendai virus. Virol 42:872, 1970

Laughlin CA, Myers MW, Risin DL, Carter BJ: Defective interfering particles of the human parvovirus adeno-associated virus. Virol 94:162, 1979

Leppert M, Kort L, Kolakofsky D: Further characterization of Sendai virus DI-RNAs: a model for their generation. Cell 12:539, 1977

Nomoto A, Jacobson A, Lee YF, Dunn J, Wimmer E: Defective interfering particles of poliovirus: mapping of the deletion and evidence that the deletions in the genomes of DI (1), (2), and (3) are located in the same region. J Mol Biol 128:179, 1979

Stark C, Kennedy SIT: The generation and propagation of defective interfering particles of Semliki forest virus in different cell types. Virol 89:285, 1978

Uchida S, Watanabe S: Transformation of mouse 3T3 cells by T antigen-forming defective SV40 virions (T particles). Virol 39:721, 1969

Weiss BE, Goran D, Cancedda R, Schlesinger S: Defective interfering passages of Sindbis virus: nature of the intracellular defective viral RNA. J Virol 14:1189, 1974

Winship TR, Thacore HR: Inhibition of vesicular stomatitis virus-defective interfering particle synthesis by shope fibroma virus. Virol 93:515, 1979

Yoshiike K: Studies on DNA from low-density particles of SV40. Virol 34:391, 402, 1968

INTERFERENCE BETWEEN VIRUSES

Aubertin A, Guir J, Kirn A: The inhibition of vaccinia virus DNA synthesis in KB cells infected with frog virus 3. J Gen Virol 8:105, 1970

Dubovi EJ, Youngner JS: Inhibition of pseudorabies virus replication by vesicular stomatitis viruses. J Virol 18:526, 534, 1976

Ehrenfeld E, Lund H: Untranslated vesicular stomatitis virus messenger RNA after poliovirus infection. Virol 80:297, 1977

Giorno R, Kates JR: Mechanism of inhibition of vaccinia virus replication in adenovirus-infected HeLa cells. J Virol 7:208, 1971

Hunt JM, Marcus PI: Mechanism of Sindbis virus-induced intrinsic interference with vesicular stomatitis virus replication. J Virol 14:99, 1974

Marcus PI, Carver DH: Intrinsic interference: a new type of viral interference. J Virol 1:334, 1967

CHAPTER 8

Antiviral Chemotherapy, Interferon, and Vaccines

The rational approach to antiviral chemotherapy

Virus multiplication consists of the synthesis of viral nucleic acids and proteins and their assembly into virions. Any rational approach to antiviral chemotherapy should examine whether and where these processes can best be interrupted without detriment to the host. The following is a relevant brief analysis.

The Replication of Viral Nucleic Acids The replication of the nucleic acids of many viruses is catalyzed by enzymes that do not exist in uninfected cells. This is true especially for all RNA-containing viruses, as well as for the pox-viruses. It should be possible to isolate and characterize these enzymes and to find specific inhibitors for them.

The Synthesis of Viral Proteins Several lines of evidence suggest that viral messenger RNAs differ in some fundamental way from host messenger RNAs. Viral messenger RNAs are as a rule translated in preference to host cell messenger RNAs. Furthermore, translation of host messenger RNAs often ceases entirely several hours after infection, when translation of viral messenger RNAs proceeds rapidly and extensively. This suggests that viral messenger RNAs differ in some recognizable manner from host messenger RNAs, and this difference should be exploitable. In fact, two of the most successful antiviral agents known, isatin-β-thiosemi-carbazone and interferon, act by preventing viral messenger RNAs from being translated under conditions when they have no detectable effect on the translation of most host cell messenger RNAs. A promising avenue of approach, therefore, is analysis of the features that differentiate viral RNAs from host cell messenger RNAs.

It is unlikely that inhibitors specific for viral protein synthesis itself will be found, since viral proteins are synthesized by the same ribosomes that synthesize host proteins and since they present no unusual features in their primary amino acid sequence.

Viral Morphogenesis Viral morphogenesis proceeds at several levels. Many viral capsid polypeptides are now known to be synthesized in the form of precursors that are cleaved to furnish the actual polypeptides used for the formation of virus particles. The nature of the enzymes that cleave these precursors is not known. It may be fruitful to characterize them and to design inhibitors for them.

No way of specifically inhibiting the assembly of virus particles exists. It is conceivable that budding could be prevented when more is known about the properties of cell membranes, but it seems that cell membranes are such important organs that it may be unwise to attempt to prevent virus particle formation at this point. By the same token, inhibition of the uptake and penetration of virus particles at the beginning of the multiplication cycle has not been attempted seriously. As will be discussed below, the drug α-adamantanamine does inhibit the uptake of certain myxoviruses, but it does not do so by directly affecting the functioning of the cell membrane. In summary, the two most promising avenues of approach for specifically inhibiting virus multiplication are inhibition of viral genome replicases and exploitation of the chemical differences between viral RNAs and host-cell messenger RNAs.

The mode of action of certain antiviral chemotherapeutic agents

No antiviral chemotherapeutic agent capable of inhibiting virus multiplication efficiently in either man or animal is yet available. However, several chemicals have been used more or less successfully in clinical trials, either as prophylactic agents or as inhibitors of virus multiplication under special conditions. In addition, a number of new agents which are now being developed show great potential; most of them act via one or another of the mechanisms that were discussed above.

Before considering individually the mode of action of successful and promising antiviral agents, it is well to examine which viral pathogens of man present the most suitable targets for chemotherapy at this time. It is reassuring to reflect that the incidence of many severe virus-caused diseases, such as smallpox, poliomyelitis, yellow fever and measles, has decreased dramatically in recent decades, owing to the success of vaccination campaigns. Other important virus-caused diseases fall into two classes. First, there are those that, although severe and life-threatening, are rare in most parts of the world; such diseases are rabies, the encephalitides, and the African and South American hemorrhagic fevers (Marburg disease, Ebola fever, Lassa fever and Machupo). Second, there are diseases that are usually no more than mild to moderate, but may become life-threatening; such diseases are caused by influenza virus and the various herpesviruses. In fact, herpesviruses may well be regarded as the primary targets for antiviral chemotherapy at this time, closely followed by influenza virus and the rare severe diseases mentioned above.

Isatin-β-thiosemicarbazone (IBT)

IBT is a potent inhibitor of poxvirus multiplication (Fig. 8-1). It also inhibits adenovirus multiplication, and some of its derivatives inhibit the multiplication of certain enteroviruses. Only its antipoxvirus activity has been investigated in detail.

At a concentration of 3 mg per liter, IBT inhibits vaccinia virus multiplication in cultured cells by over 90 percent without having any detectable effect on the host cells themselves. In the presence of IBT, the early period of the poxvirus multiplication cycle, viral DNA replication and transcription of late messenger RNA all proceed normally; however, the translation of late messenger RNA is inhibited, and the synthesis of late proteins, which include most of the viral capsid proteins, is therefore prevented, and no progeny virus particles are formed.

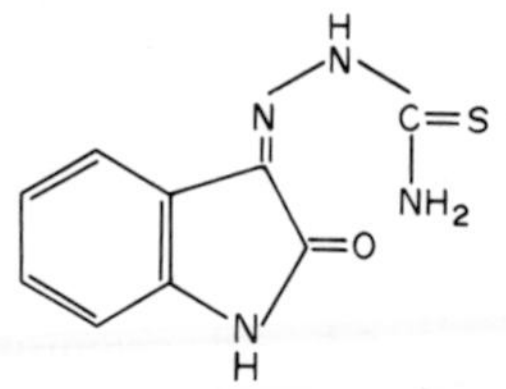

A derivative of IBT, N-methyl-IBT(Marburan), was administered by mouth to known smallpox contacts in field trials in India and Pakistan, and some beneficial results of this prophylactic treatment were observed.

2-Hydroxybenzylbenzimidazole (HBB) and Guanidine

These two reagents (Fig. 8-2) inhibit the multiplication of many picornaviruses, such as poliovirus, echoviruses, Coxsackie viruses, and foot-and-mouth disease virus. They are examples of reagents that interfere with the replication of viral RNA, either by preventing initiation of the synthesis of progeny plus strands or by preventing progeny plus strands from separating from the replicative form-replicase complex (Chap. 5). The precise manner in which this is accomplished is not known.

Although these two compounds appear to inhibit picornavirus multiplication by similar mechanisms, these mechanisms are not identical. This is shown by the fact that (1) there is little cross-resistance among the mutants to either compound (although guanidine can replace HBB in promoting the growth of HBB-dependent mutants); (2) their antiviral spectra are not identical (for example, there are guanidine-sensitive but no HBB-sensitive rhinovirus strains); and (3) simple methyl donors, such as choline and methionine, rapidly and efficiently reverse the inhibitory effect of guanidine (by an entirely unknown mechanism), but have no effect on the inhibition caused by HBB. Although both drug-resistant and drug-dependent virus mutants quickly emerge, HBB

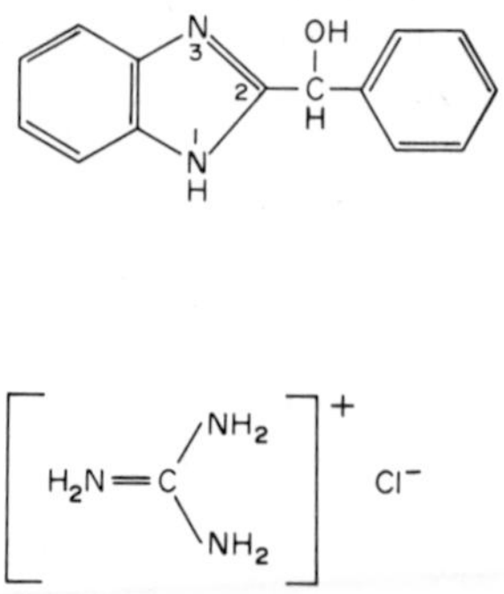

FIG. 8-1. Isatin-β-thiosemicarbazone (IBT).

FIG. 8-2. 2-Hydroxybenzylbenzimidazole (HBB) and guanidine hydrochloride.

has been used with some success in controlling enterovirus infections in animals.

Rifampicin

Rifampicin and related rifamycin derivatives bind to bacterial RNA polymerases, thereby preventing the initiation of transcription. Rifampicin does not bind to animal RNA polymerases, but it does inhibit the multiplication of poxviruses and adenoviruses. The mechanism by which it achieves this has been studied most intensively in vaccinia virus-infected cells. Inhibition of viral RNA polymerase is not involved, since both early and late messenger RNAs are transcribed normally. Rather, the mechanism involves some event in viral morphogenesis, for in the presence of the drug immature virus particles of the type illustrated in Figure 5-27 accumulate. As a result of this block in morphogenesis, the major vaccinia virus structural protein, VP4c, which arises during maturation as a result of the cleavage of a precursor, is not formed (see Fig. 2-29, band 19). The specific nature of the step that is inhibited is not known. However, it is clear that it involves a diffusible product, most probably a protein, since wild-type virus sensitive to rifampicin matures normally in its presence in cells simultaneously infected with mutants resistant to it; presumably the mutants code for a protein that is unaffected by the drug and can thus function when the normal wild-type protein cannot. As a result, resistant mutants rescue sensitive virus.

Although rifampicin does not inhibit the vaccinia DNA-dependent RNA polymerase, the RNA-dependent DNA polymerase of RNA tumor viruses (the reverse transcriptase, Chap. 9) is very sensitive to it and to certain of its derivatives.

α-Adamantanamine

α-Adamantanamine (Amantadine, Symmetrel) (Fig. 8-3), a substance with a remarkably rigid structure, inhibits an early event in the multiplication cycle of influenza virus (and also of arenaviruses). It does not inhibit adsorption and penetration, but completely inhibits primary transcription; it is thought therefore that it inhibits uncoating. Drug-resistant mutants exist; these are mutants in the M protein gene. It is

FIG. 8-3.   α-Adamantanamine (Amantadine).

conceivable therefore that the drug interacts with the M protein.

On the basis of large-scale trials that demonstrated that it had a statistically significant prophylactic effect, α-adamantanamine has been recommended for the prevention of disease caused by influenza virus type A. It has been used extensively in the Soviet Union and is presently the only drug for virus respiratory disease that is licensed in the United States. However, it is not used extensively in the United States because it seems impractical to control, through chemoprophylaxis in an open population, a disease that is generally mild and that the individual patient has a good chance of avoiding anyway. As for the problem of protecting individuals at high risk and in whom the disease may be potentially dangerous, there is a choice between this drug and the influenza vaccine (p. 157). At least at this time, the vaccine would seem to be preferred.

Phosphonoacetic acid (PAA)

Phosphonoacetic acid (Fig. 8-4) is a potent inhibitor of herpesvirus-coded DNA polymerase, but not of cellular DNA polymerases. Unfortunately it has undesirable side effects, and drug-resistant mutants emerge rapidly. It is conceivable that modification of the structure of this drug will yield compounds to which resistance

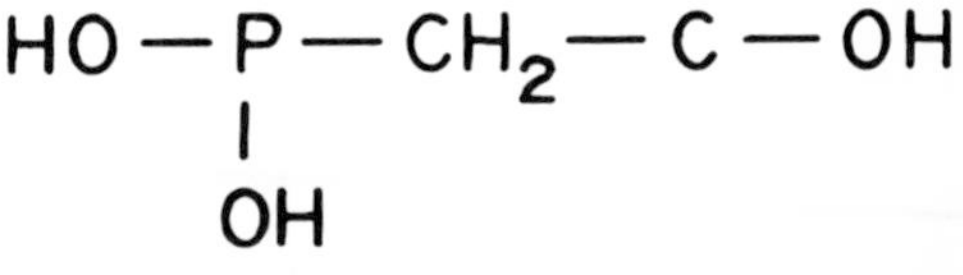

$$HO - \overset{\displaystyle O}{\underset{\displaystyle OH}{\overset{\|}{P}}} - CH_2 - \overset{\displaystyle O}{\overset{\|}{C}} - OH$$

FIG. 8-4.   Phosphonoacetic acid (PAA).

cannot develop and which will be less toxic. Recently the closely related phosphonoformic acid has been found to be an inhibitor of herpesvirus DNA polymerase and to be effective in treating cutaneous herpes simplex virus infections in guinea pigs.

Analogs of ribo- and deoxyribonucleosides

These substances consist of nucleic acid bases or derivatives of them, linked to either ribose or analogs of ribose such as arabinose. The most important antiviral substances of this type are 5'-iododeoxyuridine (IDU) (Fig. 8-5), trifluorothymidine (F_3T), cytosine arabinoside (ara-C), adenosine arabinoside (ara-A or vidarabine) and ribavirin (virazole), in which the nucleic acid base is an analog of the purine precursor 5'-aminoimidazole-4-carboxamide (Fig. 8-6).

The mechanism of action of all of these drugs, most of which were either designed or first tested as anticancer rather than antiviral agents, is similar: all are inhibitors of nucleic acid biosynthesis. All become phosphorylated to mono-, di- and triphosphates and are incorporated into nucleic acid; in the process they compete with the normal nucleic acid precursors and therefore inhibit the synthesis of cellular nucleic acids. The basis for their antiviral effect is twofold. The first, and most important, is the following: if the virus that is to be inhibited codes for an enzyme on the nucleic acid biosynthetic pathway, such as a deoxyribonucleoside kinase or a DNA polymerase, then it may be inhibited to a greater extent by the analog than by the corresponding cellular enzyme, thus causing a specific antiviral effect. In similar fashion, virazole inhibits the biosynthe-

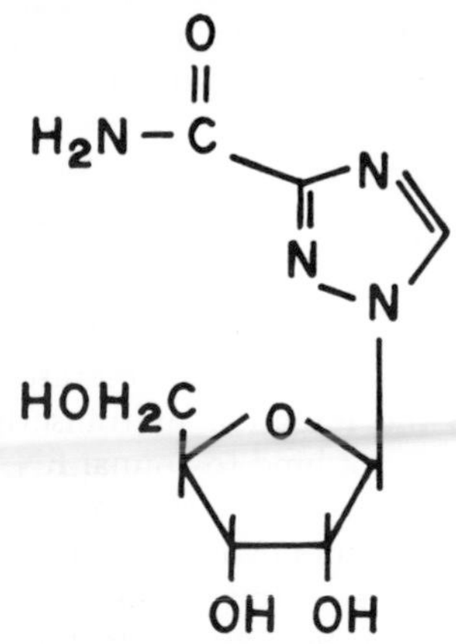

FIG. 8-6. Ribavirin (1-β-D-ribofuranosyl-1,2,4-triazole-3-carboxamide).

sis of guanine nucleotides: its antiviral target is probably virus-coded enzymes that have a higher K_m for its derivatives than normal cellular enzymes. Second, some of these analogs, particularly IDU, interfere with the functioning of the DNA into which they have been incorporated. Thus 5'-iodouracil is incorporated into DNA in place of thymine; and since it does not pair with adenine as faithfully as with thymine, mismatching occurs during both the replication and the transcription of the substituted DNA, which causes the formation of defective progeny DNA strands and defective messenger RNAs.

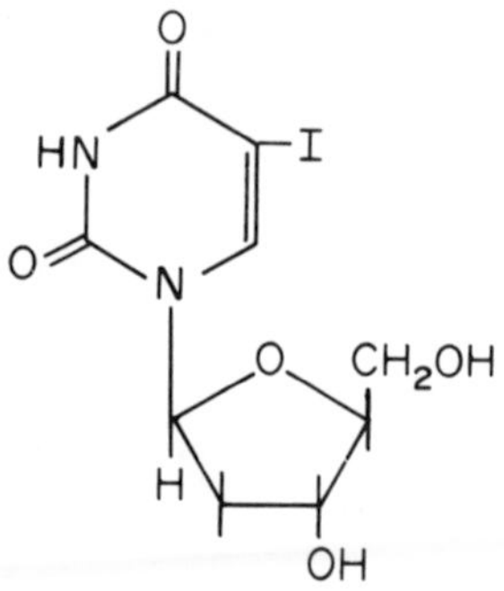

Because RNA-containing viruses do not code for enzymes on the biosynthetic pathway of nucleic acids, nucleoside analogs are more active against DNA- than RNA-containing viruses. The most promising drug of this type that is currently undergoing advanced clinical trials is vidarabine and its phosphorylated derivative, ara-AMP, which appear to be very useful in the treatment of severe herpesvirus infections of the central nervous system.

The "target cell" approach

It has long been realized that the most successful strategy in antiviral chemotherapy would be to use drugs that could enter, or would be acti-

FIG. 8-5. 5'-Iododeoxyuridine (IDU).

vated in, only virus-infected cells—in other words, a strategy that focused on the infected cell as the target cell. An approach along parallel lines is used in the treatment of herpetic keratoconjunctivitis. Here IDU is applied topically to the eye; although IDU is most probably incorporated into host cell as well as into viral DNA, the IDU reaches so few host cells that no serious damage is caused; at worst, damaged corneal cells regenerate after virus multiplication has been inhibited and infected cells have been eliminated.

Two approaches to the "target cell" strategy are as follows. The first involves the new drug acycloguanosine [9-(2-hydroxyethyoxymethyl)-guanine] (Fig. 8-7), which is phosphorylated far more extensively by the herpesvirus thymidine kinase than by cellular enzymes. It is therefore converted to its active form only in cells that are infected with the virus that it is intended to inhibit; the drug is therefore very specific and consequently not toxic. The second approach is designed to take advantage of the fact that infection with viruses often changes the permeability properties of the membranes of the cells that they infect (see Chap. 9). This provides the opportunity for introducing into infected cells antiviral agents that would not be taken up by uninfected cells; in fact, this approach should permit the administration of drugs that are quite toxic to normal cells. Both approaches promise to permit the use of very much more powerful antiviral agents than are in our current repertoire. This is likely to be particularly important in the case of RNA-containing viruses, for which, as pointed out above, effective

inhibitors are proving difficult to find. Two that have been tried are ribavirin, which inhibits influenza virus multiplication in cultured cells and for which partial success has been claimed in some clinical trials; and FANA (2-deoxy-2-3-dehydro-N-trifluoroacetyl-neuraminic acid), which is an inhibitor of the neuraminidase of influenza virus and also inhibits its multiplication, but which is too toxic for use in humans. Amantadine therefore remains the best anti-influenza virus agent. As for togaviruses, arenaviruses and rabies virus, effective chemotherapeutic agents against them remain to be devised.

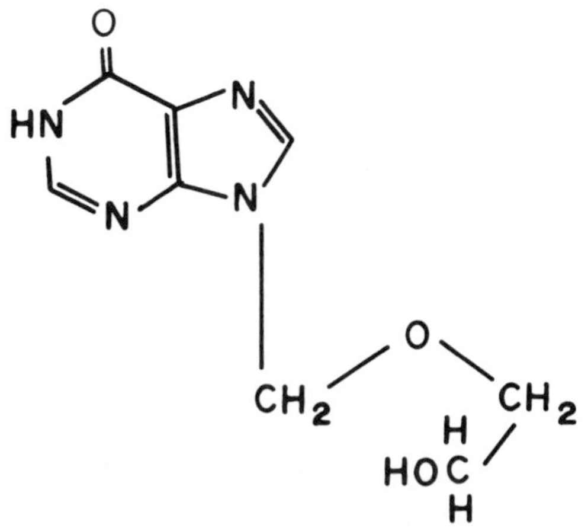

Interferon

PROPERTIES AND MODE OF ACTION OF INTERFERON

The antiviral agent on which interest is currently focused most intensely is one that is elaborated by living cells themselves. Animal cells infected with viruses very often produce a protein which, when added to uninfected cells, protects them against virus infection, or, more precisely, greatly decreases the chance that subsequent virus infection will initiate a productive multiplication cycle. This substance is called interferon. The key facts that are known concerning its nature, formation and mode of action are as follows.

1. Interferon is a host-coded protein. Normal cells do not as a rule contain interferon. Interferon released by infected cells is induced as a result of virus infection, and is generally produced most rapidly at about the time when replication of the viral genome proceeds most rapidly.

2. As a consequence, interferon is species-specific, but not virus-specific. Sometimes this specificity is very narrow: for example, chick and duck interferons exhibit little if any crossprotective ability, nor do mouse and rat interferons. However, not unexpectedly, there are exceptions; for example, human interferon protects bovine cells even better than bovine interferon. One would not, of course, expect

FIG. 8-7. Acycloguanosine.

the amino acid sequences of the various mammalian interferons to differ by more than a few residues.

3. There is not one type of human interferon, but three: fibroblast interferon and leukocyte interferon, both of which are produced in response to virus infection, and "immune" or type 2 interferon, which is produced when nonsensitized lymphocytes are stimulated by mitogens or when sensitized lymphocytes are stimulated with specific antigens. All three are distinct molecular species with no common amino acid sequences; all are glycoproteins with molecular weights of about 20,000.

4. Interferon is commonly assayed by exposing cells to preparations containing it for a period of 12 to 24 hours and then challenging the cells with a standard amount of virus known to produce a certain number of plaques. The titer of the interferon preparation is the reciprocal of that dilution which reduces this number by 50 percent. The medium of cultured cells stimulated to produce interferon commonly contains about 10^5 PRD_{50} (50 percent plaque reduction doses) of interferon per milliliter.

5. A highly distinctive feature of fibroblast and leukocyte interferons is their resistance to low pH; both are quite stable at pH 2 at 4C. Both also retain activity in the presence of sodium dodecyl sulfate—a most unusual property.

6. In spite of such stability, interferons have proved extraordinarily difficult to purify because of unaccountable losses of activity as contaminating protein is being removed. Interferon preparations have been obtained with about 10^9 PRD_{50}/mg protein, but such preparations still seem not to be absolutely pure.

7. Interferon itself is not the protein that inhibits virus multiplication; it only protects cells if RNA and protein synthesis are permitted to proceed. In effect, interferon is an inducer that causes cells to synthesize a protein which is the actual inhibitor of virus multiplication. This protein, which has not yet been isolated or characterized, confers upon cells an antiviral state that usually lasts for several days. Further evidence that interferon and the antiviral protein are different comes from genetic studies. Interferon formation in humans is controlled by genes on two chromosomes (chromosomes 2 and 5), while expression of the antiviral state depends upon a gene on chromosome 21.

8. The precise mode of action of interferon is not known, but it seems clear that the basic mechanism involves interference with the ability of parental or early viral messenger RNA molecules to be translated. As a result, no virus-specified proteins are synthesized, no progeny viral genomes are formed, and infection is aborted.

9. Finally, in addition to its antiviral activity, interferon, at much higher concentrations, also affects several cell functions: it modifies the immune response, inhibits the response to mitogenic stimuli and the expression of certain cell surface antigens and, above all, it inhibits the multiplication of both normal and tumor cells—that is, it possesses a cell multiplication inhibitory (CMI) activity. It is conceivable that all these functions are mediated by messenger RNAs that share certain sequences, and that the primary function of interferon is to regulate the expression of these *cellular* messenger RNAs rather than those of viral messenger RNAs (which may be inhibited because they are structurally similar). Thus interferon may have evolved as a cell regulatory mechanism, rather than as an antiviral defense mechanism.

The use of interferon as an anticancer agent is currently being explored intensively. Although there is an almost total lack of data concerning the molecular basis of the CMI activity of interferon in general, and of such an activity directed specifically against tumor cells in particular, clinical trials with human leukocyte as well as fibroblast interferon are proceeding at this time both in Europe and in the United States. Very promising results have already been obtained in the treatment of osteogenic sarcoma, and the effect of interferon on other forms of cancer, such as breast cancer, lung cancer, brain tumors and hepatomas is currently being explored. Unfortunately the limiting factor in such studies is still the availability of sufficient amounts of even semipurified interferon.

THE INDUCTION OF INTERFERON

The lack of toxicity of interferon, coupled with its extraordinarily high biologic activity, has placed it into the forefront of potentially useful antiviral and anti-cancer agents. There are two possible approaches to the induction of high levels of circulating interferon in humans. The first is to administer interferon by injection. Unfortunately interferon is cleared rapidly from the bloodstream, so that the amount of interferon that is necessary to protect even one person is very large. The alternative approach is to induce the formation of interferon in the individuals who are to be protected or treated. This method is receiving a good deal of attention, since it has been found that the formation of interferon can be induced by agents other than infectious virus. Among such agents are the following:

1. Viruses inactivated by heat or ultraviolet irradiation. Among these viruses are inactivated influenza virus, NDV, reovirus, and bovine enterovirus.
2. Double-stranded RNA, such as reovirus RNA, the replicative form of single-stranded viral RNAs, and the RNA present in certain fungal viruses. In fact, a crude extract of *Penicillium stoloniferum* known as "statolon" was known to be an inducer of interferon formation long before it was recognized that its active principle is the double-stranded RNA of a virus that is harbored by this mold. In addition, certain synthetic double-stranded polyribonucleotides, in particular poly I:poly C, are effective inducers of interferon formation, especially under conditions of superinduction. This is a phenomenon that is elicited when cells are treated with poly I:poly C together with the protein synthesis inhibitor, cycloheximide. If, after five hours, actinomycin D is added, followed one hour later by reversal of the inhibition of protein synthesis, about 50 times more interferon is produced than if cycloheximide had not been added initially. The basis of this phenomenon is thought to be the putative existence of an unstable regulator of the expression of the interferon gene, whatever its nature may be. During the first five-hour period both interferon and regulator messenger RNA would accumulate. After the addition of actinomycin D, there would be decay of both interferon and regulator messenger RNA, the latter more rapidly than the former; upon release of the protein synthesis block, only interferon would be formed in the presence of greatly reduced levels of regulator. Single-stranded RNA and single-stranded or double-stranded DNA are inactive as interferon inducers, as are double-stranded RNA-DNA hybrid molecules.
3. Certain synthetic polycarboxylic acids and pyran copolymers are inducers of interferon formation at relatively high concentrations, as is bacterial endotoxin. The feature common to all these substances, including the polyribonucleotides, is that they are polyanions.

The mechanism by which interferon formation is induced is unclear. Nor is it known why substances as diverse as double-stranded RNA and endotoxin can both induce interferon, whereas a substance as similar to double-stranded RNA as double-stranded DNA cannot.

Unfortunately, none of the three types of inducers fulfills the two essential requirements of an interferon inducer in humans—namely, effectiveness and low toxicity; nor are these inducers effective enough to induce the synthesis in cultured cells of the very large amounts of interferon that would be required for the treatment of large numbers of patients. Interest is therefore currently focused on the development of techniques for cloning the interferon gene. There is little doubt that the availability of large amounts of human interferon would mark a major step in the conquest of diseases caused by viruses and possibly also of cancer.

Vaccines

The only currently feasible means of preventing diseases caused by viruses is through mobilization of the immune mechanism by means of vaccines. There are two types of antiviral vaccines: inactivated virus and attenuated active virus.

INACTIVATED VIRUS VACCINES

The primary requirements for an effective vaccine of this type are complete inactivation of infectivity coupled with minimum loss of antigenicity. These requirements are not easily met simultaneously, since few reagents are available that inactivate viral genomes, the source of infectivity, without also affecting viral protein, the source of antigenicity. Ultraviolet irradiation could accomplish this best but is inapplicable because virus inactivated in this manner is capable not only of expressing the function of those genes that have not received a lethal hit, but also of undergoing multiplicity reactivation (see Chap. 7). Photodynamic inactivation (Chap. 3) inactivates viral nucleic acids efficiently and irreversibly without damaging viral proteins, which therefore retain full immunogenicity. However, it has recently been found that herpesvirus inactivated in this manner can still transform cultured cells in vitro (see Chap. 9) and may therefore be tumorigenic. Betapropiolactone is a potentially useful inactivating agent but has been used only rarely because it is a potent carcinogen. The best reagent for inactivating viral nucleic acid without compromising antigenicity is formaldehyde, but it also has drawbacks: first, it inactivates only viruses that contain single-stranded nucleic acids, and second, care must be exercised to avoid formation of a resistant virus fraction (Chap. 3).

Inactivated virus vaccines are generally made from unpurified virus preparations that contain much host cell material. Since inactivated virus cannot multiply, relatively large amounts of this type of vaccine must be administered so as to provide sufficient antigen. As a result, a great deal of nonviral material that may cause undesirable side effects is also introduced into the body.

ATTENUATED ACTIVE VIRUS VACCINES

The second method of immunizing against viral pathogens is by administering attenuated virus strains, antibody to which is capable of neutralizing the pathogen (Chap. 4). This is the principle on which Jenner's vaccination procedure against smallpox in 1798 was based; he inoculated with the essentially nonvirulent vaccinia virus to induce antibodies against the highly virulent smallpox virus. Since then many attenuated virus vaccine strains have been developed, among them Theiler's yellow fever virus vaccine strain, the attenuated Sabin poliovirus vaccine strains, and attenuated measles and rubella virus strains. The most commonly used method of producing such attenuated virus strains is by repeated passage of the human pathogen in other host species, which results in the selection of variants with drastically reduced virulence for humans.

Attenuated virus vaccines are effective in very small amounts, since the attenuated virus can multiply. This provides a powerful amplification effect; the viral progeny, rather than the virus in the inoculum, acts as the antigen. The attenuated vaccines also possess the advantage of stimulating the formation of all the correct types of antibody molecules (i.e., IgA, etc., as well as IgG). Since only small quantities of this type of vaccine need be administered, the virus is usually not purified. However, the use of purified virus would yield important benefits. In particular, it would greatly reduce the likelihood of the presence in vaccines of other, potentially dangerous, viruses, which would in turn permit vaccine viruses to be produced more efficiently than is currently advisable. Attenuated virus vaccine strains are usually grown in cultured cells derived from a variety of animal species and organs. Great care is always taken that these cells display none of the characteristics of cells transformed by tumor viruses (Chap. 9). In particular, the cells must be euploid, display a high degree of contact inhibition, and contain no particles that may be virus particles. Generally cells that have not been passaged long in vitro are preferred, so as to reduce the likelihood of transformation in vitro and the selection of transformed cells. The disadvantage of such contact-inhibited, euploid cells is that they usually do not grow rapidly and do not achieve high population densities, and that therefore the amounts of virus that they yield are relatively low. By contrast, cells of established "immortal" strains display superior growth capabilities in vitro and permit the production of larger amounts of virus, but they are often derived from cells that are preneoplastic or neoplastic in character, are usually aneuploid, and often contain particles that may be virus particles. Such cells are therefore not used for the production of attenuated virus vaccine strains. However, the potential danger of the presence of tumor viruses and passenger viruses in general in attenuated virus vaccines could be virtually eliminated by routinely purifying all such vaccines by techniques such as density gradient centrifugation. If this were done, the easily cultured, high-yielding, established cell lines could be used safely for the growth of attenuated virus vaccines.

Several potential hazards associated with the use of active attenuated virus vaccines have not caused difficulties as yet but should be kept in mind. (1) There is the danger that attenuated virus strains will, by mutation, revert to more virulent strains. This has not happened, probably because the attenuated virus strains are multistep rather than single-step variants. (2) Administration of active attenuated vaccines involves infecting humans with virus. The only desired effect of such virus is induction of the formation of antibodies, but this may not always be their only effect. For example, attenuated viruses, like their pathogenic parents, may in rare instances multiply in cells that are not their usual host cells. If such cells are those of an organ such as the brain, serious damage may result. This is in fact what happens in postvaccination encephalitis (Chap. 12). (3) The vaccine virus may activate latent viruses; models for such interactions are provided by AAV, which multiplies only in cells infected with adenovirus, and by the PARA particles (Chap. 7). Finally (4), as discussed in Chapters 6 and 8, virus strains that are less virulent than wild-type virus have been implicated in the establishment and maintenance of persistent infections. This applies particularly to temperature-sensitive variants, some of which have been shown to cause chronic infections. Proposals to develop attenuated virus vaccine strains on the basis of selecting virus strains with lower optimum growth temperatures should therefore be evaluated very carefully, and any such strains should be tested exhaustively before being certified for general use.

This discussion of the potential hazards associated with the use of inactivated and attenuated active virus vaccines is in no way intended to discourage their development and use. Indeed, it has already been emphasized that such vaccines currently provide the only practical defense against infection. However, it should serve as a warning against the indiscriminate use of antiviral vaccines, and medical practitioners should be clearly aware of the potential dangers as well as of the undoubted benefits of immunization. For example, although in abso-

lute numbers the incidence of harmful sequelae following administration of a given vaccine may be very low, a serious public health issue may develop if 100 deaths were to result from vaccinating 100 million human beings. A decision as to whether to continue using such a vaccine would then have to be weighed in terms of the situation that would arise if the vaccine were not used.

An excellent example of the type of cost-benefit analysis that must be applied to vaccination programs is provided by the national swine influenza vaccination campaign of 1976. The development of several hundred cases of Guillain-Barré syndrome was sufficient to abort this program because the expected swine influenza epidemic did not occur; the relatively small number of paralysis cases proved to be an unacceptable cost. Had there developed a swine influenza epidemic, with severe and possibly lethal disease, the vaccination program would no doubt have proceeded to completion.

FURTHER READING

Books and Reviews

Cohen SS: The lethality of aranucleotides. Med Biol: 54:299, 1976

Collins FM: Vaccines and cell-mediated immunity. Bacteriol Rev 38:371, 1974

Gresser I, Tovey MG: Antitumor effect of interferon. Biochem Biophys Acta 516:231, 1978

Ho M, Armstrong JA: Interferon. Ann Rev Microbiol 29:131, 1975

Kantoch M: Markers and vaccines. Adv Virus Res 22:259, 1978

Levy HB, Riley FL, Buckler CE: Interferon. In Nayak DP (ed): The Molecular Biology of Animal Viruses, Vol 1:41. New York, Marcel Dekker, 1977

Melnick JL: Viral vaccines. Prog Med Virol 23:158, 1977

Stewart WE II: The Interferon System. New York, Vienna, Springer-Verlag, 1979

Selected Papers

CHEMOTHERAPEUTIC AGENTS

Carrasco L: Membrane leakiness after viral infection and a new approach to the development of antiviral agents. Nature 272:694, 1978

de Clercq E, Descamps J, de Somer P, Holý A: (S)-9-(2,3-dihydroxypropyl) adenine: An aliphatic nucleoside analog with broad-spectrum antiviral activity. Science 200:563, 1978

Fyfe JA, Keller PM, Furman PA, Miller RL, Elion GB: Thymidine kinase from herpes simplex virus phosphorylates the new antiviral compound, 9-(2-hydroxy-ethoxymethyl) guanine. J Biol Chem 253:8721, 1978

Helgstrand E, et al: Trisodium phosphonoformate—a new antiviral compound. Science 201:819, 1978

Honess RW, Watson DH: Herpes simplex virus resistance and sensitivity to phosphonoacetic acid. J Virol 21:584, 1977

Katz E, Moss B: Formation of a vaccinia virus structural polypeptide from a higher molecular weight precursor: inhibition by rifampicin. Proc Natl Acad Sci USA 66:677, 1970

Korant BD: Poliovirus coat protein as the site of guanidine action. Virol 81:25, 1977

Lubeck MD, Schulman JL, Palese P: Susceptibility of influenza A viruses to amantadine as influenced by the gene coding for the M protein. J Virol 28:710, 1978

North TW, Cohen SS: Erythro-9-(2-hydroxy-3-nonyl) adenine as a specific inhibitor of herpes simplex virus replication in the presence and absence of adenosine analogues. Proc Natl Acad Sci USA 75:3684, 1978

Palese P, Compans RW: Inhibition of influenza virus replication in tissue culture by 2-deoxy-2,3-dehydro-N-trifluoroacetylneuraminic acid (FANA): mechanism of action. J Gen Virol 33:159, 1976

Schaeffer HJ, Beauchamp L, de Miranda P, Elion GP, Bauer DJ, Collins P: 9-(2-hydroxyethoxymethyl) guanine activity against viruses of the herpes group. Nature 272:583, 1978

Tamm I, Sehgal PB: Halobenzimidazole ribosides and RNA synthesis of cells and viruses. Adv Virus Res 22:188, 1978

Whitley RJ, Alford CA: Developmental aspects of selected antiviral chemotherapeutic agents. Ann Rev Microbiol 32:285, 1978

Willis RC, Carson DA, Seegmiller JE: Adenosine kinase initiates the major route of ribavirin activation in a cultured human cell line. Proc Natl Acad Sci USA 75:3042, 1978

Woodson B, Joklik WK: The inhibition of vaccinia virus multiplication by isatin-β-thiosemicarbazone. Proc Natl Acad Sci USA 54:946, 1965

INTERFERON

Brodeur BR, Merigan TC: Suppressive effect of interferon on the humoral immune response to sheep red blood cells in mice. J Immunol 113:1319, 1974

Claes P, Billiau A, de Clercq E, et al: Polyacetal carboxylic acids: a new group of antiviral polyanions. J Virol 5:313, 321, 1970

Dorner F, Scriba M, Weil R: Interferon: evidence for its glycoprotein nature. Proc Natl Acad Sci USA 70:1981, 1973

Friedman RM, Sonnabend JA: Inhibition of interferon action by p-fluorophenylalanine. Nature 203:366, 1964

Gresser I, Brouty-Boyé D, Thomas M, Macieira-Coelho A: Interferon and cell division. 1. Inhibition of the multiplication of mouse leukemia L 1210 cells in vitro by interferon preparations. Proc Natl Acad Sci USA 66:1052, 1970

Henderson DR, Joklik WK: The mechanism of interferon induction by UV-irradiated reovirus. Virol 91:389, 1978

Joklik WK, Merigan TC: Concerning the mechanism of action of interferon. Proc Natl Acad Sci USA 56:558, 1966

Kerr IM, Brown RE: pppA2'p5'A2'p5'A: an inhibitor of protein synthesis synthesized with an enzyme fraction from interferon-treated cells. Proc Natl Acad Sci USA 75:256, 1978

Repik P, Flamand A, Bishop DHL: Effect of interferon upon primary and secondary transcription of vesicular stomatitis and influenza viruses. J Virol 14:1169, 1974

Rousset S: Refractory state of cells to interferon induction. J Gen Virol 22:9, 1974

Samuel CE: Mechanism of interferon action: phosphorylation of protein synthesis initiation factor eIF-2 in interferon-treated human cells by a ribosome-associated kinase possessing site specificity similar to hemin-regulated rabbit reticulocyte kinase. Proc Natl Acad Sci USA 76:600, 1979

Stewart WE II, Gosser LB, Lockhart RZ: Priming: a nonantiviral function of interferon. J Virol 7:792, 1971

Stewart WE II, de Somer P, Edy VG, et al: Human interferons: requirements for stabilization and reactivation of human leukocyte and fibroblast interferon. J Gen Virol 26:327, 1975

Stewart WE II, Chudzio T, Lin LS, Wiranowska-Stewart M: Interferoids: in vitro and in vivo conversion of native interferons to lower molecular weight forms. Proc Natl Acad Sci USA 75:4814, 1978

Tan YH: Chromosome 21-dosage effect on inducibility of anti-viral gene(s). Nature 253:280, 1975

Tan YH, Creagan RP, Ruddle FH: The somatic cell genetics of human interferon: assignment of human interferon loci to chromosomes 2 and 5. Proc Natl Acad Sci USA 71:2251, 1974

Virelizier J, Gresser I: Role of interferon in the pathogenesis of viral diseases of mice as demonstrated by the use of anti-interferon serum. J Immunol 120:1616, 1978

Wiebe ME, Joklik WK: The mechanism of inhibition of reovirus replication by interferon. Virol 66:229, 1975

Zilberstein A, Kimichi A, Schmidt A, Revel M: Isolation of two interferon-induced translational inhibitors: a protein kinase and an oligo-isoadenylate synthetase. Proc Natl Acad Sci USA 75:4734, 1978

VACCINES

Collins FM: Vaccines and cell-mediated immunity. Bacteriol Rev 38:371, 1974

Salk J, Salk T: Control of influenza and poliomyelitis with killed virus vaccines. Science 195:834, 1977

CHAPTER 9

Tumor Viruses

Tumor viruses are considered separately from lytic viruses in order to focus attention on the role of viruses in carcinogenesis. Both RNA-containing and DNA-containing viruses can cause various types of neoplasms in animals, and evidence is mounting that they may act similarly in man. Awareness of the principles of tumor virology will therefore be important for medical practitioners in the decades ahead.

The origin of tumors

Tumor etiology, the study of the causes of tumors, has brought to light a bewildering variety of tumorigenic agents that fall into three classes: chemical substances, physical stimuli, and biologic agents. The chemical substances include compounds of the most diverse constitution, ranging from polycyclic hydrocarbons such as methylcholanthrene, benzo(a)pyrene, and dimethylbenzanthracene on the one hand, to multifunctional compounds such as dimethylnitrosamine, nitrosomethylurea, and 4-nitroquinoline-1-oxide on the other. The physical agents include both X-rays and ultraviolet irradiation, and the most important biologic agents are viruses.

The concept that infectious agents might be involved etiologically in the cancer process was advanced as early as 1908 by Ellerman and Bang, who observed that the mode of transmission of leukemia in the fowl was similar to that of an infectious disease. Shortly thereafter, Rous demonstrated that the infectious agent in avian sarcomas could pass through a filter that would not permit passage of bacteria. For many years this discovery remained an isolated finding, until in 1932 Shope discovered in wild cottontail rabbits a virus agent that transmitted a wartlike growth not only to cottontails but also to domestic rabbits. While most warts in cottontails remained benign or regressed, those in domestic rabbits sometimes developed into highly malignant carcinomas. Then, in 1938, Bittner discovered, in mammary gland tumors of mice, the milk factor, a virus which is passed from mother to offspring through suckling. The finding which, more than any other, elicited the upsurge of interest in viruses as carcino-

genic agents was the discovery by Gross in 1951 of a virus that induced leukemia in mice, a disease that is remarkably similar to leukemia in humans. Efforts to develop animal models applicable to the human disease have resulted in the isolation of a large number of viruses that cause many kinds of cancers in every major group of animals.

The characteristics of virus-transformed cells

Viruses are unique among carcinogens in that their tumorigenic activity is expressed efficiently in vitro and can therefore be studied under controlled conditions. Discussion of the virus-cell interaction up to this point has been focused on the lytic interaction, which involves multiplication of the virus and destruction of the host cell. However, certain DNA viruses can interact with cells not only by means of the lytic interaction, but also by means of an interaction in which virus multiplication is repressed and the host cell is not destroyed. The viral DNA then becomes closely associated with or integrated into the cellular genome, thereby giving rise to a new entity, the transformed or tumor cell, which can multiply indefinitely. In addition, a group of RNA-containing viruses, the RNA tumor viruses (oncornaviruses), can also transform cells and cause tumors. Until recently, the problem of how viral RNA could be integrated into the host genome presented a major conceptual hurdle. However, this has now been overcome by the discovery in RNA tumor viruses of a DNA polymerase that can transcribe RNA into DNA; this is the so-called reverse transcriptase. DNA transcribed from viral RNA by this enzyme is integrated into the cellular genome, conceivably by a mechanism analogous to that which integrates the genomes of the DNA-containing viruses.

In both cases the infected cells are altered in several fundamental characteristics and are said to be "transformed;" since these changes are at the genetic level, they are passed on to the cells' descendants. The principal properties in which virus-transformed cells differ from normal ones are as follows.

Possession of the Viral Genome

Transformed cells contain either the whole or part of the genome of the virus that causes the transformation. More often than not, this genome is integrated into the cell DNA, but in some transformed cell nuclei it may exist as a free plasmidlike entity. The extent to which viral genetic information is expressed in transformed cells varies widely, from full expression, with resultant formation of progeny virus particles, to complete silence, as judged by the absence of messenger RNA transcribed from it or proteins coded by it.

Tumorigenicity

Transformed cells generally give rise to tumors when injected into animals, particularly immunosuppressed or immunologically deficient animals, such as athymic nude mice. Like naturally occurring tumors, different lines of virus-transformed cells exhibit wide variation in invasiveness. Some transformed cells, even when injected in large numbers, produce merely benign tumors at the site of inoculation, while even small numbers of others give rise to highly invasive cancers.

Morphology

Normal and transformed cells differ in morphology. There are two major changes. First, transformed cells are usually more rounded and refractile than normal cells. Second, normal and transformed cells differ in the orientation of cells relative to each other. Normal cells usually arrange themselves in regular patterns, while transformed cells tend to orient themselves randomly (Fig. 9-1). These morphologic changes are often virus-specific. There are mutants of Rous sarcoma virus and polyoma virus that give rise to transformed cells with morphologies that differ from those induced by the corresponding wild-type viruses.

FIG. 9-1. Focus of chick embryo fibroblasts transformed with Rous sarcoma virus. Note drastically altered morphology and growth pattern of the transformed cells which are densely piled on top of one another in the center of the focus. (Courtesy of Dr. P. K. Vogt.)

Changes in Growth Patterns

Most types of untransformed cells grow in vitro to a certain cell density; cell division stops (or almost stops) when a monolayer of uniformly spread cells has formed. Under the same conditions, transformed cells continue to divide so that very much higher cell densities are attained.

This alteration of growth patterns is due primarily to two factors.

Loss of Contact Inhibition   When an untransformed cell comes into contact with another cell, there is cessation of the rapid movement of pseudopodia (ruffles) that are constantly extended and retracted, and forward movement is arrested. At the same time the cell stops dividing. This dual phenomenon is known as contact inhibition. The ability to respond to contact with other cells ensures that any given cell grows only in its appropriate location within the complete organism and does not proliferate unless it receives a signal that more of its kind are needed for the organism's orderly growth or maintenance.

While cell strains differ appreciably in their susceptibility to contact inhibition, transformed cells are always much less susceptible than their untransformed counterparts. This loss of contact inhibition represents release from one of the normal controls over multiplication; it may account in part for their unregulated growth in the organism. Some transformed cells, such as cells of benign tumors, have lost only the ability to respond to contact with each other but are still inhibited by contact with other cell types; others, such as those of metastasizing tumors, respond to contact neither with each other nor with other types of cells.

Reduction in the Requirement for Serum Contact inhibition is not the only factor that limits cell growth. The addition of serum to a nondividing contact-inhibited monolayer of untransformed cells results in further rounds of division, so that cells may actually pile up on top of one another. Virus transformation mark-

edly reduces cellular serum dependence; transformed cells require much less serum to initiate division than do untransformed cells. For example, 3T3 cells, a line of mouse fibroblasts, will not grow optimally unless the serum concentration is greater than 5 percent. By contrast, 3T3 cells transformed with SV40 can divide to a small but significant extent in serum concentrations as low as 0.5 percent. The nature of the serum factors that are required by cells and how these factors act is not known. It is hoped that studying them will hasten the deciphering of the signals that guide cells through their growth cycles.

Ability to Form Colonies in Soft Agar Suspension Untransformed cells of fibroblast origin must attach to a solid surface before they can divide; this requirement is known as "anchorage dependence of multiplication." By contrast, transformed cells will divide in suspension culture. In particular, ability to grow and form colonies when suspended in soft (0.5 percent) agar (Fig. 9-2) provides not only a very useful test for the stably transformed state but also the basis for a selective procedure that permits the isolation of transformed cells from

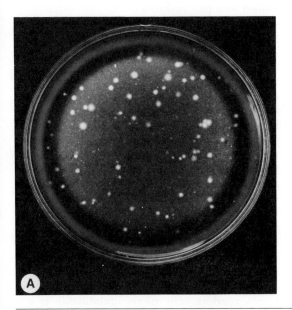
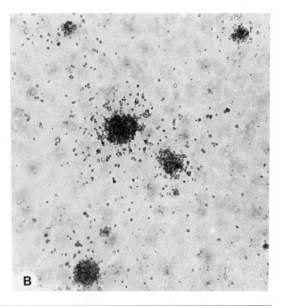

FIG. 9-2. Suspension colonies in soft agar of chick embryo fibroblasts transformed with Rous sarcoma virus. **A.** Appearance of a petri dish with numerous colonies (14 days after infection and plating). **B.** Enlargement of several colonies (8 days after infection and plating). (Courtesy of Dr. Thomas Graf.)

populations of predominantly untransformed ones.

Changes in Membrane Transport Properties

Simple sugars and other nutrients are transported several times more rapidly across the plasma membrane of transformed than of untransformed cells. This is generally the earliest observable change following transformation. It is conceivable that increased sugar transport into transformed cells is responsible, at least in part, for the increased rate of glycolysis that is generally observed in transformed cells.

Acquisition of New Surface Antigens

The surfaces of virus-transformed cells possess antigenic determinants not present on untransformed cells. These new surface antigens can be detected by immunofluorescence (Fig. 9-3) or by tests for cytotoxic and cell-mediated immunologic response. Indeed, their presence causes transformed cells to be recognized as foreign and as subject to immunologic surveillance. Most of these new antigenic determinants are virus-specified; some may be coded by the host genome. Among the mechanisms that may cause new host-specified antigens to manifest themselves following transformation are derepression of genes that do not ordinarily express themselves and unmasking or exposure of determinants that are not normally apparent.

Increased Agglutinability by Lectins

Many animal cells are agglutinated by certain glycoproteins (and proteins) that are present in plant seeds, snails, crabs, and some fish, and that are collectively known as lectins (or agglutinins). Some of these lectins preferentially agglutinate tumor cells. The two lectins that best discriminate between normal and transformed cells are the jack bean agglutinin concanavalin A (Con A) and wheat germ agglutinin (WGA). At first it was thought that transformed cells are agglutinated more readily than nontrans-

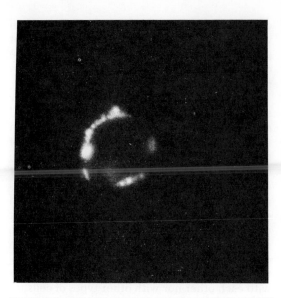

FIG. 9-3. Demonstration, by means of immunofluorescence, of the tumor-specific transplantation antigen (TSTA) on a cell derived from a tumor induced in hamsters with adenovirus type 12. The cell was treated first with a hamster antiserum to such cells and then with rabbit anti-hamster immunoglobulin conjugated with fluorescein isothiocyanate. The hamster antiserum was prepared by repeated injection of adenovirus type 12, followed by one injection of tumor cells. Note the annular pattern of specific fluorescence on the cell membrane. TSTA on cells transformed with papovaviruses can be demonstrated by analogous procedures. (From Vascencelos-Costa: J Gen Virol 8:69, 1970.)

formed ones because they bind more lectin molecules. However, it is now known that this is not so; rather the reason seems to lie in the arrangement of the lectin-binding sites which are dispersed in untransformed cells and clustered in transformed ones. Further, it is now known that both Con A and WGA can agglutinate untransformed cells under certain conditions; they will agglutinate cells in metaphase and cells exposed briefly to proteases, such as trypsin. Interestingly, contact-inhibited protease-treated cells escape from growth control for brief periods of time, probably until the proteins that are removed by proteases are regenerated. The reason treatment with proteases renders cells more agglutinable is not known; one explanation is that proteases split off surface glycopeptides, thereby increasing membrane fluidity and facilitating clustering of lectin receptor sites.

Changes in the Chemical Composition of Plasma Membrane Components

Whereas the protein composition of plasma membranes appears to be fairly constant, their lipid, glycolipid, and glycoprotein carbohydrate components are notoriously variable; they are exquisitely sensitive to changes in the medium, changes in the cell growth rate, and changes in growth conditions in general. It is difficult, therefore, to pick out changes attributable to transformation, but the following appear to be significant.

1. When untransformed cells become density (contact) inhibited, their glycolipids become larger and more complex; fucose, galactose, N-acetylgalactosamine and N-acetylneuraminic acid (NANA) are then added to them covalently. Apparently the transferases responsible for these extensions are not expressed in rapidly growing cells but are expressed in contact-inhibited cells. Transformed cells do not exhibit this cell density-dependent glycolipid extension response. Fucose-, galactose-, and NANA-containing glycolipids are present in transformed cells in smaller amounts than in untransformed cells and are smaller in size.
2. Strangely enough, changes of a similar but opposite nature occur in the oligosaccharides of glycoproteins; in transformed cells these groups are larger than in untransformed cells, and they contain more NANA (which always occupies a terminal position).
3. One of the major components of plasma membranes of normal cells is fibronectin, a large polymorphic glycoprotein (subunit M.W. 200,000–250,000); fibronectin possesses affinity for collagen, fibrin, heparin and cell surfaces. Transformed cells usually, but not always, lack this protein, which is therefore also known as LETS (large external transformation-sensitive) protein. Its absence may decrease cell-cell and cell-tissue matrix interactions; its absence may also increase membrane fluidity, thereby facilitating access to nutrients (thus lowering the requirement for serum) as well as lateral movement of membrane components (which would explain the in-creased agglutinability by lectins). Interestingly, treatment of transformed cells with LETS protein restores some morphologic features of normal cells.

Decreased Levels of Cyclic AMP

In normal cells the level of cyclic AMP increases as the cells reach confluency; lowering the concentration of serum in the medium has a similar effect. Transformed cells generally possess lower levels of cyclic AMP than normal ones, and lowering the serum concentration fails to elevate these levels.

Increased Secretion of Plasminogen Activator

Most cells produce small amounts of a protease, commonly known as plasminogen activator, which converts plasminogen to plasmin, the protease that digests fibrin. Many, but not all, transformed cells produce very much more plasminogen activator than their normal counterparts. The significance of this increased level of protease production is not clear. Since normal cells produce less plasminogen activator when contact-inhibited than when growing, and since tumor cells are less subject to contact inhibition than normal cells, production of high levels of plasminogen activator may have a role in cell growth stimulation.

Chromosomal Changes

Transformation generally results in changes in karyotype. Among the most common changes are deletion of portions of some chromosomes and duplication of portions of others. Less commonly, extra copies of entire chromosomes are incorporated in a stable manner into the cellular genome. The fact that independent transformants often exhibit the same deletions and duplications has prompted the formulation of a gene balance theory of malignancy that postulates that some genes promote malignancy, while others tend to suppress it.

Members of five virus families are now known to be tumorigenic and capable of transforming cells in culture. The manner in which transformation by these viruses is established, maintained, transmitted, and detected will now be described.

DNA tumor viruses

PAPOVAVIRUSES

Most members of the papovavirus family are oncogenic. Among them are the following:

The Papilloma Viruses Papilloma viruses cause both benign and malignant tumors in a wide variety of animals. In humans, at least five serologically distinct strains of papilloma virus cause a variety of lesions. Human papilloma virus type 1 (HPV-1) is associated with deep plantar warts; HPV-2 with common warts; HPV-3 with flat warts and with flat wart-like lesions in patients suffering from epidermodysplasia verruciformis (EV); HPV-4 with lesions of EV that exhibit a tendency to become malignant; and a recently discovered strain of HPV is associated with genital warts (condylomata acuminata). HPV-1, 2, 3 and 4 share very little, if any, DNA sequence homology; the HPV strain that is associated with genital warts shows about 10 percent sequence homology with HPV-4. Little is known concerning the interaction of these viruses with cells because they cannot be grown in cells cultured in vitro.

Polyomavirus Polyomavirus was originally isolated from mouse cell extracts used for the transmission of leukemia. It is only rarely responsible for tumors in nature. When injected into newborn mice or hamsters, it produces a wide variety of histologically distinguishable tumors, hence its name.

SV40 SV40 was first isolated from apparently normal cultures of monkey kidney cells. The only host in which it causes tumors is the baby hamster. Lymphocytic leukemia, lymphosarcoma, reticulum cell sarcoma and osteogenic sarcoma are all produced.

Human Polyomaviruses Several viruses more or less closely related to SV40 have been isolated during the last decade from patients with progressive multifocal leucoencephalopathy (PML) and from the urine of immunosuppressed patients, including renal allograft recipients. The best studied of these are BK virus, JC virus and DAR virus. These viruses are widely distributed; about 75 percent of adults have serologic evidence of infection. JC virus is highly oncogenic and produces brain tumors in over 80 percent of infected hamsters; BK virus is weakly oncogenic and produces fibrosarcomas and ependymomas in a small percentage of infected hamsters. Both readily transform cultured cells. DAR virus is very closely related to SV40 and may be regarded as a strain of SV40; BK and JC virus show 10 to 20 percent sequence homology with SV40 and with each other, the sequences with the highest homology being in the region coding for the leader sequences of late messenger RNAs. An exhaustive search for BKV in 160 human tumors failed to reveal any trace of it.

The importance of polyomaviruses (that is, members of the genus that includes polyoma, SV40, BKV and JCV) for studies of viral oncogenesis lies in the small size of their DNA, which can code for only about 1500 amino acids. Since mutants of these viruses exist that are temperature-sensitive with respect to ability to initiate and maintain the transformed cell state (see below), transformation is clearly a viral function; and since the genome of these viruses is so small, it should be possible to identify its nature.

Both polyoma and SV40 not only transform cells, but also interact with cells by means of the lytic pathway (see Chap. 5). Cells are transformed if they are nonpermissive—that is, if the virus cannot multiply in them; or, if they are permissive, they can be transformed if they are infected with defective virus particles, such as the deletion mutants that arise upon repeated passage at high multiplicity (Chap. 7), or if they are infected with inactivated virus particles [for example, by photodynamic inactivation (Chap. 3)]. In other words, if polyoma and SV40 can multiply in a given cell, that cell will generally not be transformed; if they cannot multiply, then the cell will often be transformed.

Transformation

The establishment of the transforming papovavirus-cell interaction requires high multiplicities of infection (10^6 to 10^7 particles per cell), since the cells must either be nonpermissive or at most semipermissive, or the particles defective. There is evidence that the frequency of transformation is enhanced, by some as yet unknown mechanism, by treating cells with chemical carcinogens, such as 4-nitroquinoline-1-oxide. The essential feature of the transforming polyoma virus-cell interaction is that viral DNA becomes stably integrated into the DNA of the host cell. As a result it cannot multiply, except as a cellular gene, and no progeny virus is produced. Not all infected cells are transformed in this manner; the majority are transformed abortively—that is, they escape from growth control for a few cell generations, but then revert to the normal uninfected state for reasons that are not understood. The fixation of stable transformation requires at least one cell division. The number of sites in the host genome where viral DNA can be integrated is limited, as can be shown by exposing cells to varying amounts of wild-type virus and then inquiring whether additional genomes of temperature-sensitive mutants can still be integrated. The number of integrated viral genomes generally varies from 1 to 3, but can be as high as 20. Multiple viral genomes are often inserted in a tandem head-to-tail arrangement. Transformation by one polyomavirus does not preclude transformation by another, nor by unrelated viruses. Thus, double transformants of SV40 and polyoma, as well as of SV40 and adenovirus and SV40 and various RNA tumor viruses, have been isolated and studied.

Although viral DNA in stably transformed cells does not replicate except as part of the host genome, it does express itself. In fact, substantial amounts of messenger RNA are generally transcribed from it by the α-amanitin-sensitive cellular RNA polymerase II. The nature of this RNA has been analyzed in numerous independently isolated clones of transformed cells, using the techniques outlined in Chapter 5. Although it differs in various clones, the major portion of it in all transformed cells corresponds to the early messenger RNA sequences transcribed in productively infected cells (see Chap. 5). Transformed cells almost always contain the T antigen which is easily demonstrated by immunofluorescence (Fig. 9-4). Whereas viral DNA in transformed cells never expresses itself in its entirety, it is nevertheless the entire viral DNA that is integrated, not portions or segments of it, as can be proved by several types of experiments. For example, the genome of SV40 can be rescued from transformed cells—cells that contain no infectious virus and no capsid polypeptides—by fusing them with uninfected permissive cells. Fusion may be effected by simple cocultivation or, much more efficiently, by treatment with inactivated Sendai virus (Chap. 6). Apparently only the cytoplasm of permissive cells is necessary,

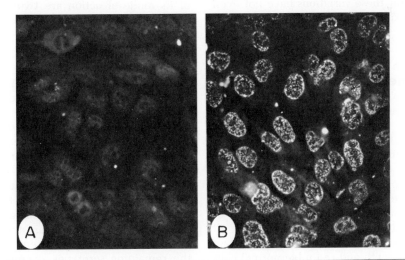

FIG. 9-4. Demonstration of the SV40 T antigen by means of immunofluorescence. Cells from a tumor induced by SV40 in a hamster were exposed to fluorescein-conjugated serum from a non-tumor-bearing hamster (**A**) and a tumor-bearing hamster (**B**). $\times 280$. (From Rapp, Butel, and Melnick: Proc Soc Exptl Biol Med 116:1131, 1964.)

since fusion with enucleated cells has been shown to rescue SV40. This rescue of virus can be effected from cells that have been transformed for many generations and many years.

The mechanism of rescue appears to involve excision of SV40 DNA from host DNA followed by normal lytic cycle type multiplication, with production of infectious progeny virus. Not all lines of cells transformed by SV40 yield virus when fused with permissive cells. The reason for this is not known; it does not depend on the number of viral genome copies that the cell harbors, nor on the extent to which they are transcribed. Sometimes virus is rescued from transformed cell lines by fusion not with uninfected permissive cells but with other lines of transformed cells (that also fail to yield virus on fusion with normal cells). The explanation here is probably that both are cells transformed by defective viral genomes, and that on fusion they complement each other and recombine to yield wild-type virus.

Another line of evidence that shows that transformed cells harbor the entire SV40 genome is that infectious SV40 DNA can be extracted from them.

While all cells transformed by polyomaviruses exhibit the characteristics just discussed, individual clones of transformed cells display wide variations in biologic properties that span the entire range from normal cells on the one hand to standard transformants on the other. The reasons for these variations have not even begun to be explored.

Revertants of Transformed Cells

Numerous attempts have been made to isolate revertants of transformed cells. Among the techniques that have led to the selection of revertant cell lines that exhibit growth control comparable to that of normal cells are selection for serum dependence, passage at high dilution, resistance to nucleic acid-base analogs at high cell density (i.e., inability to multiply at high cell density) and resistance to con A. All these methods yield cells that display contact inhibition of growth and resemble normal cells morphologically. However, almost always they still synthesize the T antigen, display unaltered virus-specific transcription patterns, and contain the entire viral genome. Reversion is,

therefore, due to a change in the cellular, rather than in the viral genome. Revertants cannot be retransformed with the virus whose genome they already harbor, but can be retransformed by other tumor viruses. Finally, revertant cell lines often contain more chromosomes than their transformed antecedents; this has suggested a hypothesis that the cell phenotype is modified as a result of changes in the relative amounts of chromosomes with genes that promote and genes that oppose transformation.

Temperature-Sensitive Polyomavirus Mutants

Several classes of ts mutants of SV40 and polyoma that are useful for studying their interaction with cells have been isolated (see Chap. 7). Among the most interesting is a class that is temperature-sensitive with respect to ability to maintain the transformed cell phenotype. Cells transformed with such mutants display the transformed phenotype at the permissive temperature (typically about 33C) and the normal phenotype at the restrictive temperature (typically about 39C); they can be switched from one temperature to the other at will, with appropriate changes in phenotype. Expression of the transformed state is thus clearly dependent on the correct functioning of a viral gene product. Identification of this protein and definition of its mode of action are two of the primary aims of cancer research.

ADENOVIRUSES

Adenoviruses also can transform animal cells in vitro, and some but not all cause tumors in animals.

Human adenoviruses comprise 31 serotypes that can be divided into three groups on the basis of their oncogenicity. Serotypes 12, 18, and 31 are highly oncogenic; when injected into newborn hamsters, they cause tumors rapidly and with high frequency. Serotypes 3, 7, 14, 16, and 21 are weakly oncogenic; they produce tumors in newborn hamsters with low frequency and after a long latent period. Most of the remaining serotypes, exemplified by serotypes 2 and 5, are nononcogenic, but even they can transform rodent cells in culture.

Adenovirus-transformed cells display many of the properties described above for cells transformed by polyomaviruses. Thus they contain adenovirus DNA that is integrated into the host cell genome and that expresses itself partially; they usually contain the adenovirus T antigen; but the virus cannot be rescued from them by fusion, which may signify either that the viral genome is trimmed during integration or that it is integrated in the form of fragments. There is evidence for both these processes with different cell lines.

Cells transformed by adenoviruses, or cells derived from tumors caused by adenoviruses, generally contain a small number, probably no more than one to three, adenovirus genomes. This corresponds to about 0.001 percent of the DNA in the cell. These integrated genomes are transcribed with very high frequency, for about 1 percent of the messenger RNA transcribed in such cells is adenovirus messenger RNA. The exact nature of transcribed sequences varies among different transformed cell lines; but they comprise almost always from 50 to 100 percent of the early sequences and none of the late ones. Interestingly, the sequences that are transcribed in cells transformed by highly oncogenic, weakly oncogenic, and nononcogenic adenoviruses are quite unrelated. The corresponding portions of the adenovirus genomes have evidently diverged very extensively during the course of the evolution.

The nature of the adenovirus transforming function is not known; it is however known that it is coded by the leftmost 7 percent of the adenovirus genome, since restriction endonuclease fragments from this region can transform cells by themselves.

The demonstrated oncogenicity of human adenoviruses in rodents raises the possibility that they may cause neoplasia in humans. This has been investigated using exquisitely sensitive hybridization techniques which have been refined to the extent that one part of viral DNA can be detected in the presence of 10^6 parts of host DNA, which corresponds to less than one adenovirus genome per cell. So far no human neoplasms have been found that contain adenovirus DNA.

HERPESVIRUSES

Numerous herpesviruses are either oncogenic, are associated with tumors, or can transform cells in vitro. Six are oncogenic in animals (Table 9-1); they include herpesviruses of primates and other mammals, birds and amphibians. As for the situation in humans, there is more evidence for a herpesvirus causing malignant disease than for any other virus; there are no fewer than three herpesviruses, Epstein-Barr virus (EBV), herpes simplex virus type 2 and cytomegalovirus, that may be involved. Direct experimental evidence for EBV and epidemiological evidence for herpes simplex virus indicate that they are present in human neoplasms; and all 3 viruses can transform cultured cells, including human ones. EBV transforms primarily lymphocytes, while the other two viruses, especially herpes simplex virus, can transform a variety of cells, particularly if the viruses are inactivated with ultraviolet light to preclude the productive, lytic virus-cell interaction. Naked herpesvirus DNA can also transform cells; by restriction endonuclease analysis the transforming sequences have been traced to the segment between map positions 0.30 and 0.45.

Epstein-Barr Virus

Burkitt lymphoma is a rather common disease in children in equatorial Africa; it also occurs with low frequency in other parts of the world. In the body, the tumor cells contain no virus,

TABLE 9-1   ONCOGENIC ANIMAL HERPESVIRUSES

Virus	Host	Malignancy
Epstein-Barr virus	Humans	Burkitt lymphoma Nasopharyngeal carcinoma
Herpesvirus ateles	Spider monkey	Lymphoma
Herpesvirus saimiri	Squirrel monkey	Lymphoma
Herpesvirus sylvilagus	Cottontail rabbit	Lymphoma
Guinea pig virus	Guinea pig	Lymphocytic leukemia
Marek's disease virus (MDV)	Chicken	Neurolymphomatosis
Frog herpesvirus	Frog	Lucké adenocarcinoma

but cell lines established from tumor tissue almost always contain some 5 to 20 percent of cells that produce virus particles with the morphology, capsid protein constitution, and genome characteristics of a herpesvirus. This virus, which is antigenically distinct from all known herpesviruses, is known as Epstein-Barr virus (EB virus), after its discoverers.

EB virus causes lymphoma in marmosets and transforms human peripheral blood leukocytes—which do not multiply in vitro—into lymphoblastlike cells with an infinite life span. Such cells do not usually produce EB virus, but they can be stimulated to do so by various means, such as treatment with bromodeoxyuridine or arginine deprivation. Such treatments will also cause the proportion of virus producer cells in established lines of Burkitt lymphoma cells to increase and will also induce the virus to multiply in some Burkitt lymphoma cell lines that do not produce it spontaneously. EB virus DNA is present in Burkitt lymphoma cells in the body, as well as in the leukocytes transformed by it in vitro, as is readily demonstrated by DNA-DNA hybridization analysis; usually, from 1 to 10 viral genome equivalents of DNA are present per cell. Interestingly enough, this DNA may be present not only as DNA integrated into the host cell genome, but it may also exist free in a circular plasmidlike form.

EB virus DNA is present not only in Burkitt lymphoma cells but also in the cells of nasopharyngeal (postnasal) carcinoma, which occurs with rather high frequency among Southern Chinese, and, to a lesser extent, among Algerians. It is not present in most cases of Burkitt lymphoma in the United States.

EB Virus and Infectious Mononucleosis

Sera from patients with Burkitt lymphoma and nasopharyngeal carcinoma contain antibodies to EB virus particles. Very surprisingly, about one-third of the normal adult population of the United States are also positive for EB virus antibodies. Conversion from the negative to the positive state occurs during the acute phase of infectious mononucleosis. Apparently, therefore, EB virus is the etiologic agent of infectious mononucleosis. Presumably, most infections with it are so mild as to go unnoticed;

some cause overt infectious mononucleosis; and, very rarely in most populations, but much more frequently in some, Burkitt lymphoma or nasopharyngeal carcinoma ensue. The reason these tumors occur much more frequently in some parts or some populations of the world than in others is not known.

Herpes Simplex Virus

Two types of herpes simplex virus infect humans: type 1, which causes primarily oral lesions, and type 2, which is associated with genital infections. Several lines of epidemiologic and serologic evidence suggest that infection with herpes simplex virus type 2 is associated with carcinoma of the cervix; in particular, women with genital herpetic infections have a higher than average incidence of cervical carcinoma, and women with cervical carcinoma have a higher than average incidence of antibodies to herpes simplex virus type 2.

Although suggestive, this evidence is not sufficient to establish a definitive causal relationship between herpes simplex virus and malignant disease in humans. In fact, such a relationship will be difficult to establish for two reasons: herpesviruses are ubiquitous, and herpesviruses tend to establish latent and persistent infections. Since experimentation with humans is not feasible, the most direct evidence is likely to be provided by a vaccine that markedly reduces the incidence of some form of cancer.

POXVIRUSES

Two poxviruses are tumorigenic: fibroma virus (including the classical Shope fibroma virus which is pathogenic for rabbits, as well as several closely related viruses pathogenic for deer, squirrels, and hares), and Yaba monkey tumor virus (pathogenic for several species of monkeys and for humans). Both produce benign tumors that soon regress.

The interest in poxvirus tumorigenicity derives from the fact that poxviruses are DNA-containing viruses that have always been thought to multiply exclusively in the cytoplasm. Recently acquired evidence suggests, however, that while vaccinia virus does not appear to have an obligatory nuclear phase in its multiplication cycle, some viral DNA and RNA is often present in the nucleus. Poxvirus DNA may therefore have the opportunity to become integrated into the host cell genome or to become otherwise stably associated with it, and when that happens, cell transformation may result. When fibroma virus transforms cells, host DNA replication is first arrested, and viral DNA replication is initiated. However, it quickly ceases, host DNA replication recommences, and the cells become transformed. They then multiply more rapidly than before, exhibit an altered morphology, are less sensitive to contact inhibition, and display new antigens on their surface. While the precise nature of the association of fibroma DNA and the host cell genome is not known, it is clear that the association is not a very stable one, since the tumors often regress and contain fibroma virions when they do so.

RNA tumor viruses

Interest in RNA tumor viruses has increased greatly in recent years, primarily for three reasons. First, in many animal species they cause leukemias and solid tumors that are often strikingly similar to certain human neoplasms; this raises the question of the role of RNA tumor viruses in causing cancer in humans. Second, it is becoming increasingly apparent that the genomes of RNA tumor viruses are part of the genetic complement of many if not all vertebrate species, with possible roles in embryological development, somatic cell mutation, and evolution, as well as in the etiology of neoplasia. Third, since cancer undoubtedly has a genetic basis, our best hope of unraveling its molecular basis lies in studying the nature of the interaction of the genomes of RNA tumor viruses with those of their host cells.

Since the term RNA tumor viruses is cumbersome, and since viruses exist which, while exhibiting the same strategy of genome structure and expression, are not tumorigenic (see Table 2-4), attempts have been made to devise a more concise and/or generally applicable designation. Among the proposed names are oncornavirus (oncoRNA), oncovirus and re-

trovirus. The latter term has found wide acceptance, particularly in biochemical and genetic studies. However, since we are here concerned primarily with those viruses of the family that are tumor viruses, we shall retain the original term "RNA tumor virus."

THE PATHOLOGY OF RNA TUMOR VIRUSES

The best studied RNA tumor viruses are the avian and mammalian ones, which are commonly referred to as the sarcoma and leukemia or leukosis viruses. These viruses cause primarily tumors of connective tissue (sarcomas) and of the hematopoietic and reticuloendothelial systems. The latter manifest themselves either as leukemias, when the blood contains large numbers of circulating tumor cells, or they may be aleukemic, the neoplasm being a solid mass of tumor cells in some organ. In chickens and cattle the leukemic and aleukemic hematopoietic neoplasms are often referred to as leukoses. In many animals, such as fowl, rodents, cattle, cats, and so on, such neoplasms are the commonest malignancies. Sarcomas and leukemias are also the commonest neoplasms in young human beings. In human adults, on the other hand, carcinomas (malignant tumors of epithelial origin) are responsible for most cancer deaths, and the same is also true of other animals that often survive into old age (such as dogs and horses). RNA tumor viruses are implicated here also. It has long been known that the mouse mammary tumor virus causes mammary carcinoma and that certain other RNA tumor viruses cause renal adenocarcinomas in chickens.

THE NATURE OF RNA TUMOR VIRUS PARTICLES

RNA tumor viruses fall into four morphologic groups. The first comprises the so-called C-type virus particles, characteristic of most leukemia and sarcoma viruses. Their nucleocapsids are never seen free within the cell except when associated with particles that are in the process of budding. They are released in the form of "immature" particles that are characterized by three distinct layers or shells surrounding an electronlucent nucleoid. During the course of

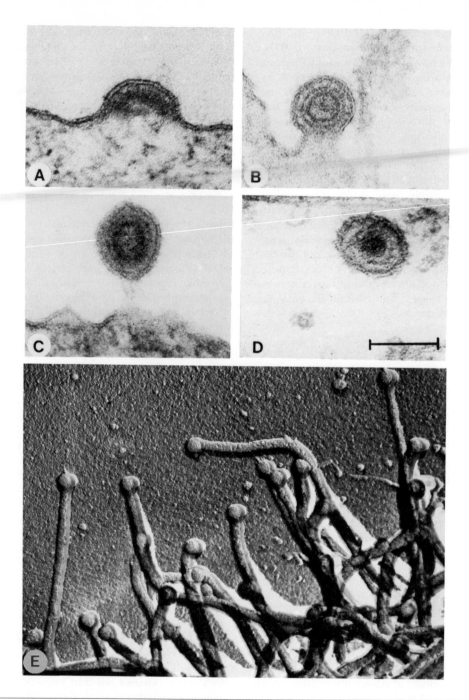

FIG. 9-5. The morphogenesis and structure of RNA tumor viruses. **A-D.** Four stages in the budding of a C-type (avian) RNA tumor virus. Note the electronlucent center or core in immature particles (**B** and **C**), and the electron-dense core in the mature particle (**D**). **E.** A micrograph of a surface replica of a GR cell that is producing murine mammary tumor virus (MMTV) particles (B-type RNA tumor virus). The micrograph shows the cell margin with microvilli, from the tips of which the virus buds. (Panels A to D, courtesy of Dr. Heinz Bauer; panel E, courtesy of Dr. J. B. Sheffield.)

the next several hours the morphology of these particles changes, apparently as the result of several of their component proteins being cleaved (see below), which results in the formation of stable, mature C-type particles that possess an electron-dense nucleoid surrounded by two shells (Fig. 9-5). The inner shell, which possesses icosohedral symmetry, is the nucleocapsid shell, while the outer consists of a lipid bilayer membrane to whose outer surface glycoprotein spikes are attached.

The second group comprises the so-called B-type particles, which are characteristic of mammary tumor viruses (MTVs). Their cores or nucleoids can often be seen in the cytoplasm of infected cells and are known as intracytoplasmic A particles. The principal feature in which they differ from C-type particles is in the eccentric rather than the central location of their nucleocapsids in thin sections and their more prominent glycoprotein spikes.

The third group comprises the D-type particles, which exhibit properties intermediate between those of C- and B-type particles; their nucleocapsids exist within cells as intracytoplasmic A-type particles, but are located centrally within mature virus particles; and their glycoprotein spikes are less prominent than those of B-type particles. D-type virus particles have been isolated from subhuman primates (rhesus monkey, squirrel monkey and langur) and from human tumor cells such as Hep-2 and HeLa cells.

The fourth group comprises the A-type particles, which fall into two classes, the intracytoplasmic and the intracisternal ones. The former are the nucleocapsids of B- and D-type particles; the latter are a distinct class of particle whose precise relation to the other groups of RNA tumor virus particles has not yet been clarified.

The RNA of RNA Tumor Viruses

The genome of RNA tumor viruses is diploid; it consists of two identical molecules (subunits) of single-stranded plus-stranded RNA that are 5000 to 10,000 nucleotides long (M.W. 1.5 to 3 $\times$ 10^6) and are hydrogen-bonded to each other via a palindromic sequence of some 50 nucleotides that is centered about 100 nucleotides from their 5′-ends (Fig. 9-6). These RNA molecules possess the following features: 1) they are capped at their 5′-termini; 2) they are polyadenylated at their 3′-termini; 3) they possess a terminally repeated sequence about 20 nucleotides long; and 4) at a distance of about 100 nucleotides from their 5′-end a $tRNA_{trp}$ molecule in the case of avian RNA tumor viruses and a $tRNA_{pro}$ molecule in the case of mammalian RNA tumor viruses is hydrogen-bonded to them. The function of these tRNA molecules is to serve as primers for the transcription of RNA tumor virus RNA into DNA (see below).

The RNA of avian sarcoma viruses encodes four genes whose order is 5′-*gag-pol-env-src-c*-3′. The *gag* gene codes for a protein that is the precursor to four structural proteins of the nucleoid which, in the early days, was known as the gebundenes antigen (German, bound antigen) to distinguish it from its unassembled monomeric or soluble protein components. The *pol* gene codes for the DNA polymerase (re-

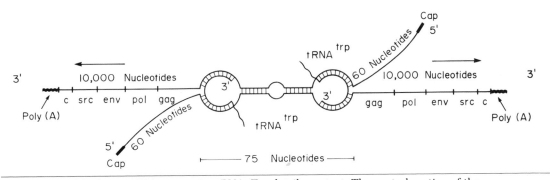

FIG. 9-6. Model of avian RNA tumor virus RNA. For details see text. The central portion of the molecule is greatly enlarged. The poly(A) sequences are about 200 residues long. ▬ is the sequence CCAUUUUACCAUUCACCACA. (Adapted from Haseltine et al: Proc Natl Acad Sci USA 74:989, 1977.)

verse transcriptase), and the *env* gene codes for the two glycoproteins that make up the spikes. The *src* genes codes for a phosphoprotein designed pp60[src] which possesses protein kinase activity and is responsible for the neoplastic transformation of host cells. In addition, there is a *c* (for "common") region, a sequence of about 1000 nucleotides that is shared by avian sarcoma and leukosis viruses. It is apparently not translated into protein and its function is unknown.

The four genes, together with the *c* region, account for very nearly the entire coding capacity of the RNA tumor virus genome.

The Proteins of RNA Tumor Viruses

RNA tumor virus particles generally consist of about six structural proteins, the exact number depending on the particular virus (Table 9-2). Two are glycoproteins that make up the envelope spikes—the larger forming the knobs, the smaller the stalks; these glycoproteins are linked covalently by —SS— bonds. The larger, gp85 in the case of avian RNA tumor viruses and gp70 in the case of mammalian ones, is the antigen that gives rise to neutralizing antibody and is responsible for both host range specificity and virus interference patterns (see below). The other structural proteins, known as p27, p19, p15, p12 and p10 in the case of avian RNA tumor viruses, and p30, p15, and p12 and p10 in the case of mammalian ones, are components of the nucleocapsid. The p27 and p30 polypeptides make up the nucleocapsid shells and are the principal virion components; p12 and p10 in avian and mammalian viruses respectively are intimately associated with the RNA. The first four of the above-mentioned nucleocapsid components of avian, and all four components of mammalian RNA tumor virus nucleocapsids are derived by cleavage of the *gag* gene product mentioned above, their order being NH$_2$-p19-p27-p12-p15-COOH and NH$_2$-p15-p12-p30-p10-COOH respectively. There is evidence that in the avian system the protease that effects this cleavage is p15. RNA tumor virus particles also contain about 10 molecules of an RNA-dependent-DNA polymerase, the reverse transcriptase, which is coded by the *pol* gene.

All these proteins exhibit a multiplicity of antigenic determinants whose characterization is of great importance for tracing genetic and evolutionary relationships. There are type-specific antigenic determinants private to individual virus strains, group-specific determinants common to groups of viruses with a common host range, and interspecies-specific determinants that are common to all mammalian or avian RNA tumor viruses. The distribution of these types of determinants among RNA tumor virus proteins is given in Table 9-2.

THE MOLECULAR BIOLOGY OF RNA TUMOR VIRUSES

There has long existed a conceptual difficulty concerning the role of RNA-containing viruses as oncogenic agents. Transformation represents a heritable alteration of cellular biosynthetic patterns that is most plausibly accounted for by a change in genetic capabilities. Conceivably there are mechanisms whereby a viral DNA genome might cause such changes after being integrated, but how could a viral RNA genome achieve this? Furthermore, many cells transformed by RNA tumor viruses always produce virus for generation after generation; what type of mechanism could ensure that host and viral genome replication would proceed in step? In 1964 Temin advanced a revolutionary hypothesis; he proposed that upon infection the viral RNA genome is transcribed into DNA by a reversal of the usual flow of information transfer, that this DNA is then integrated into the host genome, and that progeny viral genomes are transcribed from it just as all other RNA is normally transcribed. In essence, the hypothesis implied that the RNA present in RNA tumor viruses is not the genome, but rather the messenger RNA of a genome that exists only intracellularly.

The discovery of reverse transcriptase independently by Temin and Baltimore in 1970 proved Temin's hypothesis correct.

The Formation and Integration of Proviral DNA

Following successful adsorption and penetration into the cell, the RNA in parental nucleocapsids is transcribed by the reverse transcriptase into linear double-stranded DNA, often

TABLE 9-2 C-TYPE RNA TUMOR VIRUS STRUCTURAL PROTEINS[a]

Protein[b] Avian	Protein[b] Murine	Percent of Total Protein[c]	Coded by	Location and Properties	Antigenic Specificity Type	Antigenic Specificity Group (Species)	Antigenic Specificity Interspecies
p10	p12E	5,9	?	Envelope	?	?	?
p12	p10	14,6	gag	Core, RNA-associated, phosphoprotein	–	+	–
p15	p15	14,14	gag	Core, hydrophobic	+	–	–
p19	p12	12,8	gag	Core, phosphoprotein	+	–	–
p27	p30	36,50	gag	Major component, core shell	+	+	+
gp85	gp70	13,11	env	Major envelope protein, spike knob	+	+	+
gp37	gp15E	6,2	env	Minor envelope protein, spike stalk	?	?	?

[a] The structure of B-type RNA tumor viruses is similar to that of the C-type viruses. The major B-type internal *gag*-coded protein is p28; the major surface *env*-coded glycoprotein is gp52.

[b] p = protein; gp = glycoprotein. The number indicates the approximate size in daltons $\times 10^{-3}$. In addition to the seven proteins listed, each virus particle also contains about 10 molecules of the RNA-dependent DNA polymerase (reverse transcriptase).

[c] Approximate percentages in avian and murine RNA tumor viruses respectively.

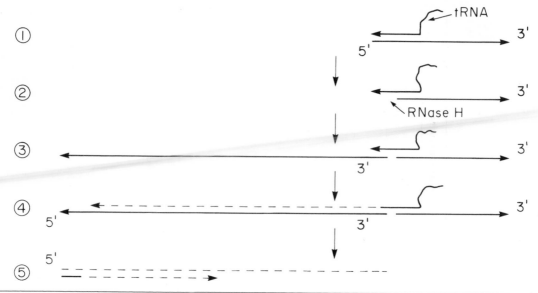

FIG. 9-7. Proposed scheme for the transcription of RNA tumor virus RNA into DNA. Step 1. Using tRNA^{trp} as primer, the 100 nucleotides at the 5'-terminus are transcribed. Step 2. The terminally redundant (T) RNA sequence is removed by the action of ribonuclease H, a nuclease activity possessed by reverse transcriptase. Step 3. The DNA transcript region of the T sequence pairs with a T sequence at the 3'-terminus of another RNA subunit; or it could also be on the same subunit that has just been transcribed at its 5'-terminus. Step 4. The new RNA subunit is transcribed into minus-strand DNA. Step 5. Transcription of minus-strand into plus-strand DNA proceeds, perhaps using viral RNA as the primer, to yield a double-stranded DNA transcript of RNA tumor virus RNA. (Adapted from Bishop: Ann Rev Biochem 47:35, 1978.)

referred to as proviral DNA. The exact mechanism by which transcription is accomplished is not known, but it is likely that it proceeds approximately as depicted in Fig. 9-7. Linear proviral DNA that is the same length as virus RNA strands can be detected in the cytoplasm within one hour of infection. These molecules then move to the nucleus and circularize (5-12 hours after infection); and they are then integrated into the genome of the host cell.

It should be noted that linear proviral DNA bears repeats of about 300 nucleotide base pairs at its termini, and that these repeats are retained in circular and integrated proviral DNA (Fig. 9-8). Presumably these repeats are essential for the alternate cycles of transcription of single-stranded RNA into/from a double-stranded DNA form that is capable of circularization (which in turn is probably essential for integration). The number of integrated viral genomes per haploid host genome may be as low as 1 or as high as 100, depending on the virus and on the cell.

Little is known concerning the sites of integration. Provisional evidence suggests that proviruses can integrate at multiple sites in the host genome, but cells chronically infected with avian reticuloendotheliosis virus apparently contain provirus integrated at a unique site. Furthermore, the genomes of endogenous viruses (see below) also presumably have defined chromosomal locations.

The Transcription of Integrated Proviral DNA

Transcription of integrated proviral DNA appears to be catalyzed by cellular RNA polymerase II. Three species of transcripts, all capped at their 5'-termini and polyadenylated at their 3'-termini, have been detected. The first are full length viral RNA molecules. They are either encapsidated into progeny virus particles, or they are translated. Usually the translation product is the *gag* gene product, but infre-

FIG. 9-8. Model showing the relationship between the sequences of avian RNA tumor virus RNA and its linear double-stranded DNA transcript (provirus) form. The DNA is longer than the RNA; the 3'-terminal 200 nucleotides of the RNA are repeated at one end of the DNA, and the 5'-terminal 100 nucleotides of the RNA are repeated at the other end of the DNA, which is therefore 300 nucleotides longer than the RNA. In double-stranded DNA transcripts of murine leukemia virus RNA the terminal repeats are about twice as long. (Adapted from Shank et al: Cell 15:1383, 1978.)

quently there is read-through into and through to the end of the *pol* gene, which gives rise to a large protein (M.W. about 180,000) which is cleaved to provide the *pol* gene product (that is, the reverse transcriptase). This mode of synthesizing the *pol* gene product (that is, without a messenger RNA of its own and by infrequent read-through from another gene) represents a convenient mechanism for ensuring that it will not be synthesized in excessive amounts; clearly far more *gag* gene products are required for viral morphogenesis than *pol* gene products. The second messenger RNA species is derived from the first by splicing out the *gag* and *pol* gene sequences. It contains a leader sequence some 300 nucleotides long derived from the 5'-terminus of the viral genome, which is linked covalently to the *env* gene and sequences downstream. This messenger RNA species is translated into the *env* gene product which is then cleaved and glycosylated to provide the two spike glycoproteins. The *src* gene product is not translated from this messenger RNA but from a third species which is derived from the first by splicing out the sequences corresponding to the *gag*, *pol* and *env* genes. It has a leader sequence similar to the second species described above.

HOST RESTRICTIONS ON VIRUS MULTIPLICATION

Two types of restrictions govern cell susceptibility to infection by RNA tumor viruses.

Host Range and Interference Patterns

The first restriction relates to the presence on plasma cell membranes of receptors specific for the major RNA tumor virus envelope glycoproteins, gp85 in avian, and gp70 in mammalian viruses. As we shall see, recombination in the *env* gene is very common, which means that there is enormous diversity with respect to the structure of these glycoproteins. The various host species respond to this diversity in different ways. In the avian system, receptors for these glycoproteins are specified in simple Mendelian manner by dominant alleles of autosomal loci. As a result, avian RNA tumor viruses fall into eight subgroups, A to H, depending on the genetic constitution of the cells that they are able to infect. Cell phenotype is designated according to the virus subgroup that is excluded; thus c/0 cells are susceptible to viruses of all subgroups, c/A cells are susceptible to all except those of subgroup A, and so on. If receptors are lacking, virus can adsorb but not penetrate, and the cell is therefore resistant.

The same grouping of viruses is obtained if the criterion is interference with ability to multiply; prior infection with a virus of the same subgroup prevents multiplication of superinfecting virus. This grouping of viruses is also upheld by their immunologic properties as judged by cross-reactions among their neutralizing antibodies. Thus host range, interference and immunologic properties are all governed by the structure of the gp85 or gp70 envelope glycoproteins.

Among the feline viruses there are three subgroups, A, B and C, as determined by host range and interference patterns. Among the murine ecotropic viruses (see below) there are two subgroups as defined by interference and antigenic properties, one of which includes the Friend, Moloney and Rauscher, and the other the Gross-AKR leukemia viruses. The xenotropic viruses (see below) constitute a third murine virus interference group.

The Fv-1 System

The second restriction system does not depend on exclusion of virus entry, but operates after penetration and before integration. Murine leukemia viruses can be classified as either N-tropic or B-tropic depending upon whether they grow preferentially in cells derived from NIH Swiss mice (N-tropic) or BALB/c mice (B-tropic). This phenotype is controlled by a cellular gene called Fv-1, located on chromosome 4, which possesses at least two alleles, Fv-1^{nn} and Fv-1^{bb}, which reciprocally restrict infection by N- and B-host range viruses. The virus determinant of this tropism is the nucleocapsid component p30.

CLASSES OF C-TYPE RNA TUMOR VIRUSES

There are two classes of RNA tumor viruses. First, there are those that are transmitted from animal to animal like other infectious agents. Examples are the Rous sarcoma virus of chickens and two primate viruses, simian sarcoma virus (SSV) and gibbon ape leukemia virus. Also in this class are viruses that were isolated some time ago from spontaneous or passaged tumors, usually leukemias, and that were then passaged in vivo and in vitro under laboratory conditions. Since such passaging provides the opportunity not only for the selection of variants, but also for recombination with the genomes of other RNA tumor viruses and with cellular DNA (see below), such strains became laboratory strains, with properties different from any virus that occurs naturally. All these viruses are known as exogenous viruses because they do not exist in provirus form in any animal.

The second class of RNA tumor viruses exist in provirus form in numerous, perhaps all, animal species. The proviruses of these viruses are transmitted vertically as dominant genetic traits, just like cellular genes, and their expression is controlled by individual specific regulatory controls which no doubt involve repression/depression at the DNA transcription level. These viruses are known as endogenous RNA tumor viruses. In feral animal populations the expression of these proviruses as virus particles is greatly repressed; breeding programs for mice and chickens, however, have produced numerous strains of inbred animals in which

the mechanisms that control endogenous provirus expression have been dissociated from the proviruses themselves, with the result that in these animals RNA tumor viruses are produced continuously and often throughout the animal's life. For example, mice of the albino AK and C58 BL strains have large numbers of leukemia virus particles in their blood stream and also develop leukemia between 6 and 12 months of age with very high frequencies; albino AK and C58 BL are high-incidence-of-leukemia strains. Mice of other strains, such as the BALB/c and DBA strains, develop leukemia only late in life and with low frequency, and are also viremic for RNA tumor viruses for some periods of their life; they are low-incidence-of-leukemia strains. Finally, mice of some strains, such as the NIH Swiss and NZB strains, do not develop leukemia at all, because the endogenous proviruses capable of producing spontaneous leukemia have been bred out of them. However, it has been discovered that in these mice still another group of RNA tumor proviruses exist, whose members never express themselves as virus particles except after being induced by treatment with one of a variety of agents.

We will now discuss in detail these various groups of viruses. C-type RNA tumor viruses have been arbitrarily divided into two groups—namely those that can transform fibroblasts in vitro, which are designated sarcoma viruses, and those that cannot, which are designated as leukemia viruses.

Avian Sarcoma Viruses

These viruses include the well-known Rous sarcoma virus (RSV) and B77 virus. They are the only C-type RNA tumor viruses that possess, and express, all four genes, namely *gag*, *pol*, *env* and *src*. They are the only viruses that are capable of transforming cultured fibroblasts and other cells and are not defective for replication. As a result, it has been possible to isolate mutants in the transforming *(src)* gene and to characterize its gene product (see above); this remains the only characterized transforming gene product of any RNA tumor virus. Interestingly, it has now been found that genes closely related to *src* are present in the genomes of all normal vertebrate cells. These genes, which are denoted *sarc*, code for one or more proteins that cross-react immunologically with pp60^{src}, but whose function is unknown. While the sequences of *src* and *sarc* are closely related, they are not identical. It seems possible, and indeed likely, that the transforming gene *src* of avian sarcoma viruses represents a cellular *sarc* gene that was acquired by the viral genome a long time ago, and that the genes resident in viral and cellular genomes have diverged to some extent since then.

Leukemia Viruses

Leukemia viruses possess only the three genes *gag*, *pol* and *env*, which are essential for their multiplication; they lack the *src* gene and therefore cannot transform fibroblasts or any other cell in vitro. Avian leukemia/leukosis viruses are widely distributed in chickens; many strains of chickens are viremic throughout life. Many of these viruses are weakly leukemogenic; late in the life of the birds that they infect, they cause a variety of neoplasms such as lymphocytic leukemia, osteopetrosis, nephroblastomas and erythroblastomas, or these neoplasms occur after a prolonged latent period when they are injected into uninfected birds.

Among the mammalian leukemia viruses, by far the most intensively studied are the murine ones. They have been broadly classified into two serological subgroups, the Gross (G)-AKR subgroup and the FMR (Friend-Moloney-Rauscher) subgroup, which are related to the extent of 8 to 10 percent as measured by nucleic acid hybridization. The naturally occurring leukemias are all of the Gross type, while the FMR complex of viruses are laboratory virus strains. Gross-type viruses typically give rise to thymic lymphosarcomas as well as myeloid leukemias after a latent period of several months; FMR viruses cause erythroblastic leukemias after a latency of only a few weeks.

Leukemia viruses have also been isolated from other animal species, including other rodents (rats and hamsters), cats and monkeys. Feline leukemia virus (FeLV) infects cats horizontally and causes generalized lymphosarcoma, one of the most common cancers in cats. It readily crosses species barriers; it causes leukemia also in dogs and grows readily in human cells. Several strains of gibbon ape leukemia virus have been isolated.

All these viruses multiply without helper viruses. In view of the fact that their genomes do not contain a transforming gene, the mystery is how they cause leukemia. Recent evidence indicates that they do not do so directly, but that they recombine with other viral genomes, or possibly even host genetic material, to generate the actually transforming viruses (see below).

Defective Viruses: Acute Leukemia Viruses and Mammalian Sarcoma Viruses

The viruses in this group possess several fascinating properties. They comprise several avian leukemia viruses such as the avian carcinoma virus MH2, avian erythroblastosis virus (AEV) strain ES4, avian myeloblastosis virus (AMV) strain BAI-A, and avian myelocytomatosis virus strain MC29; the mammalian sarcoma viruses (such as the Moloney, Kirsten and Harvey murine sarcoma viruses (Mo-MSV, Ki-MSV and Ha-MSV) and feline sarcoma virus (FeSV)), and two murine leukemia viruses (the Abelson and Friend murine leukemia viruses). All these viruses possess high oncogenic potential. The chick leukemia viruses rapidly cause erythroblastosis, myelocytomatosis, neurolymphomatosis, endotheliomas and sarcomas; the Friend murine leukemia viruses rapidly induce splenomegaly, granulocytic myelogenous leukemia and erythroleukemia; Abelson virus quickly induces reticulum cell sarcomas and thymus-independent lymphomas; and the mammalian sarcoma viruses rapidly cause a variety of sarcomas. In addition, most of these defective viruses can transform not only hematopoietic cell cultures, but also cultured fibroblasts. Yet none possesses the *src* gene, or any gene even remotely related to it. What, then, is the nature of the genomes of these highly oncogenic viruses?

It turns out that all these viruses possess genomes grossly deficient in the genes that code for structural viral components. These viruses code for no more than two of their *gag* proteins, since they possess only the 5′-terminal portion of the *gag* gene; and they possess neither the *pol* nor the *env* gene. As a result, these viruses are replication-defective and need helper viruses. For example, cells infected with Moloney murine or feline sarcoma virus become transformed, but fail to yield progeny virus particles; cells transformed by these viruses are nonproducer cells. The viruses that act as helpers for all these viruses are standard leukemia viruses such as were described above, which supply the full range of nucleocapsid components, the reverse transcriptase and the envelopes for these defective viruses, which therefore exist in the form of pseudotype particles that are denoted thus: Mo-MSV (Mo-MLV) for Moloney murine sarcoma virus particles, the virus in brackets denoting the helper virus. Strains of all these viruses consist of mixtures of virus particles, the defective pseudotype particles that cause cell transformation, and the helper virus particles that contain the information for virus particle formation. The helpers are generally present in large excess, although elaborate techniques for removing them have been worked out in most cases. Often, but not always, a wide variety of leukemia viruses can act as helpers; for example Mo-MSV can use FeLV as the helper virus, and the resulting pseudotype particles, Mo-MSV (FeLV), have the host range, interference patterns and antigenic properties of FeLV, but interact with cells as Mo-MSV. Usually, however, one type of helper is preferred. Thus the viruses that are associated with AMV and AEV in nature are known as myeloblastosis-associated virus (MAV) and erythroblastosis-associated virus (EAV) respectively. The Friend leukemia virus strain, which was isolated from a Swiss mouse that was inoculated at birth with an extract of the Ehrlich transplantable tumor and developed an erythroid leukemia, consists of a defective component [the spleen focus-forming virus (SFFV), that transforms erythroid cells] and a helper leukemia virus [the lymphoid leukemia virus (LLV)] which causes lymphoid leukemia.

The genome of all these defective viruses is very small; it is only about 5,700 nucleotides long (M.W. about 1.7×10^6), only slightly more than half the size of RSV RNA (10,000 nucleotides). As pointed out above, the genome lacks the 3'-terminal half of the *gag* gene, the *pol* gene, the *env* gene, and the *src* gene. Instead it contains sequences that a) are unrelated to those of standard leukemia viruses [since the overall degree of relatedness of defective and standard leukemia virus genomes is only 50 to 60 percent (Fig. 9-9)], and b) code for a polyprotein that ranges in size from 75,000 to 130,000 daltons, depending on the virus; the only thus far identified antigenic determinants

in this polyprotein are the two *gag* proteins coded by the 5'-terminal half of the *gag* gene. For some viruses, such as the Abelson murine leukemia virus and the avian adenocarcinoma virus MH2, this protein appears not to be cleaved; for other viruses it is cleaved. Thus the 130,000 daltons protein coded by FeSV is cleaved to a protein that contains the two *gag* proteins p15 and p12, and to a protein with a molecular weight of 65,000 that is designated FOCMA (*f*eline *o*ncornavirus-associated *c*ell *m*embrane *a*ntigen) and is present on the surface of nonproducer cells transformed by FeSV. This protein plays a central role in immunosurveillance against naturally occurring tumors, since resistance to tumor development correlates with the presence of antibody to FOCMA. Similar, but not entirely analogous cleavage products of the polyprotein have been found on the surfaces of cells transformed by murine sarcoma viruses; examples are the 95,000 dalton Gross cell surface antigens (GCSA) and a 60,000 dalton protein specified by Mo-MSV; both are glycosylated N-terminal portions of the polyprotein (since they possess *gag* protein antigenic determinants, which FOCMA does not). The glycosylated polyprotein of the Abelson murine leukemia virus is also found on the surface of transformed cells. Interestingly, phosphorylated portions of the polyprotein are components of the pseudotype particles of many of these virus strains; thus FeSV pseudotype particles contain the phosphorylated 85,000 dalton N-terminal portion of the polyprotein, Mo-MSV particles contain a phosphorylated 60,000 dalton portion, and Ki-MSV and SSV particles contain a phosphorylated 45,000 dalton portion.

In addition to its important roles as cell-surface marker and virus particle constituent, the polyprotein or its cleavage products may also function as the transformation protein for these highly oncogenic viruses. Of course, another transformation candidate protein for these viruses is the protein that may be coded by that portion of the defective viral genome that does not code for the polyprotein; even if the smallest genome coded for the largest polyprotein, there would still be room to code for another protein at least 50,000 daltons in size. However, no such protein has yet been detected.

The nucleic acid sequences that are present

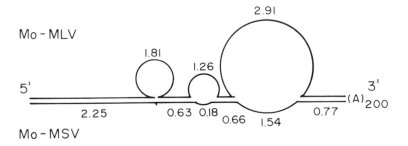

FIG. 9-9. Scheme of the regions of homology between the RNAs of Moloney murine leukemia and Moloney murine sarcoma virus RNA. The scheme was constructed on the basis of heteroduplex analysis. Parallel sections denote regions of homology, looped-out sections denote regions of nonhomology. Lengths are in kilobases. It is seen that only about 50% of the sequences of Mo-MLV RNA are shared with Mo-SMV RNA. (Adapted from Hu, Davidson and Verma: Cell 10:469, 1977.)

in the genomes of these defective viruses, but not in those of standard leukemia viruses, are most probably derived from cellular genetic material; and it is likely that they include portions of the proviruses of endogenous xenotropic RNA tumor viruses (see below). The defective viruses are therefore probably recombinants between leukemia viruses and host genetic material. Indeed many of these viruses were isolated from systems in which such recombination would have been favored. Many were originally derived from animals through which homologous or heterologous leukemia viruses had been passaged, or from cells that had been infected with high multiplicities of leukemia viruses; both sets of circumstances would favor recombination between the genetic material of the leukemia virus and that of the host. Thus Ki-MSV and Ha-MSV were both isolated from rats infected with Mo-MLV and contain large amounts of rat genetic material; Abelson leukemia virus was isolated from a tumor arising in a prednisolone-treated BALB/c mouse inoculated with Mo-MLV; the two-component Friend leukemia virus strain arose upon repeated mouse-to-mouse passage of uncloned Friend leukemia virus; and Mo-MSV was isolated from a rhabdomyosarcoma induced in BALB/c mice by infection with high doses of Mo-MLV. Indeed, it has recently been shown that new mouse sarcoma viruses that rapidly produce sarcomas in animals can be generated readily by infecting cultured mouse cells with high multiplicities of non-transforming murine RNA tumor viruses.

ENDOGENOUS C-TYPE RNA TUMOR VIRUSES

Mice and chickens of numerous inbred strains are throughout their lives viremic for C-type RNA tumor viruses of the leukemia type. Although these viruses are first produced in animals that are completely healthy, they usually develop leukemia at some later stage of their life (see above). These viruses are endogenous viruses that can express themselves spontaneously; their proviruses exist in the mouse or chick genome in from 1 to about 100 copies.

In addition to these viruses, mice and chickens, as well as animals of other species such as other rodents, cats, monkeys and primates, contain endogenous viruses that do not normally express themselves but which can be induced to do so by treatment with a variety of substances, including inhibitors of protein synthesis, inhibitors of nucleic acid synthesis, halogenated deoxyribonucleosides, inhibitors of glycosylation and B-lymphocyte mitogens. Different inducers induce different viruses, which indicates that the nature of the cellular controls that regulate their ability to express themselves are diverse. It also suggests that the endogenous viruses which constantly express themselves to high titer may be those whose proviruses have been separated by animal breeding programs from the mechanisms that regulate their ability to be transcribed. That proviruses are indeed under the control of cis-acting control mechanisms is suggested by transfection experiments. Proviral DNA is infectious, even when integrated; but proviruses have been found that are not infectious unless the cellular DNA in which they reside is reduced in size to approximate provirus size, a procedure that would be expected to remove control elements linked to proviruses.

It should be noted that the control mechanisms themselves seem to be programmed. Some proviruses express themselves during embryonic life, even to the extent of virus particle formation, without causing disease. The proviruses then usually become repressed, and virus is usually undetectable in most young adults. Later in life, however, the proviruses often begin to express themselves again, virus particles begin to be formed, and leukemia may result.

Ecotropic, Xenotropic, and Amphotropic Endogenous Viruses

There are three groups of endogenous viruses that differ in host range, interference patterns and antigenic determinants. The first comprises viruses that multiply readily in cells of the species in which they were induced; these are the ectropic viruses. The second comprises viruses that cannot infect cells of the species in which they were induced, but can infect cells of other species; these are the xenotropic viruses. The third are the viruses that can multiply either in cells of the species in which they were induced, or in cells of other species; these are the amphotropic viruses.

Numerous ecotropic, xenotropic and amphotropic viruses have now been isolated from numerous mammalian species, including primates. Mice, in particular, have been found to contain numerous proviruses in their genomes; it is probably conservative to estimate that wild-type mice, that is, mice that have not been bred for provirus repression or derepression, harbor as many as ten different proviruses of all three types, each present to the extent of an average of ten copies. Further, different species of mice from different parts of the world harbor different populations of proviruses. It also seems, especially in birds, that incomplete proviruses, or provirus fragments, are prevalent. Table 9-3 shows that pheasants, Japanese quail, ducks and pigeons all contain small numbers (0.5 to 10 per cell) of provirus fragments of the endogenous chick virus RAV-0.

A great deal of work has been done to assess the genetic relatedness of all these newly discovered endogenous viruses to one another and to exogenous leukemia virus strains. The criteria used in this type of analysis are comparison of nucleic acid sequences on the one hand (by hybridization analysis, oligonucleotide mapping or direct sequencing), and comparison of the sequences of the various virus-coded proteins on the other (determination of antigenic relatedness, peptide mapping and direct sequencing). An example of the former type of comparison (endogenous versus exogenous virus) is provided by the following analysis of the genomes of RAV-0, the most important endogenous virus of chickens, and Rous sarcoma virus (Fig. 9-10). The two genomes are virtually identical in their *gag* and *pol* genes. However, there are substantial sequence differences in their *env* genes, most of them due to single base changes, and these differences are more pronounced in the central portion of their *env* genes than in their terminal regions, which

TABLE 9-3   **NATURE OF THE ENDOGENOUS RNA TUMOR VIRUS GENOMES PRESENT IN THE DNAS OF VARIOUS BIRDS**

Species	Extent of Hybridization*	Copies per Haploid Genome
Chicken	100	1–2
Japanese quail	8	3
Pigeon	12	0.5
Peking duck	10	6
Ring-necked pheasant	50	10

Adapted from Tereba et al: Virol *65*:524, 1975.
* The DNA of various birds was hybridized to DNA transcribed with reverse transcriptase from the RNA of RAV-0, the endogenous RNA tumor virus of chickens. The extent of hybridization measures the fractions (in percent) of RAV-0 sequences that are present in the various cellular genomes.

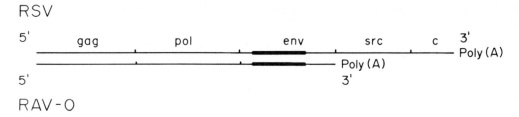

FIG. 9-10. Relationship map of the RSV and RAV-0 genomes. For details see text. Substantial nonhomology is confined to a central portion of the *env* gene, which may contain the sequences that code for subgroup specificity (━). This portion is flanked by two regions of closer homology. In the *gag* and *pol* genes the two genomes are virtually identical, with only scattered single nucleotide differences. In spite of the fact that the RAV-0 genome lacks the *src* gene and the *c* region, it does contain the same terminally redundant sequence of 20 nucleotides at each end that is characteristic of most if not all avian RNA tumor virus RNAs.

may therefore be a hypervariable region. Further evidence for this view will be discussed below.

Evolutionary Relationships Among Endogenous C-type Viruses

The most fascinating insight provided by studies such as have just been described is into the evolutionary origin of proviruses; for it has proved possible to determine when and whence particular proviruses were introduced into their present day host. This is possible because the proviruses are generally highly conserved evolutionarily. Among the conclusions that have been reached are (1) that the avian reticuloendotheliosis virus is much more closely related to murine viruses than to endogenous or exogenous chicken viruses, which suggests that murine viruses were transferred from mice to birds many millions of years ago; (2) that the endogenous proviruses of pigs were acquired from mice after the mouse had separated from the rat (10 million years ago); (3) that the exogenous primate simian sarcoma virus and gibbon ape leukemia virus are derived from a murine xenotropic virus; (4) that the xenotropic cat virus RD114 was acquired in the Pliocene from a monkey; and (5) that old-world monkeys as well as humans possess gene sequences related to the xenotropic baboon virus M7. These gene sequences can be divided into two groups: those in monkeys and apes that evolved in Asia, and those in monkeys and apes that evolved in Africa. Among the apes, the go-

rilla and the chimpanzee seem, by this criterion, to be African in origin, whereas the gibbon and the orangutan are Asian in origin; and humans fall into the Asian category.

The Expression of the *env* Gene of Endogenous Viruses

While endogenous viruses usually do not express themselves to the extent of virus particle formation, they do, however, frequently express their *env* gene. The first indication in this direction was the discovery that a strain of RSV whose RNA contains a deletion in the *env* gene (the Bryan strain) is incapable of growing in some chick cells, but is capable of growing in others, which were therefore said to possess a "chick-helper-factor" (chf); this factor turned out to be the spontaneously expressed glycoprotein coded by the *env* gene of the endogenous chick RNA tumor virus RAV-0. Thus RAV-0 does not express itself at all in chick cells that are chf⁻, it expresses its *env* gene in cells that are chf⁺, and there are one or two inbred strains of chickens in which it expresses itself completely, to the extent of infectious virus particle formation.

The situation is similar in the mouse. It has long been known that a differentiation alloantigen designated G_{IX} (since it appeared to be specified by a gene in linkage group IX of the mouse) is present in the thymocytes of certain mouse strains. It is also present in the serum of some mouse strains (particularly that of strain 12A), in epithelia associated with the digestive tract, and in the epithelial lining of the male reproductive tract and therefore in seminal fluid. There is one locus, Gv-1, that regulates the expression of this antigen in all sites in strain 129 mice. It was then found that the G_{IX} antigen could be induced in rats by infection with murine leukemia virus; and that it represents, in fact, the gp70 glycoprotein specified by the *env* gene of a provirus which can express itself irrespective of *gag* and *pol* gene expression, just like chf above. However, unlike in the chicken, which harbors only one provirus, mice harbor numerous proviruses, all of which may express their *env* genes independently of each other in the form of gp70 molecules located on the surface of cells of various organs. Thus in mouse strains such as the 129/J mouse, from which no C-type virus has ever been isolated, the gp70s of four different endogenous viruses are expressed in different developmental compartments: one in the serum, one on thymus and spleen cells, one on bone marrow cells, and one on cells lining the male genital tract. All these gp70s are related, more or less closely, both to each other and to the gp70s coded by ecotropic, xenotropic and amphotropic endogenous viruses isolated from a variety of inbred mouse strains, as well as from feral mouse populations. It seems therefore that there is a compatible relationship between the mouse and an extensive family of endogenous C-type RNA tumor viruses that exists in proviral form in all somatic and germ line cells. The gp70s of these viruses exhibit a diversity that rivals that of immunoglobulins or the H-2 histocompatibility locus. At the present time it

is not clear whether this polymorphism benefits only the virus (by modifying its host range, thereby promoting its adaptability), or whether it also benefits the mouse, through the various gp70s acting as true differentiation antigens. However, as specified above, the extent of provirus expression varies enormously with age of the mouse, and both the formation of serum G_{IX}-gp70, as well as B-type RNA tumor virus expression (see below), is profoundly hormone-dependent. Furthermore, it seems that in certain cases the compatible provirus-mouse relationship breaks down and oncogenic transformation results in neoplasia. How this may occur will be considered next.

HOW LEUKEMIA VIRUSES CAUSE LEUKEMIA

The central mystery of RNA tumor viruses has long been how leukemia viruses cause leukemia.

Among avian sarcoma viruses such RSV, it is clearly the *src* gene product that transforms cells. Recently it has been found that even mutants of RSV that contain substantial deletions in the *src* gene can cause tumors in newly hatched chickens, and that such tumors yield viruses that once again contain a full-sized *src* gene. However this *src* gene is not identical to that in the originally injected RSV; it is probably derived by recombination between the partially deleted parental RSV *src* gene and cellular *sarc* genes.

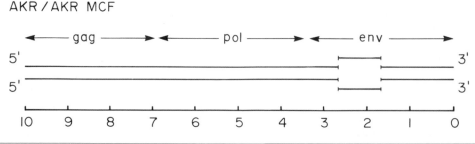

FIG. 9-11. The sequence homology relationships, as measured by heteroduplex analysis, of the AKR murine leukemia virus and an MCF virus derived from it (MCF 247). The complementary DNA of AKR virus DNA was hybridized to the RNA of AKR MCF, and the resulting structures were examined by electron microscopy. For further details, see text. Note that there is one sequence, which amounts to about 8 percent of the RNA, which is completely nonhomologous. Presumably it represents part of the *env* gene of an endogenous RNA tumor virus. (Adapted from Chien et al: J Virol 28:352, 1978.)

In the highly oncogenic defective leukemia viruses the transforming function resides either in the polyprotein, in its cleavage products, or in an as yet undetected virus-coded protein (see p. 179). The point is that for both the avian sarcoma viruses and the defective leukemia viruses there are candidates for the role of transforming protein. However, nondefective leukemia viruses cause oncogenic transformation and yet contain only the *gag, pol,* and *env* genes; what is their transforming protein?

It has recently been found that the development of spontaneous lymphomas is associated with the generation of a new class of leukemia viruses; such new viruses have been isolated from the thymuses of late preleukemic or leukemic AKR and preleukemic C58 mice, and from the lymphomas that develop in NIH Swiss mice. These viruses, which induce focal morphologic alterations in a mink lung cell line and which have therefore become known as mink cell focus (MCF)-inducing viruses, have the following properties: (1) they grow not only in murine, but also in nonmurine cells, and are therefore said to be dual-tropic; (2) they are interfered with by ecotropic as well as by xenotropic viruses; and (3) they are neutralized by antisera to both ecotropic and xenotropic viruses. All these properties suggest that MCF viruses are recombinants in the *env* gene between ecotropic and xenotropic viruses, and tryptic peptide analysis of their gp70s, as well as heteroduplex analysis of their RNAs (Fig. 9-11) has confirmed this conclusion. All available evidence indicates that MCF viruses possess the *gag* and *pol* genes characteristic of ecotropic viruses, but that their *env* gene region comprises sequences that are characteristic of both ecotropic and xenotropic endogenous viruses. It is not known whether there are restrictions on where within the *env* gene recombination may or may not occur; conceivably a very large number of MCF viruses may exist.

The emergence of MCF viruses from lymphoid tissue in the preleukemic and leukemic state indicates that either they, or recombination in the *env* gene, may be involved in the pathogenesis of leukemia; it remains to be established whether the recombinant *env* glycoproteins themselves are the transforming proteins. It is interesting to recall in this regard that the SSFV component of Friend leukemia virus, which rapidly causes splenomegaly and leukemia (see p. 179), is a recombinant between Friend ecotropic leukemia virus and an *env* gene portion of a xenotropic virus.

MOUSE MAMMARY TUMOR VIRUSES (MMTV)

Mouse mammary tumor virus, the prototype of B-type RNA tumor viruses, exists in the form of numerous proviruses that express themselves to varying degrees in different strains of mice. The most familiar is the milk factor of Bittner, a virus strain that is resident in C3H and A strain mice (high-incidence strains), expresses itself in all cells of its hosts, and is present in particularly large amounts in lactating mammary tissue and, therefore, in milk. As a result, it is passed on readily to progeny animals, in which it produces mammary adenocarcinomas with high frequency early in life (generally between age 6 and 12 months). It is also passed on to animals of low-incidence strains when they are reared by foster mothers of the C3H and A strains and with high frequency produces tumors in them also.

The other MMTV strains express themselves far less readily; they are not present in large amounts in milk and are transmitted vertically through the gametes. They are also much less oncogenic; they cause tumors with low frequency late in life. They are resident in both low- and high-incidence mouse strains; their presence in the latter is demonstrated by the fact that they develop adenocarcinomas late in life even when they are freed of the milk factor by being nursed by foster mothers of low-incidence strains.

Until very recently, no cultured cell line was known in which MMTV could be grown. As a result, it had to be assayed by injecting virus into newborn mice and waiting for the 6 to 12 months before tumors developed. Recently it has been found that MMTV will multiply in a feline kidney cell line, and it should now be possible to develop a much more efficient and rapid assay in these cells.

Mice of high- and low-incidence strains contain similar numbers of MMTV proviruses (2 to 10) in their genomes. However, they are transcribed to very differing degrees. In virus-producing tumors and lactating mammary tissue of high-incidence mouse strains there are from 100 to 1000 genome equivalents of MMTV RNA per cell; in tumors of low-incidence mouse strains only 1 to 2, and in livers and spleens from either high- or low-incidence mouse strains (that is, in cells that never become transformed and never produce MMTV), there is only 0.1 to 1 genome equivalent of viral RNA per cell. These levels of transcription correlate with the extent of virus production and indicate that the primary control of MMTV gene expression occurs at the level of transcription. The factors that actually regulate the transcription of MMTV provirus are both host genetic and hormonal. A Mendelian gene with a dominant allele for MMTV expression is known; estrogens enhance the development of mammary cancer (which normally occurs only in female mice, but can also be induced by estrogens in male mice of high-incidence strains); and glucocorticoids greatly increase the degree of MMTV provirus transcription and the level of MMTV formation in lactating mammary tissue cells and mammary tumor cells, but are unable to effect significant increases in other mouse cells.

TYPE D RNA TUMOR VIRUSES

Type D RNA tumor viruses are distinguished from type C viruses by the fact that type D immature forms exist within the cytoplasm of infected cells, and from type B viruses by the fact that type D "intracytoplasmic A particles" (see below) are smaller (diameters of 90 and 75 nm, respectively) and that type D glycoprotein spikes are shorter.

The first D-type RNA tumor virus to be isolated was the Mason-Pfizer monkey virus (MPMV) which was isolated from a breast carcinoma of a female rhesus monkey, in which it is horizontally transmitted as an exogenous virus. Subsequently two other type D viruses were isolated, one from a squirrel monkey and the other from a langur; both are endogenous viruses in these species. All these viruses share interspecies-specific determinants among the major internal proteins and the glycoproteins, and they also show cross-reactivity with the gp70 of the baboon xenotropic M7 virus. Since the squirrel monkey is a new-world primate whereas the langur and the rhesus monkey are old-world primates, a progenitor of type D

RNA tumor viruses appears to have become genetically associated with primates at a very early stage of their evolution, at least 50 million years ago.

A-TYPE PARTICLES

Mouse cells often contain particles some 70-90 nm in diameter that consist of toroidal (ring-shaped) nucleoids enclosed by a membrane (Fig. 9-12). These particles, which are referred to as A-type particles, belong to two classes.

Particles of the first class occur primarily in the cytoplasm of mouse mammary tumor cells; these intracytoplasmic A-type particles are the precursors of B-type particles, that is, mouse mammary tumor virus particles. The evidence for this view includes the following facts: (1) their structure is similar to that of MMTV cores; (2) their RNA is homologous to the RNA of MMTV; (3) they possess an RNA-dependent DNA polymerase with properties similar to that of MMTV; (4) they share antigenic determinants with MMTV; and (5) although their component proteins differ in size from those of MMTV cores, proteolytic enzymes such as trypsin cleave them into proteins of the correct size. The evidence is strong, therefore, that intracytoplasmic A-type particles are precursors of mature MMTV. Not surprisingly, they are noninfectious.

Similar but smaller intracytoplasmic A-type particles occur in cells infected with or transformed by type D viruses (see above).

A-type particles of the second class are located within the cisternae of the endoplasmic reticulum and are therefore known as intracisternal A-type particles. They appear regularly in the early stages of mouse embryogenesis as well as in certain kinds of tumors, particularly myeloma, rhabdomyosarcoma and neuroblastoma. Their appearance is not correlated with the formation of either B- or C-type virus particles; not only is this so under natural conditions, but induction of C-type particle formation with substances such as iododeoxyuridine or dexamethasone does not induce the formation of increased amounts of intracisternal A-type particles. Like intracytoplasmic A-type particles, they possess a DNA-dependent RNA polymerase and are noninfectious. Their RNA, which has a genetic complexity of about 2.5×10^6, is present in proviral form in mouse DNA to the

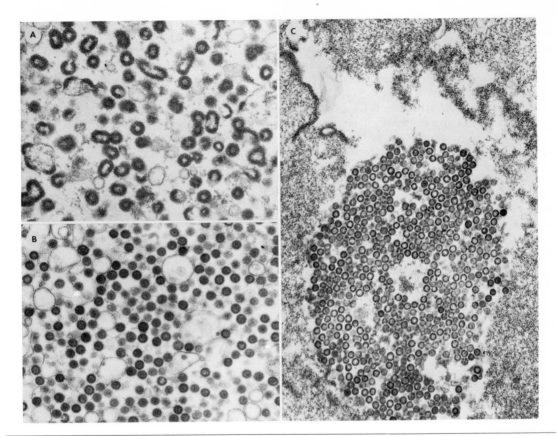

FIG. 9-12.   A-type particles. **A.** Intracisternal A particles, extracted from mouse plasma cell tumor MOPC 104E and partially purified by sucrose density gradient centrifugation. ×70,000. **B.** Intracytoplasmic A particles from a mouse mammary tumor, partially purified by sucrose density gradient centrifugation. ×53,000. **C.** Intracytoplasmic A particle inclusion in the nuclear pellet fraction of a homogenate of a mouse mammary tumor. ×35,000. (Courtesy of Dr. N. A. Wivell.)

extent of about 1000 copies per haploid genome, and this reiteration frequency is not higher in the DNA of A-type particle-rich tumor cells than in normal cells. The nucleic acid from intracisternal A-type particles isolated from *Mus musculus* (the species to which laboratory mouse strains belong) will not hybridize to the nucleic acid of C-type RNA tumor viruses of *Mus musculus* but will hybridize to a very significant extent not only to the nucleic acid of C-type RNA tumor virus of *Mus cervicolor*, an Asian mouse species, but also to the DNA of this mouse.

Intracisternal A-type particles may be defective forms of endogenous RNA tumor viruses that fail to achieve egress from cells. Conceivably their genomes became associated with *env* gene sequences to produce a new class of replication-competent virus in the ancestor of

Mus cervicolor, but not of *Mus musculus*. Their function in the latter species is obscure; but it is striking that their expression is regulated by normal developmental mechanisms. They may thus represent an intermediate stage of virus evolution, with the potential for exchange of information with endogenous viral genes or other cellular genetic elements.

Do viruses cause cancer in humans?

What assessment can be made at this time of the role of viruses in carcinogenesis in humans?

No causal relationship has been demonstrated, and it may well be that the most direct evidence concerning the etiologic role of some virus will be the demonstration that vaccination against it lowers the incidence of some form of cancer.

In the meantime, attention is focused on attempts to detect the presence of viral genetic material or of its expression in human tumor cells. As pointed out above (see p. 169), the evidence is best concerning the herpesvirus EB virus, whose genome is certainly present in Burkitt lymphoma and nasopharyngeal carcinoma cells. By contrast, adenoviruses and papovaviruses are very unlikely candidates for human tumor viruses: numerous human neoplasms have been examined for traces of infection with viruses of these two families, and all have been negative.

There have been numerous reports that RNA tumor virus particles, their nucleic acids and antigens, as well as reverse transcriptase, are present in normal and malignant human tissue. Since the question of whether RNA tumor viruses play a role in causing cancer in humans is potentially very important, the evidence that is available at this time will now be reviewed in some detail.

The Search for Infectious Endogenous RNA Tumor Viruses

In spite of intensive search, no endogenous RNA tumor virus has yet been found in humans. There have been several reports of the isolation of such viruses from a variety of human tumors, but after careful study none of the candidate viruses has passed the fundamental test—namely, that its proviral DNA be demonstrable in the human genome (that is, in the DNA of normal individuals). Yet it seems almost certain that the human genome does contain proviral DNA. The reason is simply that primates are known to contain proviruses—three xenotropic type C viruses and two type D viruses have been isolated from primates so far—and there seems to be no good reason why they should not also be present in the human genome. In fact, it has recently been found that all primates including humans contain sequences in their genomes that are partially ho-

mologous to those of the baboon xenotropic M7 virus. It does seem, however, that the expression of primate endogenous viruses is tightly controlled; only the baboon xenotropic M7 virus and one of the type D viruses express themselves readily to the extent of infectious particle formation. If the human genome does harbor proviruses, as seems likely, they are clearly in the tightly controlled group.

The Presence of RNA Tumor Virus Particles in Human Tissue

Particles that resemble RNA tumor virus particles have been observed repeatedly by electron microscopy in normal and malignant human tissues such as placenta and melanoma. The frequency of these particles is always very low, and their morphology generally differs somewhat from that of C-type particles. Such particles have also been seen in cells cultured in vitro from malignant human tissue, and particles that possess the enzyme reverse transcriptase and RNA of the size of RNA tumor virus genomes have been isolated from them by density gradient centrifugation. However, the status of these particles is uncertain, since not only does in vitro culture always entail the risk of contamination, but it has not yet been possible to isolate these particles in amounts sufficient for detailed characterization.

The Search for Nucleic Acid Homologous to RNA Tumor Virus RNA

As pointed out above, normal human DNA contains sequences that are at least partially homologous to those of the baboon xenotropic virus M7. It has been reported that leukemic tissue DNA from about 20 percent of patients hybridizes with M7 RNA both more extensively and more faithfully than normal human DNA. DNA sequences distantly related to Rauscher murine leukemia virus, Mo-MLV and SSV RNA have also been reported to be present in the DNA of leukocytes of some leukemia patients. These data have been interpreted as evidence that all these viruses may play a role as exogenous viruses in causing leukemia in humans.

The Search for Antibodies Cross-Reactive with C- and B-type RNA Tumor Virus Proteins

Healthy humans contain antibodies to the gp70 of the xenotropic baboon virus M7, as well as to that of SSV, and, to a lesser extent, of the Friend leukemia complex. It has also been found that women develop selective cell-mediated reactivity against human fibroblasts infected with M7 virus, as well as temporarily elevated antibody levels to M7 virus during the second and third trimesters of pregnancy. Antibodies to the D-type primate virus, Mason-Pfizer monkey virus, have also been observed in sera from patients with renal disease; antibodies to the p30, the principal internal protein, of several mammalian RNA tumor viruses, including especially the xenotropic feline virus RD114, have been found in lupus glomerular immune deposits; and sera from patients with chronic myelogenous leukemia have been found to contain antibodies to the reverse transcriptase of FeLV, while some normal sera have been observed to neutralize the reverse transcriptase of the SSV-gibbon ape leukemia virus group. All these data suggest that infection of humans with exogenous RNA tumor viruses may not be uncommon.

The Nature of the Particles Present in Human Breast Cancer Tissue

Particles have been found in human breast cancer tissue that exhibit many of the properties characteristic of the B-type MMTV and the D-type MPMV, the viruses that cause mammary tumors in mice and monkeys respectively. Among the similarities are morphologic resemblance, possession of similar size nucleic acid and partial nucleic acid homology, similarity of reverse transcriptases, and antigenic cross-reactivity with gp52, the spike glycoprotein of MMTV.

None of this evidence is conclusive. It does however suggest that the human genome does harbor endogenous RNA tumor viruses, just like those of all other animal species that have been examined so far, and that humans may occasionally become infected with horizontally transmitted exogenous RNA tumor viruses. Continued intensive effort will be required to isolate and identify these viruses and to assess their tumorigenic potential.

FURTHER READING

Books and Reviews

GENERAL

Cairns HJF: Mutation, selection and the natural history of cancer. Nature 255:197, 1975

Eckhart W: Genetics of DNA tumor viruses. Ann Rev Genetics 8:301, 1974

Gillespie D, Gallo RC: RNA processing and RNA tumor virus origin and evolution. Science 188:802, 1975

Gross L: Facts and theories on viruses causing cancer and leukemia. Proc Natl Acad Sci USA 71:2013, 1974

Hanafusa H: Cell transformation by RNA tumor viruses. In Fraenkel-Conrat H, Wagner RR (eds): Comprehensive Virology, Vol 10, p 401. New York and London, Plenum, 1977

Sivak A: Cocarcinogenesis. BBA 560:67, 1979

Tooze J (ed): The Molecular Biology of Tumor Viruses. Cold Spring Harbor Lab, 1973

Tooze J, Sambrook J (eds): Selected Papers in Tumor Virology. Cold Spring Harbor Lab, 1974

CELL TRANSFORMATION

Hakomori S: Structures and organization of cell surface glycolipids. Dependency on cell growth and malignant transformation. Biochim Biophys Acta 417:55, 1975

Hynes RO: Cell surface proteins and malignant transformation. Biochim Biophys Acta 458:73, 1976

Nicholson GL: Trans-membrane control of the receptors on normal and tumor cells. II. Surface changes associated with transformation and malignancy. Biochim Biophys Acta 458:1, 1976

Rafferty KA Jr: Epithelial cells: growth in culture of normal and neoplastic forms. Adv Cancer Res 21:249, 1975

Vaheri A, Mosher DF: High molecular weight, cell-surface-associated glycoprotein (fibronectin) lost in malignant transformation. Biochim Biophys Acta 516:1, 1978

Warren L, Buck CA, Tuszynski GP: Glycopeptide changes and malignant transformation. A possible role for carbohydrate and malignant behavior. Biochim Biophys Acta 516:97, 1978

DNA TUMOR VIRUSES

Deinhardt F, Falk LA, Wolfe LG: Simian herpesviruses and neoplasia. Adv Cancer Res 19:167, 1974

Doerfler W: Integration of viral DNA into the host genome. Curr Top Microbiol Immunol 71:1, 1975

Epstein MA, Achong BG: Recent progress in Epstein-Barr virus research. Ann Rev Microbiol 31:421, 1977

Levine AJ: SV40 and adenovirus early functions involved in DNA replication and transformation. Biochim Biophys Acta 458:213, 1976

Rapp F, Westmoreland D: Cell transformation by DNA-containing viruses. Biochim Biophys Acta 458:167, 1976

Rawls WE, Bacchetti S, Graham FL: Relation of herpes simplex viruses to human malignancies. Curr Top Microbiol Immunol 77:71, 1977

Sambrook J: The molecular biology of the papovaviruses. In Nayak DP (ed): The Molecular Biology of Animal Viruses. Vol 2, p 578. New York, Marcel Dekker, 1978

Weil R: Viral tumor antigens: a novel type of mammalian regulator proteins. Biochim Biophys Acta 516:301, 1978

Weinberg RA: How does T antigen transform cells? Cell 11:243, 1977

zur Hausen H: .Oncogenic herpesviruses. Biochim Biophys Acta 417:25, 1975

zur Hausen H: Human papilloma viruses and their possible role in squamous cell carcinomas. Curr Top Microbiol Immunol 78:1, 1977

RNA TUMOR VIRUSES

Bishop JM: Retroviruses. Ann Rev Biochem 47:35, 1978

Coffin JM: Structure, replication and recombination of retrovirus genomes: some unifying hypotheses. J Gen Virol 42:1, 1979

Eisenman RN, Vogt VM: The biosynthesis of oncovirus proteins. Biochim Biophys Acta 473:188, 1978

Elder JH, Gautsch JW, Jensen FC, Lerner RA: Multigene family of endogenous retroviruses: recombinant origin of diversity. J Natl Cancer Inst 61:625, 1978

Fan H: Expression of RNA tumor viruses at translation and transcription levels. Curr Top Microbiol Immunol 79:1, 1978

Fine D, Schochetman G: Type D primate retroviruses: a review. Cancer Res 38:3123, 1978

Friis RR: Temperature-sensitive mutants of avian RNA tumor viruses: a review. Curr Top Microbiol Immunol 79:261, 1978

Gardner MB: Type C viruses of wild mice: characterization and natural history of amphotropic, ecotropic and xenotropic MuLV. Curr Top Microbiol Immunol 79:215, 1978

Gilden RV: Biology of RNA tumor viruses. In Nayak DP (ed): The Molecular Biology of Animal Viruses, Vol 1:435. New York, Marcel Dekker, 1977

Graf T, Beug H: Avian leukemia viruses: interaction

with their target cells in vivo and in vitro. Biochim Biophys Acta 516:269, 1978

Levy JA: Xenotropic type C viruses. Curr Top Microbiol Immunol 79:111, 1978

Montelaro RC, Bolognesi DP: Structure and morphogenesis of type C retroviruses. Adv Cancer Res 28:63, 1978

Moore DH: Evidence in favor of the existence of human breast cancer virus. Cancer Res 34:2322, 1974

Temin HM: On the origin of RNA tumor viruses. Ann Rev Genetics 8:155, 1974

Van Zaane D, Bloemers HJ: The genome of the mammalian sarcoma viruses. Biochim Biophys Acta 516:249, 1978

Wang L: The gene order of avian RNA tumor viruses derived from biochemical analyses of deletion mutants and viral recombinants. Ann Rev Microbiol 32:561, 1978

Weinberg RA: Structure of the intermediates leading to the integrated provirus. Biochim Biophys Acta 473:39, 1977

Selected Papers

DNA TUMOR VIRUSES

Anzai T, Dreesman GR, Courtney RJ, et al: Antibody to herpes simplex virus type 2-induced nonstructural proteins in women with cervical cancer and in control groups. J Natl Cancer Inst 54:1051, 1975

Campo MS, Cameron IR, Rogers ME: Tandem integration of complete and defective SV40 genomes in mouse-human somatic cell hybrids. Cell 15:1411, 1978

Doerfler W: Integration of the deoxyribonucleic acid of adenovirus type 12 into the deoxyribonucleic acid of baby hamster kidney cells. J Virol 6:652, 1970

Duff R, Rapp F: Oncogenic transformation of hamster embryo cells after exposure to inactivated herpes simplex virus type 1. J Virol 12:209, 1973

Gallimore PH, Sharp PA, Sambrook J: Viral DNA in transformed cells. II. A study of the sequences of adenovirus 2 DNA in nine lines of transformed rat cells using specific fragments of the viral genome. J Mol Biol 89:49, 1974

Graham FL, van der Eb AJ, Heijneker HJ: Size and location of the transforming region in human adenovirus type 5 DNA. Nature 251:687, 1974

Henle G, Henle W, Diehl V: Relation of Burkitt's tumor-associated herpes-type virus to infectious mononucleosis. Proc Natl Acad Sci USA 59:94, 1968

Hinze HC, Walker DL: Response of cultured rabbit cells to infection with the Shope fibroma virus. I. Proliferation and morphological alteration in the infected cells. J Bacteriol 88:1185, 1964

Jones KW, Kinross J, Maitland N, Norval M: Normal human tissues contain RNA and antigens related to infectious adenovirus type 2. Nature 277:274, 1979

Klein G, Giovanella BC, Lindahl T, et al: Direct evidence for the presence of Epstein-Barr virus DNA and nuclear antigen in malignant epithelial cells from patients with poorly differentiated carcinoma of the nasopharynx. Proc Natl Acad Sci USA 71:4737, 1974

Koprowski H, Jensen FC, Steplewski Z: Activation of production of infectious tumor virus SV40 in heterokaryon cultures. Proc Natl Acad Sci USA 58:127, 1978

Lancaster WD, Meinke W: Persistence of viral DNA in human cell cultures infected with human papilloma virus. Nature 256:434, 1975

Morris AG, Lavialle C, Suárez H, Cassingena R: The induction of SV40-transformed Chinese hamster and mouse kidney cells by mitomycin. Intervirol 5:305, 1975

Orth G, Favre M, Jablonska S, Brylak K, Croissant O: Viral sequences related to a human skin papillomavirus in genital warts. Nature 275:334, 1978

Risser R, Pollack R: A nonselective analysis of SV40 transformation of mouse 3T3 cells. Virol 59:477, 1974

Sambrook J, Westphal H, Srinivasan PR, Dulbecco R: The integrated state of viral DNA in SV40-transformed cells. Proc Natl Acad Sci USA 60:1288, 1968

Schlegel R, Benjamin TL: Cellular alterations dependent upon the polyoma virus Hr-t function: separation of mitogenic from transforming capacities. Cell 14:587, 1978

zur Hausen H, Schulte-Holthausen H, Klein G, et al: EBV DNA in biopsies of Burkitt tumor and anaplastic carcinomas of the nasopharynx. Nature 228:1056, 1970

RNA TUMOR VIRUSES
VIRUS COMPONENTS

Bender W, Chien Y-H, Chattopadhyay S, Vogt PK, Gardner MB, Davidson N: High-molecular-weight RNAs of AKR, NZB, and wild mouse viruses and avian reticuloendotheliosis virus all have similar dimer structures. J Virol 25:888, 1978

Bolognesi DP, Montelaro RC, Frank H, Schafer W: Assembly of type C oncornaviruses: a model. Science 199:183, 1978

Cheung K-S, Smith RE, Stone MP, Joklik WK: Comparison of immature (rapid harvest) and mature Rous sarcoma virus particles. Virol 50:851, 1972

Dittmar KJ, Moelling K: Biochemical properties of p15-associated protease in an avian RNA tumor virus. J Virol 28:106, 1978

Hizi A, Joklik WK: RNA-dependent DNA polymerase of avian sarcoma virus B77. I. Isolation and partial characterization of the α, β_2, and $\alpha\beta$ forms of the enzyme. J Biol Chem 252:2281, 1977

Witte ON, Baltimore D: Relationship of retrovirus polyprotein cleavages to virion maturation studied with temperature-sensitive murine leukemia virus mutants. J Virol 26:750, 1978

MOLECULAR BIOLOGY AND GENE EXPRESSION

Brugge JS, Erikson RL: Identification of a transformation-specific antigen induced by an avian sarcoma virus. Nature 269:346, 1977

Coffin JM, Champion M, Chabot F: Nucleotide sequence relationships between the genomes of an endogenous and an exogenous avian tumor virus. J Virol 28:972, 1978

Collett MS, Brugge JS, Erikson RL: Characterization of a normal avian cell protein related to the avian sarcoma virus transforming gene product. Cell 15:1363, 1978

Collett MS, Faras AJ: Avian retrovirus RNA-directed DNA synthesis: transcription at the 5′ terminus of the viral genome and the functional role for the viral terminal redundancy. Virol 86:297, 1978

Copeland NG, Cooper GM: Transfection by exogenous and endogenous murine retrovirus DNAs. Cell 16:347, 1979

Cordell B: At least 104 nucleotides are transposed from the 5′ terminus of the avian sarcoma virus genome to the 5′ termini of smaller viral mRNAs. Cell 15:79, 1978

Halpern MS, Bolognesi DP, Friis RR, Mason WS: Expression of the major viral glycoprotein of avian tumor virus in cells of chf(+) chicken embryos. J Virol 15:1131, 1975

Hu SSF, Lai MMC, Vogt PK: Characterization of the env gene in avian oncoviruses by heteroduplex mapping. J Virol 27:667, 1978

Jamjoom GA, Naso RB, Arlinghaus RB: Further characterization of intracellular precursor polyproteins of Rauscher leukemia virus. Virol 78:11, 1977

Krzyzek RA, Collett MS, Lau AF, Perdue ML, Leis JP, Faras AJ: Evidence for splicing of avian sarcoma virus 5′-terminal genomic sequences onto viral-specific RNA in infected cells. Proc Natl Acad Sci USA 75:1284, 1978

Levinson AD, Oppermann H, Levintow L, Varmus HE, Bishop JM: Evidence that the transforming gene of avian sarcoma virus encodes a protein kinase associated with a phosphoprotein. Cell 15:561, 1978

Murphy EC Jr, Arlinghaus RB: Tryptic peptide analyses of polypeptides generated by premature termination of cell-free protein synthesis allow a determination of the Rauscher leukemia virus gag gene order. J Virol 28:929, 1978

Oppermann H, Bishop JM, Varmus HE, Levintow L: A joint product of the genes gag and pol of avian sarcoma virus: a possible precursor of reverse transcriptase. Cell 12:993, 1977

Padgett TG, Stubblefield E, Varmus HE: Chicken macrochromosomes contain an endogenous provirus and microchromosomes contain sequences related to the transforming gene of ASV. Cell 10:649, 1977

Rothenberg E, Donoghue DJ, Baltimore D: Analysis of a 5′ leader sequence on murine leukemia virus 21S RNA: heteroduplex mapping with long reverse transcriptase products. Cell 13:435, 1978

Schincariol AL, Joklik WK: Early synthesis of virus-specific RNA and DNA in cells rapidly transformed with Rous sarcoma virus. Virol 56:532, 1973

Shank PR, Hughes SH, Kung H-J, Majors JE, Quintrell N, Guntaka RV, Bishop JM, Varmus HE: Mapping unintegrated avian sarcoma virus DNA: termini of linear DNA bear 300 nucleotides present once or twice in two species of circular DNA. Cell 15:1383, 1978

Shields A, Witte ON, Rothenberg E, Baltimore D: High frequency of aberrant expression of Moloney murine leukemia virus in clonal infections. Cell 14:601, 1978

Spector DH, Varmus HE, Bishop JM: Nucleotide sequences related to the transforming gene of avian sarcoma virus are present in DNA of uninfected vertebrates. Proc Natl Acad Sci USA 75:4102, 1978

Steffen D, Weinberg RA: The integrated genome of murine leukemia virus. Cell 15:1003, 1978

Stehelin D, Guntaka RV, Varmus HE, Bishop JM: Purification of DNA complementary to nucleotide sequences required for neoplastic transformation of fibroblasts by avian sarcoma viruses. J Mol Biol 101:349, 1976

Tal J, Fujita DJ, Kawai S, Varmus HE, Bishop JM: Purification of DNA complementary to the *env* gene of avian sarcoma virus and analysis of relationships among the *env* genes of avian leukosis-sarcoma viruses. J Virol 21:497, 1977

Tereba A, Skoog L, Vogt PK: RNA tumor virus-specific sequences in nuclear DNA of several avian species. Virol 65:524, 1975

Weiss SR, Hackett PB, Oppermann H, Ullrich A, Levintow L, Bishop JM: Cell-free translation of avian sarcoma virus RNA: suppression of the *gag* termination codon does not augment synthesis of the joint *gag/pol* product. Cell 15:607, 1978

ACUTE LEUKEMIA AND SIMILAR VIRUSES

Duesberg PH, Bister K, Vogt PK: The RNA of avian acute leukemia virus MC29. Proc Natl Acad Sci USA 74:4320, 1977

Graf T, Royer-Pokora B, Meyer-Glauner W, Claviez M, Götz E, Beug H: In vitro transformation with avian myelocytomatosis virus strain CMII: characterization of the virus and its target cells. Virol 83:96, 1977

Hayman MJ, Royer-Pokora B, Graf T: Defectiveness of avian erythroblastosis virus: synthesis of a 75K *gag*-related protein. Virol 92:31, 1979

Hu S, Davidson N: A heteroduplex study of the sequence relationships between the RNAs of M-MSV and M-MLV. Cell 10:469, 1977

Hu SSF, Vogt PK: Avian oncovirus MH2 is defective in *gag, pol,* and *env*. Virol 92:278, 1979

Kahn AS, Stephenson JR: FeSV-coded polyprotein: enzymatic cleavage by a type C virus-coded structural protein. J Virol 29:649, 1979

Royer-Pokora B, Beug H, Claviez M, Winkhardt H-J, Friis RR, Graf T: Transformation parameters in chicken fibroblasts transformed by AEV and MC29 avian leukemia viruses. Cell 13:751, 1978

Witte ON, Rosenberg N, Paskind M, Shields A, Baltimore D: Identification of an Abelson murine leukemia virus-coded protein present in transformed fibroblast and lymphoid cells. Proc Natl Acad Sci USA 75:2488, 1978

ENDOGENOUS AND EXOGENOUS MURINE
LEUKEMIA VIRUSES

Albino A, Korngold L, Mellors RC: Tryptic peptide analysis of *gag* gene proteins of endogenous mouse type C viruses. J Virol 29:102, 1979

Barbacid M, Robbins KC, Aaronson SA: Wild mouse RNA tumor viruses. A nongenetically transmitted virus group closely related to exogenous leukemia viruses of laboratory mouse strains. J Exp Med 149:254, 1979

Barbacid M, Stephenson JR, Aaronson SA: Evolutionary relationships between *gag* gene-coded proteins of murine and primate endogenous type C RNA viruses. Cell 10:641, 1977

Bryant ML, Roy-Burman P, Gardner MB, Pal BK: Genetic relationship of wild mouse amphotropic virus to murine ecotropic and xenotropic viruses. Virol 88:389, 1978

Fischinger PJ, Nomura S: Efficient release of murine xenotropic oncornavirus after murine leukemia virus infection of mouse cells. Virol 65:304, 1975

Hartley NW, Rowe WP: Naturally occurring murine leukemia viruses in wild mice: characterization of a new "amphotropic " class. J Virol 19:19, 1976

GENERATION OF HIGHLY ONCOGENIC VIRUSES BY
RECOMBINATION

Chien Y-H, Verma IM, Shih TY, Scolnick EM, Davidson N: Heteroduplex analysis of the sequence relations between the RNAs of mink cell focus-inducing and murine leukemia viruses. J Virol 28:352, 1978

Halpern CC, Hayward WS, Hanafusa H: Characterization of some isolates of newly recovered ASV. J Virol 29:91, 1979

Hartley NW, Wolford NK, Old LJ, Rowe WP: A new class of murine leukemia virus associated with development of spontaneous lymphomas. Proc Natl Acad Sci USA 74:789, 1977

Ihle JN, Fischinger P, Bolognesi D, Elder J, Gautsch JW: B-MuX: a unique murine C-type virus containing the *"env"* gene of xenotropic viruses and the *"gag"* gene of the ecotropic virus. Virol 90:255, 1978

Rapp UR, Todaro GJ: Generation of new mouse sarcoma viruses in cell culture. Science 201:821, 1978

Rapp UR, Todaro GJ: Generation of oncogenic type C viruses: rapidly leukemogenic viruses derived from C3H mouse cells in vivo and in vitro. Proc Natl Acad Sci USA 75:2468, 1978

Shih TY, Williams DR, Weeks MO, Maryak JM, Vass WC, Scolnick EM: Comparison of the genomic organization of Kirsten and Harvey sarcoma viruses. J Virol 27:45, 1978

Troxler DH, Lowy D, Howk R, Young H, Scolnick EM: Friend strain of spleen focus-forming virus is a recombinant between ecotropic murine type C virus and the *env* gene region of xenotropic type C virus. Proc Natl Acad Sci USA 74:4671, 1977

Vigne R, Breitman ML, Moscovici C, Vogt PK: Restitution of fibroblast-transforming ability in *src* deletion mutants of avian sarcoma virus during animal passage. Virol 93:413, 1979

VIRUS-CODED CELL SURFACE PROTEINS, INCLUDING THE G$_{IX}$ ANTIGEN

Bryant ML, Pal BK, Gardner MB, Elder JH, Jensen FC, Lerner RA: Structural analysis of the major envelope glycoprotein (gp70) of the amphotropic and ecotropic type C viruses of wild mice. Virol 84:348, 1978

del Villano BC, Lerner RA: Relationship between the oncornavirus gene product gp70 and a major protein secretion of the mouse genital tract. Nature 259:497, 1976

Elder JH, Jensen FC, Bryant ML, Lerner RA: Polymorphism of the major envelope glycoprotein (gp70) of murine C-type viruses: virion associated and differentiation antigens encoded by a multigene family. Nature 267:23, 1977

Ledbetter J, Nowinski RC, Emery S: Viral proteins expressed on the surface of murine leukemia cells. J Virol 22:65, 1977

McLellan WL, Ihle JN: Purification and characterization of a murine tumor cell surface glycoprotein of 75,000 daltons that is related to the major envelope glycoprotein of murine leukemia virus. Virol 89:547, 1978

O'Donnell PV, Stockert E: Induction of G$_{IX}$ antigen and Gross cell surface antigen after infection by ecotropic and xenotropic murine leukemia viruses in vitro. J Virol 20:545, 1976

PRIMATE C-TYPE RNA TUMOR VIRUSES

Benveniste RE, Todaro GJ: Evolution of type C viral genes: evidence for an Asian origin of man. Nature 261:101, 1976

Hu S, Davidson N, Nicolson MO, McAllister RM: Heteroduplex study of the sequence relations between RD-114 and baboon viral RNAs. J Virol 23:345, 1977

Todaro GJ, Sherr CJ, Benveniste RE: Baboons and their close relatives are unusual among primates in their ability to release nondefective endogenous type C viruses. Virol 72:278, 1976

B-TYPE RNA TUMOR VIRUSES: MAMMARY TUMOR VIRUSES

Cohen JC, Shank PR, Morris VL, Cardiff R, Varmus HE: Integration of the DNA of mouse mammary tumor virus in virus-infected normal and neoplastic tissue of the mouse. Cell 16:333, 1979

Morris VL, Kozak C, Cohen JC, Shank PR, Jolicoeur P, Ruddle F, Varmus HE: Endogenous mouse mammary tumor virus DNA is distributed among multiple mouse chromosomes. Virol 92:46, 1979

Morris VL, Medeiros E, Ringold GM, Bishop JM, Varmus HE: Comparison of mouse mammary tumor virus-specific DNA in inbred, wild and Asian mice, and in tumors and normal organs from inbred mice. J Mol Biol 114:73, 1977

Parks WP, Ransom JC, Young HA, Scolnick EM: Mammary tumor virus induction by glucocorticoids. Characterization of specific transcriptional regulation. J Biol Chem 250:3330, 1975

Ringold GM, Yamamoto KR, Bishop JM, Varmus HE: Glucocorticoid-stimulated accumulation of mouse mammary tumor virus RNA: increased rate of synthesis of viral RNA. Proc Natl Acad Sci USA 74:2879, 1977

Schochetman G, Oroszlan S, Arthur L, Fine D: Gene order of the mouse mammary tumor virus glycoproteins. Virol 83:72, 1977

Yagi MJ, Compans RW: Structural components of mouse mammary tumor virus. I. Polypeptides of the virion. Virol 76:751, 1977

TYPE D VIRUSES AND A-TYPE PARTICLES

Colcher D, Drohan W, Schlom J: Mason-Pfizer Virus RNA genome: relationship to the RNA of morphologically similar isolates and other oncornaviruses. J Virol 17:705, 1976

Kuff EL, Lueders KK, Scolnick EM: Nucleotide sequence relationship between intracisternal type A particles of *Mus musculus* and an endogenous retrovirus (M432) of *Mus cervicolor*. J Virol 28:66, 1978

Lueders KK, Kuff EL: Sequences associated with intracisternal A particles are reiterated in the mouse genome. Cell 12:963, 1977

Lueders KK, Segal S, Kuff EL: RNA sequences specifically associated with mouse intracisternal A particles. Cell 11:83, 1977

Smith GH: Evidence for a precursor-product relationship between intracytoplasmic A particles and mouse mammary tumor virus cores. J Gen Virol 41:193, 1978

DO RNA TUMOR VIRUSES CAUSE CANCER IN MAN?

Aulakh GS, Gallo RC: Rauscher-leukemia-virus-related sequences in human DNA: presence in some tissues of some patients with hematopoietic neoplasias and absence in DNA from other tissues. Proc Natl Acad Sci USA 74:353, 1977

Balda B-R, Hehlmann R, Cho J-R, Spiegelman S: Oncornavirus-like particles in human skin cancers. Proc Natl Acad Sci USA 72:3697, 1975

Kalter SS, Helmke RJ, Heberling RL, et al: C-type particles in normal human placentas. J Natl Cancer Inst 50:1081, 1973

Kurth R, Mikschy U: Human antibodies reactive with purified envelope antigens of primate type C tumor viruses. Proc Natl Acad Sci USA 75:5692, 1978

Ohno T, Spiegelman S: Antigenic relatedness of the DNA polymerase of human breast cancer particles to the enzyme of the Mason-Pfizer monkey virus. Proc Natl Acad Sci USA 74:2144, 1977

Reitz MS, Miller NR, Wong-Staal F, Gallagher RE, Gallo RC, Gillespie DH: Primate type-C virus nucleic acid sequences (woolly monkey and baboon types) in tissues from a patient with acute myelogenous leukemia and in viruses isolated from cultured cells of the same patient. Proc Natl Acad Sci USA 73:2113, 1976

Stephenson JR, Aaronson SA: Search for antigens and antibodies crossreactive with type C viruses of the woolly monkey and gibbon ape in animal models and in humans. Proc Natl Acad Sci USA 73:1725, 1976

Thiry L, Sprecher-Goldberger S, Bossens M, Neuray F: Cell-mediated immune response to simian oncornavirus antigens in pregnant women. J Natl Cancer Inst 60:527, 1978

Wong-Staal F, Gillespie D, Gallo RC: Proviral sequences of baboon endogenous type C RNA virus in DNA of human leukemic tissues. Nature 262:190, 1976

MISCELLANEOUS

Huebner RJ, Todaro GJ: Oncogenes of RNA tumor viruses as determinants of cancer. Proc Natl Acad Sci USA 64:1087, 1969

Igel HJ, Huebner RJ, Turner RC, Kotin P, Falk HL: Mouse leukemia virus activation by chemical carcinogens. Science 166:1624, 1969

Proffitt MR, Hirsch MS, Black PH: Murine leukemia: a virus-induced autoimmune disease? Science 182:821, 1973

Strand M, August JT: Structural proteins of mammalian oncogenic RNA viruses: multiple antigenic determinants of the major internal protein and envelope glycoprotein. J Virol 13:171, 1974

Weiss RA, Boettiger D, Murphy HM: Pseudotypes of avian sarcoma viruses with the envelope properties of vesicular stomatitis virus. Virol 76:808, 1977

Weiss RA, Friis PR, Vogt PK: Induction of avian tumor viruses in normal cells by physical and chemical carcinogens. Virol 46:920, 1971

Weiss RA, Wong AL: Phenotypic mixing between avian and mammalian RNA tumor viruses: I. Envelope pseudotypes of Rous sarcoma virus. Virol 76:826, 1977

CHAPTER 10

The Bacteriophages

In 1915 Twort published a note describing the infectious destruction of micrococcal colonies and offered three possible explanations for this phenomenon: (1) the destroyed colonies represented a stage in the bacterial life cycle that induced normal colonies to undergo a glassy transformation; (2) the lytic agent was an enzyme that caused both cell destruction and the production of more enzyme; and (3) the causative agent was a virus that grew in bacteria and destroyed them. All of these explanations could be reconciled with the original findings that the agent would not grow on any medium, passed through bacteria-proof filters, and was inactivated by heating to 60C for one hour. d'Hérelle, who discovered this phenomenon independently, soon demonstrated the particulate nature of what he called "bacteriophages." The elegant pioneering work of Burnet and of Schlesinger in the 1930s confirmed the viral nature of the Twort-d'Hérelle agents.

Early hopes of using the action of phage on susceptible bacteria as a means of preventing and treating infectious diseases were not fulfilled (Chap. 1). However, in the early 1940s, Delbrück and a group of investigators around him realized that the availability of viruses multiplying in cloned populations of rapidly growing host cells provided an ideal tool for gaining an insight into the mechanisms of biologic self-replication. Their expectation was amply borne out. The bacteriophage-bacterium system proved to be highly amenable to experimentation, and intensive investigation of it over the last three decades has provided many of the fundamental concepts concerning molecular genetics, nucleic acid replication, and the transcription and translation of genetic information. These concepts are applicable not only to bacterial viruses but also to animal viruses and to cells in general. Thus, although bacteriophages were a failure as therapeutic agents, they have proved invaluable for the elucidation of the reactions most basic to life.

The purpose of this chapter is to review the mechanisms involved in the multiplication of several classes of bacteriophages, emphasizing those that already have been shown to operate also during animal virus multiplication, or may in the future be shown to do so.

The structure of bacteriophages

There are bacteriophages for every bacterial species. Very few have been investigated in detail, but intense concentration on a small number of them has permitted rapid progress. Many of the well-studied phages are active on *Escherichia coli* or on *Bacillus* or *Pseudomonas* species (Table 10-1).

THE STRUCTURE OF PHAGE CAPSIDS

The structure of bacteriophages is governed by the same principles as are described in Chapter 2. Some of the smaller phages, such as ϕX174 and MS2, have icosahedral capsids (Fig. 10-1); others, such as phage f1, are filamentous and possess helical symmetry (Fig. 10-2). Larger phages generally consist of a head that comprises the phage genome enclosed within a single capsid shell that is usually hexagonal and may or may not be elongated, and a tail that serves both as the cell attachment organ and as a tube through which phage DNA passes into the host cell. The complexity of this tail varies

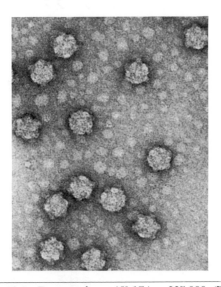

FIG. 10-1. Bacteriophage ϕX 174. $\times$225,000. (Electron micrograph by Dr. R. B. Luftig.)

TABLE 10-1 CHARACTERISTICS OF SOME WELL-STUDIED BACTERIOPHAGES

Phage	Host	Particle Dimensions (nm)		Structure	Nucleic Acid			
		Head	Tail Length		Type	Strandedness	M.W. $\times 10^6$	Structure
T1	E. coli	50	150	Hexagonal head, simple tail	DNA	DS	27	Linear
T2,T4,T6	E. coli	80 × 110	110	Prolate icosahedral head, complex tail with fibers	DNA	DS	130	Linear; circularly permuted and terminally redundant; contains glucosylated 5-hydroxymethylcytosine
T3,T7	E. coli	60	15	Hexagonal head, short tail	DNA	DS	25	Linear
T5	E. coli	65	170	Hexagonal head, simple tail	DNA	DS	76	Linear, one strand segmented
Lambda	E. coli	54	140	Hexagonal head, simple tail	DNA	DS	31	Linear, cohesive ends
P22	S. typhimurium	55	20	Hexagonal head, complex tail	DNA	DS	27	Linear, circularly permuted, terminally redundant
SPO1	B. subtilis	100	210	Hexagonal head, complex tail	DNA	DS	100	Linear, contains 5-hydroxymethyluracil
φ29	B. amyloliquefaciens	30 × 40	30	Prolate icosahedral head with attached fibers, complex tail with collars and neck appendges	DNA	DS	11	Linear
PM2	Pseudomonas BAL-31	60	None	Hexagonal head, envelope contains lipid	DNA	DS	6	Circular
φX174,S13, M12,G4	E. coli	27	None	Icosahedral	DNA	SS	1.7	Circular
f1,fd,M13	E. coli	5–10 × 800	None	Filamentous	DNA	SS	1.3	Circular
φ6	Pseudomonas phaseolica	65	None	Polyhedral head, envelope contains lipid	RNA	DS	9.5	3 linear segments (2.2, 2.8 and 4.5 × 10^6)
MS2,f2,fr,Qβ	E. coli	24	None	Icosahedral	RNA	SS	1.2	Linear

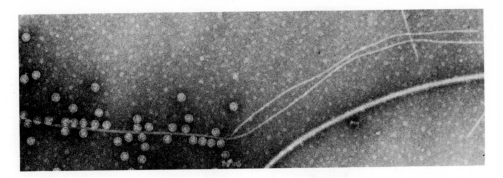

FIG. 10-2. An F-pilus with two types of male-specific phage attached: icosahedral RNA-containing MS2, and filamentous DNA-containing fl. The former are attached along the entire length of the F-pilus; the latter are adsorbed by their ends to the tip of the pilus. (From Caro and Schnos: Proc Natl Acad Sci USA 56:128, 1968.)

greatly from phage to phage, but in most phages it conforms to one of three morphologic patterns. In the first, exemplified by coliphages T3 and T7, the tail is very short (Fig. 10-3). In the second, exemplified by coliphages T1 and λ, the tail is long but rather simple in construction; it consists essentially of a noncontractile flexible tube that may or may not possess either a knob or one or several spikes or fibers at its distal end (Fig. 10-4). The third morphologic tail pattern, exhibited by the coliphages T2, T4, and T6, the so-called T-even phages, is almost staggeringly intricate (Fig. 10-5). These tails

consist of a hollow core which is attached at one end to the head and bears at its distal end a hexagonal base plate to which are attached six pins and six long tail fibers bent in the middle. A thin collar with several (possibly six) whiskers is attached to the core close to the head and it is surrounded for most of its length by a sheath composed of 24 rings of helically arranged capsomers. These tails, which consist of more than 20 different protein species, serve as syringes by means of which phage DNA is injected into the cell (p. 202).

Very few phages are enveloped. Among

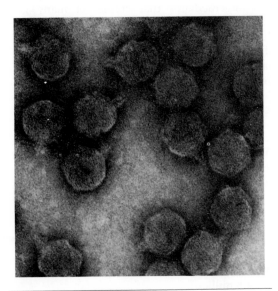

FIG. 10-3. Bacteriophage T7. Magnification: ×225,000. (Electron micrograph by Dr. R. B. Luftig.)

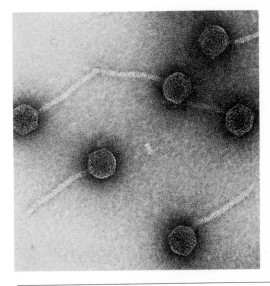

FIG. 10-4. Bacteriophage λ. ×135,000. (Electron micrograph by Dr. R. B. Luftig.)

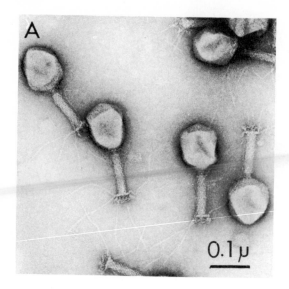

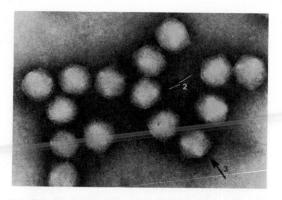

FIG. 10-6. The enveloped phage PM2 fixed with glutaraldehyde and negatively stained with phosphotungstate. The numbered arrows indicate axes of 2-fold and 3-fold symmetry. ×120,000. (From Silbert, Salditt, and Franklin: Virol 39:666, 1969.)

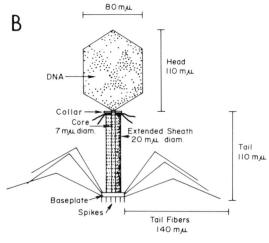

FIG. 10-5. **A.** Bacteriophage T4. ×100,000. **B.** Model of phage T4. (Electron micrograph by Dr. R. B. Luftig.)

those that are is the *Pseudomonas* phage PM2 (Fig. 10-6).

THE STRUCTURE OF PHAGE NUCLEIC ACIDS

Bacteriophage genomes take several forms (Table 10-1). Most of them consist of single, linear, double-stranded DNA molecules that vary in molecular weight from about 1 million to more than 100 million. Some of them, in particular

those of the T-even phages, exist in a form which is both circularly permuted and terminally redundant. In them the nucleotide sequence of some molecules is A, B, C, D, . . . W, X, Y, Z, A, B; of others B, C, D, . . . W, X, Y, Z, A, B, C; of others still C, D, . . . W, X, Y, Z, A, B, C, D; and so on. In other words, all molecules contain the same nucleotides, but the terminal nucleotide sequences differ in individual genomes and a few nucleotides are reiterated at the ends of the molecules. This is due to the fact that their replication proceeds via replicative intermediates that consist of many genomes linked covalently end-to-end which are then cut at random so as to yield molecules of identical size (p. 211). The reason for the terminal redundancy is to avoid the loss of the gene that would otherwise be cut at each end.

Other phage genomes, such as that of PM2, consist of circular double-stranded DNA molecules; others, such as those of phages φX174, G4 and fd, consist of single-stranded circular DNA molecules with plus polarity; still others are linear single-stranded nonpermuted RNA molecules with plus polarity, analogous to poliovirus RNA. There is also a phage of *Pseudomonas phaseolica*, φ6, that contains double-stranded RNA. Like the double-stranded RNA-containing genomes of animal viruses, φ6 RNA is segmented; it consists of three unique segments, whose molecular weights are 2.2, 2.9, and 4.8 million.

Phage nucleic acids are infectious, just as are the nucleic acids of animal viruses. The bacte-

TABLE 10-2 EXTENT OF GLUCOSYLATION OF 5-HYDROXYMETHYLCYTOSINE
IN THE DNA OF T-EVEN PHAGES*

Phage	Not Glucosylated	α-Glucosyl	β-Glucosyl	β-1:6-Glucosyl-α-Glucosyl
T2	25	70	0	5
T4	0	70	30	0
T6	25	3	0	72

From Lehman: J Biol Chem 235:3254, 1960.
* Figures represent percentages.

rial cell wall, however, presents a more formidable barrier to the entry of nucleic acids than is presented by the plasma membrane of animal cells, and infection by naked phage nucleic acids, termed transfection, can only be demonstrated under special circumstances. Transfection of whole bacterial cells is rare; cells of *B. subtilis* that are competent for transformation can be transfected, and so can cells of *E. coli* treated with Ca^{2+} (only with λ DNA). The more readily transfectible bacterial forms are spheroplasts and frozen-thawed cells, which can be transfected not only with the small single-stranded phage RNAs and DNAs but also with double-stranded phage DNAs as large as those of the T-even phages.

Unusual Nucleic Acid Bases Whereas the genomes of animal viruses contain only the normal nucleic acid bases adenine, guanine, cytosine, and uracil or thymine, phage genomes sometimes contain unusual or substituted bases. For example, the DNA of the T-even bacteriophages contains not cytosine but 5-hydroxymethylcytosine; the DNA of certain *B. subtilis* phages contain not thymine but uracil, 5-hydroxymethyluracil, or 5-(4',5'-dihydroxypentyl) uracil, and in the DNA of a *Pseudomonas acidovorus* phage, thymine is partially replaced by 5-(4'-aminobutylaminomethyl) uracil (also known as α-putrescinylthymine).

Glucosylation and Methylation The bases of bacteriophage DNA are often glucosylated and/or methylated in highly characteristic patterns. Both 5-hydroxymethyluracil and particularly 5-hydroxymethylcytosine are frequently linked by phage-coded enzymes to either one or two glucose residues via α or β bonds. The situation concerning the T-even phages is summarized in Table 10-2. In addition, a small proportion of the bases in phage DNAs is generally methylated by host-specified enzymes. Both types of substitution confer resistance to degradation by nucleases. Bacteria guard themselves against foreign DNAs by degrading them, and in order to prevent their own DNA from being destroyed, they cause methyl groups to be attached to their own DNA in specific locations, thereby rendering it resistant to their own restriction endonucleases. Since phage nucleic acids also become methylated, highly specific restriction patterns, known as "host-induced modifications," exist among bacteriophages. For example, the DNA of a phage grown in host A will be methylated by that host's enzymes in such a manner that it will be resistant to that host's restriction endonucleases, but most of the DNA of this phage will be degraded by the restriction endonucleases of many other bacteria. However, any DNA molecules that escape degradation in such hosts can replicate, and their progeny is then modified by the methylating enzymes of the new hosts, acquiring new specificities in the process.

The physiology of lytic phage infection

Like animal viruses, some bacteriophages lyse their hosts, while others integrate their genomes into the host genome. The former are known as virulent phages, the latter as temperate or lysogenic phages. We will consider first the lytic phage one-step multiplication cycle which is formally analogous to that of animal viruses (Chap. 5) but proceeds much more rapidly. For example, whereas the length of the latent period (the interval between infection and release of progeny virus) is at least 4 hours for poliovirus and 12 hours for vaccinia virus, the latent period is 13 minutes for phages T1, T3, T7, and ϕX174, and 21 to 25 minutes for the T-even phages and MS2.

ADSORPTION

The initial step of the phage multiplication cycle illustrates with particular clarity the existence of virus-specific receptors on host cells. Phages are usually highly specific for a limited number of bacterial host strains. The basis of this specificity is the complementarity of molecular configurations on phage attachment organs and of receptor molecules on the bacterial surface. The nature of the phage attachment organ varies from phage to phage: small icosahedral phages, such as ϕX174 and MS2, have multiple attachment sites; filamentous phages, such as f1, adsorb with their ends (Fig. 10-2); tailed phages adsorb with the knobs, spikes, or fibers that are located at the tips of their tails. Phage receptors on the bacterial surface are sometimes on lipopolysaccharide, at other times on lipoprotein, and sometimes on F pili, which causes the phages that adsorb to them, typically-single-stranded DNA-containing and RNA-containing phages, to be male-specific (Fig. 10-2).

Mutations in bacteria that destroy the complementarity of the phage-receptor interaction result in resistance. Phage populations usually contain mutants that are themselves altered in such a way as to restore the necessary complementarity. These are known as "host-range mutants" and can grow in the mutated bacteria.

Since phage receptors are surface components, bacteria capable of adsorbing the same phage are often antigenically related. This observation has found diagnostic application in the practice of phage typing. In this method the sensitivity/resistance patterns of bacterial strains to a series of bacteriophages are determined. Since these patterns are both readily determined and highly characteristic, they are useful tests for identification. For example, some phages of *Salmonella typhi* are specific for strains possessing the Vi antigen, which is characteristic of virulent strains, and these phages are used for the identification of viru-lent strains of typhoid bacilli. Several other phage-typing systems exist, the best known being that for staphylococci.

INJECTION AND UNCOATING

The events immediately following phage adsorption are involved with the injection of phage nucleic acid into the cell; these events provide an excellent illustration of the general principle that uncoating of viral genomes involves the physical separation of genome and capsid. This was first shown in 1952 by Hershey and Chase in an experiment that represents one of the milestones of virology. They infected bacteria with phage T2 labeled in the protein with the radioisotope^{35}S. Following an incubation period of several minutes, the mixture of phage and bacteria was sheared by blending in a Waring blender; after blending, only a small amount of radioactive label was associated with the bacteria. When this experiment was repeated with phage in which the DNA has been labeled with the radioisotope^{32}P, the converse was true; that is, the radioactive label was now associated with the cells. This experiment was correctly interpreted as signifying that the DNA had passed into the interior of the cell, while the protein coat had remained attached to the outer cell surface, from which it could be dislodged by shearing forces. Since the bacteria from which phage coats were removed by shearing yielded normal phage progeny, this experiment also clearly showed that viral DNA itself contained all the information necessary for phage multiplication.

The actual infection process has been best studied for the T-even phages. Following adsorption by means of the tail fiber-receptor interaction, the six tail pins make contact with the host surface and firmly anchor the phage tail plate to it. The conformation of the tail plate (an extremely complex structure consisting of at least 14 different polypeptide species) then changes. The cause of the change is apparently the interaction of one of the tail plate components, the enzyme dihydrofolate reductase, with pyridine nucleotides (such as NADPH) that temporarily leak from cells in response to infection (Fig. 10-7). The tail plate conformation change in turn triggers a change in the manner in which the protein subunits of the tail sheath are arranged, as a result of

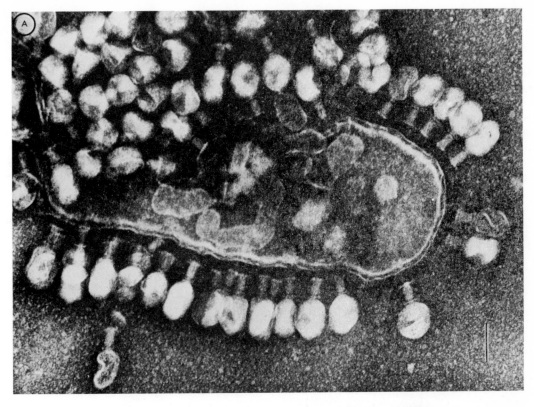

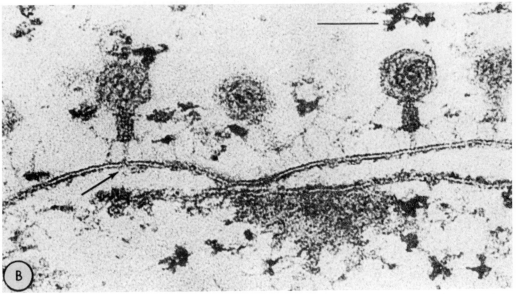

FIG. 10-7. Bacteriophage T4 adsorbed to *Escherichia coli.* A. The phages' sheaths are con-tracted, and their base plates are 30-40 nm from the cell wall. B. The phages visible in this section are seen to be attached to the cell wall by the tail fibers; the cell wall appears as two continuous electron-dense lines separated by an electron-lucent space. The arrow indicates where the tail core of an adsorbed phage may have penetrated the cell wall. (From Simon and Anderson: Virol 32:279, 1967.)

which the sheath contracts, thereby driving the tail core about 12 nm into the bacterial cell (Fig. 10-7). The phage DNA then passes through the tail core into the bacterial cell. The role of the host cell in DNA injection is obscure; on the one hand, it is known that if T2 phages are mixed with cell wall fragments, the DNA is ejected into the medium, suggesting a passive role for the host (Fig. 10-8), but, on the other hand, there is evidence that host protein synthesis is necessary for the transfer of the entire genome from the phage head into the cell.

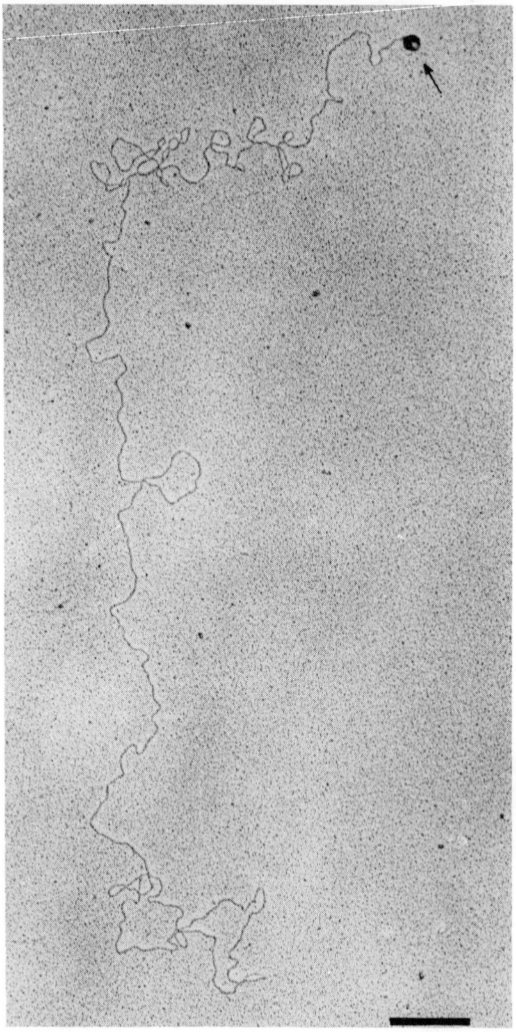

FIG. 10-8. A T7 phage particle ejecting its DNA. Almost the entire DNA molecule (12nm long) has been ejected. The DNA was caused to eject by treatment with formamide. The bar is 0.3 nm long. (Courtesy of Dr. Kaoru Saigo.)

Most probably, cooperation between virus and host is required. The holes created in the cell surface by the phage tail cores are normally quickly sealed by cell wall material newly synthesized under the control of a phage gene called "spackle." Clearly, it would not be to the advantage of the phage to permit these holes to remain, since cell contents would leak out through them. In fact, if the multiplicity of infection is too high, excessive leakage does occur, resulting in a phenomenon known as "lysis from without," and the multiplication cycle is aborted.

As pointed out previously, most phages do not possess tails with contractile sheaths, and still others possess no tails at all. Nevertheless, here also the phage protein coat remains on the outside, and only the nucleic acid is introduced into the cell.

The multiplication cycle of phages containing double-stranded DNA

THE NATURE OF THE INFORMATION ENCODED IN THE PHAGE GENOME

During the phage multiplication cycle the genetic information encoded in the viral genome expresses itself. The nature of this information is much more completely defined for several phage systems than for most animal viruses. The principal reason for this is the ease with which phage mutants deficient or recognizably altered in a variety of genes can be selected, characterized, and mapped. Most useful for this purpose are the conditional lethal mutants— mutants that can function under one set of conditions but not under others. The nature of one class of such mutants, the temperature-sensitive mutants, has already been described (Chap. 7). Another class of conditional lethal mutants are the nonsense mutants, such as the amber, ochre and opal mutants. In these mutants, codons coding for an amino acid are mutated to termination codons, such as UAG, the amber termination codon. Only protein fragments, which are rarely functional, can be specified by genes containing such codons. However, bacterial mutants exist in which transfer RNA molecules are mutated in such a way that they recognize UAG not as chain-terminating but as specifying some amino acid. In such cells the mutated genes can be translated into complete and often at least partially functional protein molecules, and as a result the amber mutants can multiply. Since they suppress the consequences of nonsense mutations, such mutant bacterial strains are known as "suppressor strains." A third type of mutant that has found wide application in the analysis of the phage multiplication cycle are deletion mutants, which simply lack portions of phage DNA.

Thousands of phage mutants have by now been obtained and characterized with respect to the function in which they are defective. For example, there are mutants that fail to synthesize early proteins, mutants that fail to replicate phage DNA, mutants that fail to form mature phage particles, and so on. Over 100 such functions, each corresponding to a specific gene, have so far been discovered for bacteriophage T4, which is probably about one-half of the total number of T4 genes; and for smaller, less complicated phages, this proportion is considerably higher. Most of the known T4 genes have been mapped; all lie on a circular linkage map, which agrees with the finding that although T4 DNA is linear, it is circularly permuted (Fig. 10-9). Similar maps have been prepared for several other phages, notably T7, λ, and the small phages that contain single-stranded nucleic acid.

The proteins specified by the genes of the large complex phages, such as the T-even phages, have a wide variety of functions. Most of them can be grouped into eight classes:

1. Proteins that repair the bacterial cell membrane early in infection
2. Enzymes that degrade host DNA
3. Enzymes that synthesize nucleic acid precursors
4. Proteins that function in phage DNA replication
5. Proteins that program transcription of the phage genome

6. Proteins that are structural phage components
7. Proteins that function catalytically in phage morphogenesis
8. Enzymes that degrade the bacterial cell wall and cell membrane during the late stages of the multiplication cycle.

In addition, the T-even phages code for eight species of transfer RNA, the genes for which are clustered in a small region of the phage genome where they may form a single transcription unit or operon. Mutants lacking these genes can multiply, but their burst size, or yield, is smaller than that of wild-type phage. This suggests that the phage-coded transfer RNAs supplement the reading capacity of those codons that are used more commonly by the virus than by the host, thereby ensuring optimal rates of protein synthesis.

We will now consider each of the eight categories of proteins and examine their role in phage multiplication.

THE FUNCTIONS OF VARIOUS CATEGORIES OF PHAGE GENE PRODUCTS

Proteins that Repair the Bacterial Cell Membrane Early in Infection

Infection causes the bacterial cell membrane to become leaky, so that cellular contents begin to escape; but within several minutes sealing occurs and the preinfection integrity of the membrane is restored. The precise mechanism by which this is achieved is not clear, but it is known that at least five phage-coded proteins are synthesized early during the infection cycle and incorporated into the cell membrane.

Enzymes that Degrade Host DNA

Among the earliest effects of phage T4 infection is the breakdown of host DNA (Fig. 10-10), which has several consequences. First, it provides a source of nucleic acid precursors, which may be limiting under certain conditions. More important, transcription of host-specified mes-

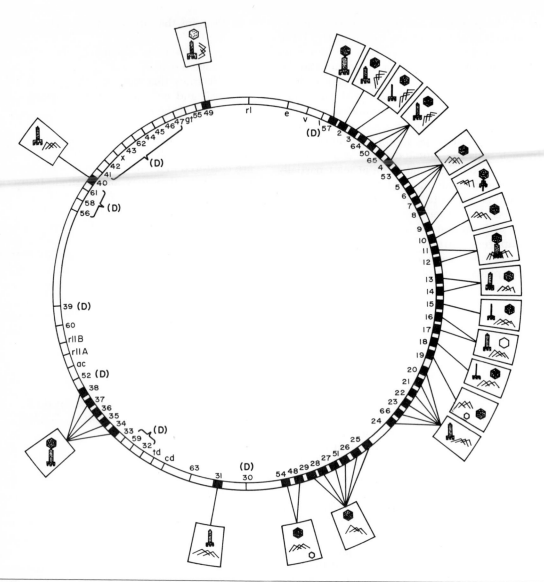

FIG. 10-9. The genetic map of T4. The solid segments indicate late genes with morphogenetic functions, while the short bars represent early genes with enzymatic or regulatory functions; genes with functions in DNA replication are identified by "D." The boxes indicate viral components such as heads, tails and tail fibers that are present in extracts of cells infected with conditional lethal mutants under restrictive conditions. (Adapted from Edgar and Wood: Proc Natl Acad Sci USA 55:498, 1966.)

senger RNA ceases abruptly; since most bacterial messenger RNAs are very short-lived, this not only leads to the sudden cessation of host-specified protein synthesis but also quickly provides ribosomes for the translation of phage-specified messenger RNA. The following is a graphic example of the consequences of the cessation of host protein synthesis. If a culture of *E. coli* grown in glucose-containing medium and infected with T4 is transferred to medium containing lactose, phage infection is aborted. The reason is that lactose utilization requires the induction of the synthesis of certain proteins, among them a permease and the enzyme β-galactosidase. T4 infection halts host protein synthesis. Therefore, lactose cannot be metabo-

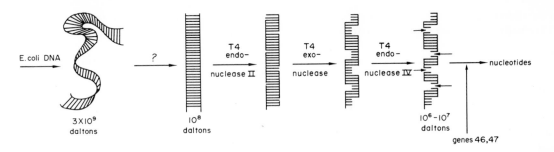

FIG. 10-10. A tentative scheme to account for the degradation of *E. coli* DNA after infection with phage T4. *E. coli* DNA is first degraded to fragments about one-tenth its size, which are then nicked by endonuclease II. The nicks are then widened by an exonuclease and the single-stranded stretches so produced are then cut by endonuclease IV. The products are duplex DNA molecules which are more than 100 times smaller than the original *E. coli* genome. Since fragments of this size accumulate in cells infected with phage mutants carrying lesions in genes 46 and 47 but not in cells infected with wild-type phage, these genes presumably code for enzymes that function in their further degradation. (Adapted from Warner et al: J Virol 5:700, 1970.)

lized, energy production is prevented, and no phage progeny can be formed.

Enzymes that Synthesize Nucleic Acid Precursors

Numerous enzymes function in the synthesis of nucleic acid precursors, particularly when the phage DNA contains both unusual bases and substituted bases. For example, in the case of T4, which contains glycosylated 5-hydroxymethylcytosine instead of cytosine, the following enzymes are required: (1) enzymes that destroy deoxycytidine triphosphate (so that cytosine is not incorporated into phage DNA), (2) an enzyme that hydroxymethylates deoxycytidylic acid, and (3) enzymes that glucosylate 5-hydroxymethylcytosine residues in polymerized DNA. In addition, numerous enzymes, among them enzymes that function in the synthesis of the other three deoxyribonucleoside triphosphates, are specified by the T-even genome. They are not absolutely essential but increase the yield of progeny phage.

Proteins that Function in Phage DNA Replication

The replication mechanism of double-stranded phage DNA is exceedingly complex, and although enormous effort has been devoted to its

elucidation, no clear picture has yet emerged. The following facts are pertinent.

1. Phage DNA replicates by a semiconservative mechanism.
2. During the early stages of the replication of some phage DNAs, extensive recombination occurs. This results in the dispersal of the parental genome among numerous progeny genomes.
3. Phage DNA most probably replicates while associated with membrane in some as yet unspecified manner.
4. The replication of some double-stranded DNA genomes, notably those of the T-even phages, proceeds via the formation of giant intermediates that consist of up to 100 repeats of the mature phage DNA genome.
5. Progeny phage genomes are cut from this intermediate in "headful" units that consist of circularly permuted and terminally redundant genomes (see above).

Among the enzymes known to function in the replication of T4 DNA are a DNA polymerase (coded by gene 43), a polynucleotide ligase (coded by gene 30) and, most probably, several nicking enzymes (nucleases). A protein that facilitates unwinding of DNA may well participate also.

Proteins that Function in Transcription Programming

Phage genomes, like all other genomes, express themselves via messenger RNA. In fact, it is worth noting that messenger RNA was discovered in phage-infected cells. In 1956 Volkin and Astrachan noted that the base composition of newly synthesized RNA in T2-infected *E. coli* resembled that of T2 DNA rather than that of host DNA; in 1959 Brenner, Jacob, and Meselson firmly established the concept of messenger RNA by showing that this RNA combines with ribosomes and is responsible for the synthesis of phage proteins.

Phage-specified proteins, like those specified by animal virus genomes (Chap. 5), are synthesized in a strictly ordered sequence. The necessity for this is obvious; for example, enzymes breaking down host DNA should be synthesized first, and viral structural proteins should be synthesized last. There are two ways of achieving programmed protein synthesis: (1) either the entire viral genome is transcribed continuously and translation is programmed, or (2) transcription is programmed, and messenger RNAs are translated as they are formed. Although the former mechanism may apply to a few species of messenger RNA, it is primarily transcription that is programmed in phage-infected and animal virus-infected cells. This is shown most readily by hybridization experiments of the following type. At various times after infection, cells are incubated for brief periods, say for two minutes, with RNA precursors labeled with some radioisotope. On extraction such cells yield labeled messenger RNA populations synthesized from 0 to 2, 2 to 4, 4 to 6, and so on, minutes after infection. These messenger RNA populations are then allowed to hybridize with denatured phage DNA in the presence of excess unlabeled messenger RNA extracted from cells at various stages of infection. If the labeled and unlabeled RNA populations are identical, the large excess of unlabeled RNA will, by simple competition, prevent the labeled RNA from hybridizing; if the two populations are different, the presence of unlabeled RNA will have no effect on the ability of labeled RNA to hybridize to DNA.

Another type of analysis that can be carried out is to hybridize the messenger RNA molecules that are formed at various times after infection to the series of fragments that are generated when phage DNA is digested with restriction endonucleases. As described in Chap. 5, such fragments can be ordered to provide restriction endonuclease fragment maps. With this technique it can readily be demonstrated that different regions of the phage genome are transcribed at different stages of the multiplication cycle. The situation here is quite analogous to that described in Chap. 5 concerning the sequential expression of viral genetic information during the multiplication of double-stranded DNA-containing animal viruses.

How are such complex transcription programs managed? The complete answer is not yet evident, but it seems that one of the factors of critical importance is the specificity of the transcribing enzymes.

Since phages do not contain RNA polymerases, the first phage genes are always transcribed by the host RNA polymerase. The *E. coli* RNA polymerase is a complex enzyme that consists of four subunits—two α subunits, one β subunit, and one β' subunit—and has associated with it one protein molecule, the σ factor, which controls its specificity. The portion of the phage genome that is transcribed by this enzyme is usually not large and is specified by the presence of initiation and termination signals that it can recognize. The messenger RNAs transcribed by it are known as early or immediate early (in the case of very complex programs).

The remainder of the transcription program is managed by proteins specified by early phage messenger RNAs. There are two principal means by which this is accomplished. The first, employed by the simpler phages such as T7, involves the synthesis of a new RNA polymerase; this new enzyme consists of only a single polypeptide chain, and its specificity is such that it transcribes its homologous DNA more efficiently than any other. The second means, employed by the more complex phages such as T4 and SPO1, involves the synthesis of a series of polypeptides that modify the host RNA polymerase in a series of successive steps. Among these modifications are adenylation of the α subunits, proteolytic cleavage of small portions from the β and β' subunits, and the replacement of the σ factor by phage-coded proteins. If the ability to recognize and react with promoters and transcription termination signals of each of these modified RNA polymerase species is different, they could readily effect transcription programs such as those described above.

Structural Phage Components

The messenger RNAs that code for the proteins described so far are generally transcribed only from parental DNA. Those that code for the proteins to be described now are usually transcribed only from progeny DNA. These proteins are known as the "late proteins," and the quantitatively most important late proteins are the structural components of progeny phage particles (compare Chap. 5). Tailed phage particles are complex structures composed of a relatively large number of protein species; for example, T4 contains at least 30 protein species, and even simple phages like T7 and λ contain 10 to 15 different species.

Proteins that Function Catalytically in Morphogenesis

Phage genomes usually code for several proteins that function in a catalytic capacity during morphogenesis. For example, in addition to the 30-odd structural protein species, T4 DNA codes for at least 17 additional proteins that are also essential for the formation of mature phage particles. Their nature and the precise manner in which they function is largely unknown (but see below for the scaffolding protein). The genes for tail proteins are clustered into three sets in the T4 genome (Fig. 10-9): one for proteins that are needed in very few copies, one for proteins needed in intermediate numbers of copies, and one for proteins needed in very many copies per cell. The genetic apparatus for the formation of phage tails is thus very highly refined. The situation seems to be similar for several other phages.

Enzymes Necessary for Progeny Phage Liberation: Cell Lysis

Most virulent phages are liberated from their hosts by a mechanism that differs fundamentally from any employed by animal viruses (Chap. 5); the bacterial cell lyses suddenly, thereby liberating the entire progeny. This is known as cell lysis and, in the case of the T-even phages, is the result of the function of two late phage-coded proteins. One is a lipase coded by the *t* gene which attacks the cell membrane, thereby halting metabolic processes; the other is the enzyme lysozyme, coded by the *e* gene, which then hydrolyzes the cell wall.

Phage Morphogenesis

Phage morphogenesis has been studied most intensively for the T-even coliphages, in particular T4, and for the lysogenic coliphage λ. The most successful approach has been to study the products of infection of a large number of conditional lethal mutants, both temperature-sensitive and amber, under restrictive conditions. Cells infected with structural or catalytic gene mutants under restrictive conditions accumulate not mature phage particles but phage components, such as partially completed or complete heads, tail fibers, tail cores, tail sheaths, partially assembled tails, and so on. As a result, it has been possible to link defects in both structural and catalytic proteins to specific genes whose location on the genetic map of the phage DNA can then be identified.

Phage morphogenesis often involves the processing of capsid proteins. Two kinds of processing have been observed: protein fusion (during λ head assembly) and, more commonly, proteolytic cleavage (during T4 and P2 head assembly, as well as during T5 and λ tail assembly). Possible reasons why viral capsid proteins are often synthesized in the form of precursors have been discussed in Chap. 5.

As for the study of the morphogenetic pathway itself, one of the most useful approaches has been to examine the ability of extracts of cells infected with phage mutants to complement each other. In other words, pairs of such extracts are mixed, and one determines whether infectious phage particles are formed from the components present in each. By such means, the scheme for T4 tail assembly shown in Figure 10-11 has been worked out.

As for the assembly of phage heads and the insertion of DNA into them, current evidence indicates that these processes are achieved with the aid of a catalytic head assembly or scaffolding protein. Heads of phage P22 (a *Salmonella typhimurium* phage that resembles phage λ in many respects) do not self-assemble. Instead,

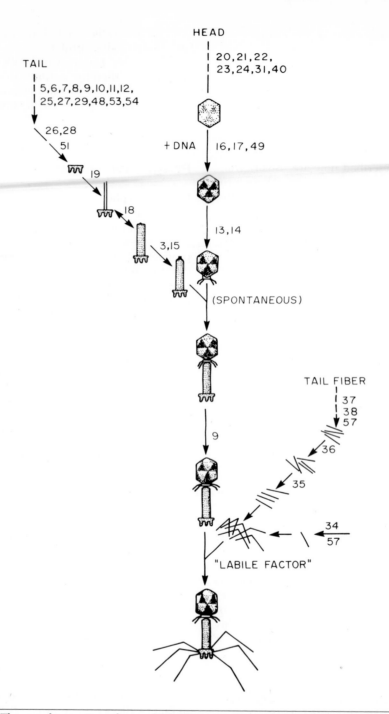

FIG. 10-11. The morphogenetic pathway of T4. There are three principal branches leading independently to the formation of heads, tails and tail fibers, which then combine to form complete virus particles. The numbers refer to the gene product(s) involved in each step. (Adapted from Wood and Edgar: Sci Am 217:74, 1967. Courtesy of Freeman and Co.)

TABLE 10-3   GENES AND GENE FUNCTIONS OF SINGLE-STRANDED DNA-CONTAINING BACTERIOPHAGES LIKE ϕX174

Gene Designation	Molecular weight of the protein product	Function
A	60,000	RF replication, ss DNA synthesis
B	20,000	Morphogenesis, SS DNA synthesis
C	12,000	SS DNA synthesis
D	14,000	SS DNA synthesis, morphogenesis (?)
E	(10,000?)	Lysis
F	50,000	Major capsid protein, SS RNA synthesis
G	20,000	Spike (vertex) protein, SS DNA synthesis
H	37,000	Spike (vertex) protein

Adapted from Denhardt DT: In Fraenkel-Conrat H, Wagner RR (eds): Comprehensive Virology, Vol 7, p 1. Plenum, 1977.

200 to 300 molecules of a scaffolding protein catalyze the polymerization of the coat protein into a head precursor shell that contains both proteins. The scaffolding protein then exits and is replaced by DNA; possibly the exit of the scaffolding protein exposes charged sites that serve to collapse the DNA. The actual amount of DNA that is inserted into the phage head is determined by the so-called headful cutting mechanism; that is, the DNA, in the form of large replicative intermediates, is inserted until the head is full, and the DNA is then cut.

The multiplication cycle of phages containing single-stranded DNA

Several *E. coli* phages contain not double-stranded but single-stranded DNA (Table 10-1). The most completely studied of these phages is ϕX174, whose genome is a plus-stranded single-stranded DNA molecule 5370 nucleotides long (M.W. 1.7×10^6), which has been sequenced. It contains 8 genes which have been mapped (Table 10-3). Three of the proteins specified by ϕX174 DNA are structural proteins, five are involved in DNA synthesis, and one causes cell lysis.

The principal problems posed by the discovery of these phages were (1) how does single-stranded DNA replicate and still obey the rules

of base pairing, and (2) since messenger RNAs are transcribed from double-stranded DNA, how can single-stranded DNA serve as their template? These questions were answered when it was found that upon entering the host cell, single-stranded DNA is converted to a double-stranded replicative form.

A scheme describing the mode of replication of ϕX174 DNA is presented in Figure 10-12. Although several of the steps depicted may be greatly oversimplified, most of the scheme's key elements have been demonstrated experimentally. The parental DNA (V) first attaches to a host cell membrane component (M), and a complementary strand is synthesized by bacterial enzymes (presumably a polymerase and a ligase) so as to yield a supercoiled circular double-stranded replicative form known as RF1, which is analogous to papovavirus DNA (Fig. 2-25). This is then nicked by an endonuclease to yield RF2, a relaxed circular molecule (step 2). In step 3 the V strand is elongated, using the C strand as template, and a new C strand is synthesized with the elongated portion of the V strand serving as its template. This process is a modification of the rolling circle mode of DNA replication. In step 4 this intermediate is split into a membrane-attached RF1 identical to the original one and a free RF1. The original one then repeats steps 2 to 4, forming new RFs, while the free RF1 has two options. If membrane sites are available, it can attach to them (step 5) and then also generate progeny RFs, or it may be converted to RF2 (step 6), on which the V strand is again elongated (step 7). This intermediate is then split to yield another RF2 (possibly via RF1) and a V strand, which is encapsidated (step 8). The net result is that double-stranded RF molecules first replicate and then serve as templates for the synthesis of

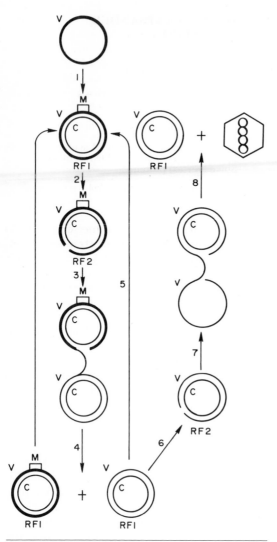

FIG. 10-12. A scheme to account for the replication of φX 174 DNA. V represents the DNA strand present in phage; C represents the complementary strand. The heavy V strand is that in the parental particle.

Messenger RNAs are transcribed from membrane-bound RFs. Among the proteins into which they are translated are the structural phage components, one species of which attaches to the loose end of progeny V strands arising in step 7, thereby initiating encapsidation.

The multiplication cycle of the filamentous phages f1, fd, and M13 differs somewhat from that of φX174 DNA. Whereas nascent φX174 DNA is immediately encapsidated, forming progeny that accumulates within the cell and is eventually liberated when the cells lyse, large amounts of free viral DNA accumulate intracellularly in the case of the filamentous phages. These molecules are then encapsidated at the cell membrane, and virus particles are released without cell lysis in a process that is in part analogous to that by which enveloped animal viruses are released from their host cells.

The multiplication cycle of phages containing single-stranded RNA

Single-stranded RNA-containing phages are small icosahedral viruses that resemble picornaviruses. Like them, their capsids consist of 180 protein molecules, but whereas there are 60 molecules each of three different protein species present in picornaviruses, all 180 protein molecules are identical in the case of the RNA phages. In addition, each phage particle contains one molecule of an additional protein, the maturation or A protein, which is necessary both for proper assembly and for ability to adsorb. The RNA of these phages has a molecular weight of just over 1 million, which is less than one-half the size of the picornavirus genome.

Most of the known RNA phages are coliphages. They fall into three or four serologic groups. Most of the well-studied RNA phages (f2, fr, MS2, R17, and so on) belong to one group, while one, Qβ, belongs to another. All RNA phages are male-specific because their receptor sites are on F pili.

large numbers of single-stranded progeny DNA molecules. It is worth noting that infectious φX174 DNA has been synthesized in vitro with purified enzymes. It was the first functioning DNA genome to be synthesized outside a living cell.

The enzymology of all these steps is very complex. Even step 1, the synthesis of a C strand on a template V strand, requires at least eight host and phage enzymes and protein factors.

The study of the multiplication cycle of these phages has provided important clues for investigations of animal viruses that contain RNA. In addition, their extraordinary simplicity—they are the simplest viruses known—has made them fascinating objects for the elucidation of fundamental biologic processes.

The genomes of RNA phages comprise only three genes. These are, from the 5'-terminus of the RNA, the genes for the A protein, the coat protein, and the RNA polymerase. The latter is actually only one of four polypeptides that together make up the functional RNA polymerase, the others being three bacterial polypeptides that are normally located on ribosomes, namely, factor i, elongation factor Tu, and elongation factor Ts. The precise function of these host-coded polypeptides as components of the viral RNA polymerase is not known, but it has been suggested that they somehow control specificity and affinity for initiation sites and that they are essential for maintaining the enzyme complex in its active conformation.

The genome of RNA phages possesses plus polarity and starts functioning as a messenger very soon after it enters the cell. After coding for the RNA polymerase, it replicates by a mechanism similar to that described for poliovirus (Fig. 5-15). This involves, first, the formation of complementary minus strands and then their successive transcription into progeny plus strands. Coat protein then begins to be synthesized in large amounts together with some A protein; as soon as their concentrations are sufficiently high, encapsidation of RNA commences and mature progeny phage accumulates (Fig. 10-13). These events are quite analogous to those that occur during the picornavirus multiplication cycle (Chap. 5).

The Programming of the RNA Phage Multiplication Cycle

The multiplication cycle of RNA phages, like that of other viruses, is programmed, with the various reactions necessary for morphopoiesis taking place in a regularly progressing sequence. Yet, the genome of RNA phage is too small to code for proteins with a purely regulatory function. How then does it happen that coat protein, which is present in virions in 180-fold excess over A protein, is in fact synthesized in vastly greater amounts than A protein? How

is it arranged that the RNA polymerase polypeptide is synthesized early during the infection cycle, when it is required, and not throughout the cycle? And since parental RNA genomes must be both translated and transcribed, what mechanism prevents collisions between ribosomes, which traverse messenger RNA from the 5'- to the 3'-terminus, and RNA polymerase molecules, which traverse templates in the opposite direction?

The genomes of RNA phages are polycistronic messenger RNAs; there are three ribosome attachment sites, one at the start of each cistron. Like other RNAs, they possess a pronounced secondary structure that is determined by base pairing between homologous nucleotide sequences, probably rather like that exhibited by tRNA molecules, and, as a result, there is a marked difference in accessibility of individual sequences to approaching molecules or particles. In the naked phage RNA the most accessible ribosome binding site is at the beginning of the cistron that codes for the coat protein, which would therefore be formed in the greatest amount. However, attachment of ribo-

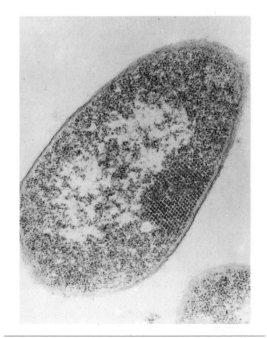

FIG. 10-13. An E. coli cell 50 minutes after infection with MS2. Note the paracrystalline virus particle array. The virus is slightly larger than the ribosomes that are present throughout the cell. This cell is about to lyse. ×37,500. (Courtesy of Dr. M. Van Montagu.)

somes there renders the site at the beginning of. the polymerase cistron also accessible to ribosomes, so that during the initial stages of infection, both coat protein and polymerase are formed freely. Coat protein itself, however, can bind to the ribosome binding sites at the beginning of the A protein and polymerase cistrons, thereby preventing ribosome attachment, so that as coat protein accumulates, it gradually inhibits translation of these two cistrons more and more strongly. The net result is that the A protein cistron is translated only infrequently throughout most of the infection cycle, and the polymerase cistron is translated frequently during the early stages of the infection cycle but only very infrequently during the later stages.

The collision problem is solved in the following manner. When a polymerase molecule attaches to the 3′-terminus of the RNA strand, it also binds to the coat protein cistron ribosome attachment site, thereby preventing further ribosome attachment. When ribosomes have cleared from the region between these two points, the polymerase molecule can then progress along the RNA strand unimpeded.

It is clear therefore that, although it is very simple and small, the RNA phage genome can nevertheless regulate a sophisticated transcription and translation program. This is achieved by dual functioning of several of the proteins that it encodes.

Lysogeny

Not all phage infections result in lysis of the host cell. There are some phages which, upon entering a sensitive cell, either elicit a lytic response (that is, a typical multiplication cycle such as already described) or repress the expression of almost all their genetic information and integrate their genome into that of the host cell. The integrated phage genome, the prophage, behaves from then on like any other portion of the bacterial chromosome. This phenomenon is known as lysogeny. Bacteria that harbor prophages in their chromosomes are known as lysogens, and phages capable of integrating their genomes into those of their host

cells are known as lysogenic, or temperate, phages.

When it was discovered that some animal viruses could enter into a similar relationship with their hosts (Chap. 9), the analogy between these viruses and temperate phages was quickly recognized; and although there are differences in the manner in which tumor viruses and temperate phages interact with their host cells, the concepts which have evolved from temperate phage studies have profoundly influenced our thinking concerning tumor viruses.

THE PROPHAGE

The frequency of lysogenization varies both with the phage-host system and with the multiplicity of infection. It is greater the higher the multiplicity; this suggests that each phage genome has a finite probability of lysogenization. The number of any particular species of prophage per lysogen is generally the same as the number of bacterial genomes, which suggests that the prophage attachment site is restricted to a particular locus on the bacterial chromosome. That a unique location generally exists can be demonstrated by genetic studies, such as the interrupted mating technique. However, occasionally there is more than one attachment site for the same prophage, and all may be occupied simultaneously. Superinfection may result in the integration of two prophages at the same site, inserted either into each other or adjacent to each other. Further, bacterial chromosomes generally contain prophage attachment sites for several different unrelated prophages, and all of these may be occupied at the same time. Finally, one temperate phage, phage mu, has no specific integration site but can insert its genome anywhere in the bacterial chromosome; it is, in essence, a transposon.

THE CONCEPT OF IMMUNITY

Under normal conditions prophage is stable. Termination of the lysogenic state does, however, occur spontaneously with low frequency (of the order of 10^{-5}). The phage genome is then excised from the host chromosome (see

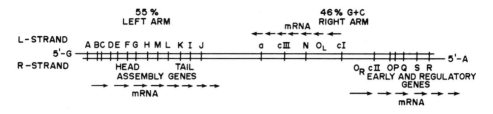

FIG. 10-14.   A simplified genetic map of phage lambda. (Adapted from Taylor, Hradecna, and Szybalski: Proc Natl Acad Sci USA 57:1618, 1967.)

below), and a normal lytic multiplication cycle ensues. Cultures of lysogenic bacteria therefore generally contain low titers of infectious virus.

Termination of the lysogenic state and formation of phage progeny is known as induction. The frequency of induction can be increased enormously by agents that inhibit DNA replication, such as irradiation with ultraviolet light, mitomycin C and agents capable of alkylating DNA. Under appropriate conditions treatment with such agents is capable of "inducing" all cells in cultures of lysogenic bacteria at the same time, with lysis and progeny phage liberation occurring some 20 minutes later.

Although cultures of lysogenic bacteria always contain infectious phage, they are not lysed by it. They are immune. This is not because the phage cannot adsorb and inject its DNA, but because the injected DNA cannot express itself and replicate. The reason for this is that one of the prophage genes codes for a repressor that prevents transcription of all other phage genes, both those on prophage and those on superinfecting phage genomes. Immunity is highly specific and bacteria lysogenic for one phage are not immune to other temperate phages unless they are closely related.

THE NATURE OF PHAGE LAMBDA

By far the best known of the many temperate bacteriophages is the coliphage λ, probably the most intensively studied of all viruses. The exquisite and intricate mechanisms that control its interaction with the host and its multiplication represent the system par excellence with which to study biologic control. We will therefore discuss the mechanisms operating in lysog-

eny in terms of the λ system.

Phage λ possesses a hexagonal head and a simple noncontractile tail with one terminal tail fiber; it comprises about 10 protein species. Its DNA, which contains no unusual bases, is linear and has a molecular weight of 31×10^6, representing about 50 genes. Figure 10-14 shows a simplified genetic map of λ. About 20 genes code for its structural proteins and for proteins that function in its morphogenesis. These genes, which include genes *A* to *J*, are in the left arm of the DNA, which contains 55 percent guanine plus cytosine. Some 30 other genes have been identified as regulating in one way or another the interaction of the λ genome with that of the host cell and its transcription and translation. These genes are located in the right arm of the DNA, which contains 46 percent guanine plus cytosine. Not all genes are transcribed from the same DNA strand: as shown in Figure 10-14, some genes are transcribed from one strand, while others are transcribed in the opposite direction from the other strand.

The ends of the two strands of λ DNA are not equal in length. Each 5'-terminus has an unpaired single-stranded region 12 nucleotides long. The base sequence at the left arm 5'-terminus is GGGCGGCGACCT, and that at the right arm 5'-terminus is complementary to this. Phage λ DNA cyclizes readily because these two sequences, known as "cohesive" or "sticky" ends, tend to hybridize. The origin of the sticky ends is as follows: just like the DNA of T4, λ DNA replicates via intermediates that are longer than it is itself. Whereas the T4 intermediate is cut into "headful" portions of DNA with a "headful" portion being slightly larger than the phage genome, causing T4 DNA to be circularly permuted, λ DNA is cut at a specific location, and the scissions in the two strands do not coincide, but are 12 nucleotides apart.

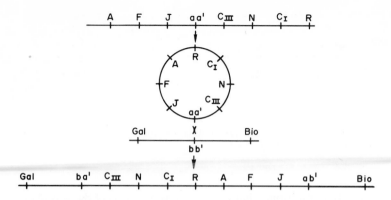

FIG. 10-15. The Campbell model for prophage integration. After circularization of lambda DNA by joining of cohesive ends, reciprocal recombination between phage aa' and bacterial bb' recognition sites provides for linear insertion of the viral DNA into that of the host. (Adapted from Echols and Joyner: In Fraenkel-Conrat H., (ed): Molecular Basis of Virology, 1968. Courtesy of Reinhold Book Corp.)

EVENTS LEADING TO LYSOGENY: INTEGRATION

When λ DNA enters the cell, it cyclizes, and the two nicks are sealed by polynucleotide ligase, thus forming covalently closed circles. Within a brief critical period of time the decision is made whether a lytic multiplication cycle is to be initiated or whether the phage genome is to become integrated. The decision appears to be influenced principally by the relative rates at which several viral gene products reach critical concentrations and by the ability of these products to function. Two of these gene products are the proteins specified by genes N and cI. The N gene product, which apparently modifies the host RNA polymerase, is essential for the switching on all early λ genes. If the N gene product is not synthesized, or if it is defective, phage λ cannot multiply, and the lytic response to λ infection is precluded. The protein specified by gene cI is the λ repressor, a protein with a molecular weight of about 30,000, which binds to two operators, O_R and O_L, on either side of gene cI (Fig. 10-14), thereby preventing the transcription of all λ genes, including that of gene N. While the cI and N gene products are of critical importance in determining the outcome of the phage-host interaction, they are by no means the only ones that function in this manner. In fact, the mechanism that is involved is extraordinarily complex. The reason for this is that phage-coded proteins are required for integration, and the

phage must therefore not impose repression before these proteins are synthesized, but must shut off viral development before an irreversible commitment to lytic growth is made. Among several other proteins that are also important in determining the outcome of infection are those specified by gene Q, which provides for the rapid synthesis of head, tail, and lysis proteins and therefore promotes lytic development, and those specified by genes cII and cIII, which provide for rapid synthesis of the repressor.

If sufficient N and Q gene products are formed, the λ genome can express itself, and a lytic infection cycle ensues. This also happens in the case of mutants with mutations in gene cI that render the repressor defective and which are therefore obligatorily virulent. If, however, the repressor accumulates, the λ genome cannot express itself. In that case the λ genome has two options. (1) It is integrated into the host genome by a mechanism that is illustrated schematically in Figure 10-15. The region aa' between genes J and cIII then aligns itself with region bb', the prophage attachment site on the bacterial chromosome, which is situated between the galactose and biotin genes. Enzymes coded by the phage int system then make scissions in all four nucleotide strands, the phage DNA is inserted by means of double crossing over, and covalent bonds are then reformed. The λ genome is now integrated. (2) If the aa' and bb' regions do not match, most commonly because of a deletion, the phage genome cannot become integrated. It then multiplies in the host cell as an episome. In either event, the only λ gene that is transcribed is gene *cI.*

RELEASE FROM REPRESSION: INDUCTION

The termination of the lysogenic state, or induction, requires both release from repression and excision of the phage DNA from the host DNA. The cellular signal for release from repression is inhibition of host DNA synthesis, which apparently causes generation of an inducing substance that prevents the repressor from functioning. As a result, enzymes specified by the *int* and *xis* phage genes are synthesized; these enzymes excise the phage DNA from the host DNA. Excision proceeds essentially via reversal of the integration process and yields a circular phage DNA molecule which then replicates in the normal manner characteristic of virulent λ infection.

The juxtaposition of bacterial and phage genes in the chromosome of the lysogen sometimes leads to excision of some bacterial DNA together with phage DNA. This provides an opportunity for the transfer of bacterial genetic material from one bacterium to another via phage; this is known as transduction.

THE SIGNIFICANCE OF LYSOGENY

Every bacterium may well carry one or more prophages. There are numerous defective or cryptic prophages—that is, prophages that lack the complete set of information necessary for phage multiplication, so that even when induced and therefore excised, no infectious phage particles are formed. Bacteria harboring such prophages are analogous in some respects to cells transformed by defective SV40 (Chap. 7). Even if they possess complete genomes, lysogenic phage cannot be detected unless there are available nonimmune bacterial indicator strains that lack prophage attachment sites and that can be infected by the lytic pathway.

Lysogenic Phage Conversion Although in most cases the phenotype of bacteria that carry prophage is indistinguishable from that of bacteria that do not, there are several interesting examples where there is a clear difference. These bacteria exhibit the phenomenon known as lysogenic conversion.

One case of lysogenic conversion involves the phage-mediated conversion of the somatic O antigen of *Salmonella.* This antigen forms part of the lipopolysaccharide structure of the cell wall and is composed of highly esterified lipid (lipid A) linked to a polysaccharide core to which are attached side-chains consisting of repeating sequences of a variety of sugars. The antigenic specificity resides in the sugars and can be modified by altering their nature.

A wide variety of temperature phages of *Salmonella* can modify the antigenic properties of the somatic O antigen. For example, the O antigen of *Salmonella anatum* normally has the antigenic formula 3,10 (based on its response pattern to various antisera) and carries terminal O-acetyl-D-galactosyl groups linked by means of α-1:6 linkages to D-mannosyl residues. When lysogenized by the temperature phage ϵ^{15}, antigenic specificity 10 is lost, and specificity 15 is acquired. This is due to loss of the O-acetyl group and replacement of the α-1:6 bond by a β-1:6 bond. Cells lysogenized by phage ϵ^{15} can themselves be lysogenized by phage ϵ^{34}, which causes a further change in antigenic specificity, this time due to the addition of D-glucosyl residues by means of β-1:4 linkages. There are numerous other examples.

There is no doubt that the chemical alterations in *Salmonella* polysaccharides are phage-induced, since ϵ^{15} causes repression or inhibition of the α1:6-galactosyl transferase and acetylase and codes for a β-1:6-galactosyl transferase. These converting functions are, however, not essential to the phage, since mutants defective in them grow well. On the other hand, it is significant that ϵ^{15} cannot adsorb to cells with antigenic specificity 3,15; that is, cells carrying ϵ^{15} cannot be infected by it. It is likely that the significance of this type of lysogenic conversion lies in the specification of host range patterns, perhaps to augment phage immunity systems.

A second example of lysogenic conversion concerns *Corynebacterium diphtheriae.* Toxigenic strains carry the prophage β, and nontoxigenic strains can be made toxigenic by lysogenization. It is clear that the structural gene for toxin is located on the phage genome, since phage mutants coding for defective toxin have been isolated. However, it is not clear what advantage the phage gains by coding for a polypeptide with the remarkable biologic and enzymatic properties of the toxin. Most probably the advantage relates not to the phage but to the bacterial lysogen. In other words, corynebacteria that can synthesize toxin most probably have a survival advantage over those that can not do so.

A third example involves *Clostridium botulinum* types C and D, which produce toxin only when demonstrably infected with (and yielding) phages CE β and DE β respectively. However, with types C and D, the relation between host cell and phage may not be true lysogeny, since infected *C. botulinum* becomes nontoxigenic (and nonvirogenic) when treated with antiserum to phage.

FURTHER READING

Books and Reviews

Ackerman H, Audurier A, Berthiaume L, Jones LA, Mayo JA, Vidaver AK: Guidelines for bacteriophage characterization. Adv Virus Res 23:2, 1978

Bradley DE: Ultrastructure of bacteriophages and bacteriocins. Bacteriol Rev 31:230, 1967

Cairns J, Stent GS, Watson JD: Phage and the origins of molecular biology. New York, Cold Spring Harbor Lab of Quantitative Biol, 1966

Calendar R, Geisseloder J, Sunshine MG, Six EW, Lindquist B: The P2-P4 transactivation system. In Fraenkel-Conrat H, Wagner RR (eds): Comprehensive Virology, Vol 8, p 329. New York and London, Plenum, 1977

Campbell A: Defective bacteriophages and incomplete prophages. In Fraenkel-Conrat H, Wagner RR (eds): Comprehensive Virology. Vol 8, p 259. New York and London, Plenum, 1977

Casjens S, King J: Virus assembly. Ann Rev Biochem 44:555, 1975

Denhardt DT: The isometric single-stranded DNA phages. In Fraenkel-Conrat H, Wagner RR (eds): Comprehensive Virology, Vol 7, p 1. New York and London, Plenum, 1977

Denhardt DT, Dressler D, Ray DS (eds): The single-stranded DNA phages. New York, Cold Spring Harbor Lab, 1978

Echols H, Murialdo H: A genetic map of λ. Micro Rev. 42:577, 1978

Eoyang L, August JT: Reproduction of RNA bacteriophages. In Fraenkel-Conrat H, Wagner RR (eds): Comprehensive Virology, Vol 2, p 1. New York and London, Plenum, 1974

Hausmann R: Bacteriophage T7 genetics. Curr Top Microbiol Immunol 75:77, 1976

Hohn T, Katsura I: Structure and assembly of bacteriophage lambda. Curr Top Microbiol Immunol 78:69, 1977

Lemke PA, Nash CH: Fungal viruses. Bacteriol Rev 38:29, 1974

Mathews CK; Reproduction of large virulent bacteriophages. In Fraenkel-Conrat H, Wagner RR (eds): Comprehensive Virology, Vol 8, p 179. New York and London, Plenum, 1977

Miller RC Jr: Replication and molecular recombination of T-phage. Ann Rev Microbiol 29:355, 1975

Murialdo H, Becker A: Head morphogenesis of double-stranded DNA containing bacteriophages. Micro Rev 42:529, 1978

Nash HA: Integration and excision of bacteriophage lambda. Curr Top Microbiol Immunol 78:171, 1977

Pirotta V: The lambda repressor and its action. Curr Top Microbiol Immunol 74:21, 1976

Rabussay D, Geiduschek EP: Regulation of gene action in the development of lytic bacteriophages. In Fraenkel-Conrat H, Wagner RR (eds): Comprehensive Virology, Vol 8, p 1. New York and London, Plenum, 1977

Ray DS: Replication of filamentous bacteriophages. In Fraenkel-Conrat H, Wagner RR (eds): Comprehensive Virology, Vol 8, p 105. New York and London, Plenum, 1977

Skalka AM: DNA replication—bacteriophage lambda. Curr Top Microbiol Immunol 78:201, 1977

Weisberg RA, Gottesman S, Gottesman ME: Bacteriophage lambda: the lysogenic pathway. In Fraenkel-Conrat H, Wagner RR (eds): Comprehensive Virology, Vol 8, p 197. New York and London, Plenum, 1977

Weissman C, Billeter MA, Goodman HM, Hindley J, Weber H: Structure and function of phage RNA. Ann Rev Biochem 42:303, 1973

Selected Papers

SINGLE-STRANDED DNA-CONTAINING PHAGES

Godson GN: G4 DNA replication. III. Synthesis of replicative form. J Mol Biol 117:353, 1977

Godson GN, Barrell BG, Staden R, Fiddes JC: Nucleotide sequence of bacteriophage G4 DNA. Nature 276:236, 1978

Ikeda J-E, Yudelevich A, Hurwitz J: Isolation and characterization of the protein coded by gene *A* of bacteriophage ϕX174 DNA. Proc Natl Acad Sci USA 73:2669, 1976

Ohsumi M, Vovis GF, Zinder ND: The isolation and characterization of an in vivo recombinant between the filamentous bacteriophage fl and the plasmid pSC101. Virol 89:438, 1978

Sanger F, Coulson AR, Friedmann T, Air GM, Barrell

BG, Brown NL, Fiddes JC, Hutchison CA III, Slocombe PM, Smith M: The nucleotide sequence of bacteriophage φX174. J Mol Biol 125:225, 1978

Ueda K, McMacken R, Kornberg A: *dna* B Protein of *Escherichia coli:* purification and role in the purification of φX174 DNA. J Biol Chem 253:261, 1978

SINGLE-STRANDED RNA CONTAINING PHAGES

Domingo E, Sabo D, Taniguchi T, Weissman C: Nucleotide sequence heterogeneity of an RNA phage population. Cell 13:735, 1978

Edlind TD, Bassel AR: Secondary structure of RNA from bacteriophages f2, Qβ, and PP7. J Virol 24:135, 1977

Schaffner W, Ruegg KJ, Weissman C: Nonvariant RNAs: Nucleotide sequence and interaction with bacteriophage Qβ replicase. J Mol Biol 117:877, 1977

Taniguchi T, Weissman C: Site-directed mutations in the initiator region of the bacteriophage Qβ coat cistron and their effect on ribosome binding. J Mol Biol 118:533, 1978

THE LYTIC MULTIPLICATION CYCLE OF
DOUBLE-STRANDED DNA-CONTAINING PHAGES

de Wyngaert MA, Hinkle DC: Involvement of DNA gyrase in replication and transcription of bacteriophage T7 DNA. J Virol 29:529, 1979

Fox TD, Pero J: New phage-SPO1-induced polypeptides associated with *Bacillus subtilis* RNA polymerase. Proc Natl Acad Sci USA 71:2761, 1974

Hiatt WR, Whiteley HR: Translation of RNAs synthesized in vivo and in vitro from bacteriophage SP82 DNA. J Virol 25:616, 1978

Müller UR, Marchin GL: Purification and properties of a T4 bacteriophage factor that modifies valyl-tRNA synthetase of *Escherichia coli.* J Biol Chem 252:6640, 1977

Majumder HK, Bishayee S, Chakraborty PR, Maitra U: Ribonuclease III cleavage of bacteriophage T3 RNA polymerase transcripts to late T3 mRNAs. Proc Natl Acad Sci USA 74:4891, 1977

Spiegelman GB, Whiteley HR: Bacteriophage SP82 induced modifications of *Bacillus subtilis* RNA polymerase result in the recognition of additional RNA synthesis initiation sites on phage DNA. Biochem Biophys Res Commun 81:1058, 1978

Talkington C, Pero J: Restriction fragment analysis of the temporal program of bacteriophage SP01 transcription and its control by phage-modified RNA polymerases. Virol 83:365, 1977

von Gabain A, Bujard H: Interaction of *Escherichia coli* RNA polymerase with promoters of several

coliphage and plasmid DNAs. Proc Natl Acad Sci USA 76:189, 1979

PHAGE MORPHOGENESIS

Berget PB, King J: Antigenic gene products of bacteriophage T4 baseplates. Virol 86:312, 1978

Black LW, Silverman DJ: Model for DNA packaging into bacteriophage T4 heads. J Virol 28:643, 1978

Coombs DH, Eiserling FA: Studies on the structure, protein composition and assembly of the neck of bacteriophage T4. J Mol Biol 116:375, 1977

Crowther RA, Lenk EV, Kirkuchi Y, King J: Molecular reorganization in the hexagon to star transition of the base plate of T4. J Mol Biol 116:489, 1977

Feiss M, Siegele DA: Packaging of the bacteriophage lambda chromosome: dependence of *cos* cleavage on chromosome length. Virol 92:190, 1979

King J, Hall C, Casjens S: Control of the synthesis of phage P22 scaffolding protein is coupled to capsid assembly. Cell 15:551, 1978

Kozloff LM, Lute M, Crosby LK: Bacteriophage T4 virion baseplate thymidylate synthetase and dihydrofolate reductase. J Virol 23:637, 1977

Poteete AR, King J: Functions of two new genes in *Salmonella* phage P22 assembly. Virol 76:725, 1977

Showe MK, Onorato L: Kinetic factors and form determination of the head of bacteriophage T4. Proc Natl Acad Sci USA 75:4165, 1978

Sternberg N, Weisberg Y: Packaging of coliphage lambda DNA. I. The role of the cohesive end site and the gene *A* protein. II. The role of the gene *D* protein. J Mol Biol 117:717, 733, 1977

Wood WB, Conley MP: Attachment of tail fibers in bacteriophage T4 assembly: role of the phage whiskers. J Mol Biol 127:15, 1979

LYSOGENY

Kikuchi Y, Nash HA: The bacteriophage λ *int* gene product. J Biol Chem 253:7149, 1978

Shaw JE, Jones BB, Pearson ML: Identification of the *N* gene protein of bacteriophage λ. Proc Natl Acad Sci USA 75:2225, 1978

Takeda Y: Specific repression of in vitro transcription by the cro repressor of bacteriophage lambda. J Mol Biol 127:177, 1979

Takeda Y, Folkmanis A, Echols H: Cro regulatory protein specified by bacteriophage λ: structure, DNA-binding and repression of RNA synthesis. J Biol Chem 252:6177, 1977

DIPHTHERIA TOXIN PRODUCTION

Uchida T, Kanei C, Yoneda M: Mutations in Corynephage β that affect the yield of diphtheria toxin. Virol 77:876, 1977

SECTION II

CLINICAL VIROLOGY

CHAPTER 11

Diagnostic Virology

Diagnosis of specific viral infections is useful to the practicing physician by delineating an etiologic agent for the observed illness. At the present time, this information may alter unnecessary antimicrobial therapy and allow earlier discharge from the hospital, or provide informative research data. The time required for viral diagnosis by standard inoculation of tissue culture, embryonated egg, or animals often exceeds the duration of the illness in question. Increasingly rapid diagnosis will provide far more information for the responsible physician. The development of specific and rapid diagnosis will be fostered by the development of antiviral drugs. Virus-specific compounds demand that quick and accurate diagnosis be an integral part of patient management in order to use the drugs safely and in appropriate situations.

Viral diagnostic facilities vary greatly, and the clinician must determine what specific resources are available in his hospital or community. The facilities of the municipal or state health departments are accessible without direct patient costs, and through these laboratories referral of clinical materials can be made to the Center for Disease Control in Atlanta, Georgia. In addition, laboratories within community hospitals and associated with medical schools are accessible in some areas.

There are viral illnesses where the accurate diagnosis is of extreme importance to the patient, his family, or his community. For example, the World Health Organization's global program for eradication of smallpox has resulted in a dramatic decline in disease, so that no naturally occurring cases of the disease have been reported since the end of 1977. Many physicians have had no personal experience with smallpox but must know when they are confronted with a possible case and understand how the laboratory can efficiently assist in the diagnosis. Similarly, the correct diagnosis of an initial case or paralytic poliomyelitis can allow effective immunization to be instituted in a susceptible community, thereby preventing a sizable outbreak of disease. This requires isolation, identification, and typing of the virus in the laboratory. When rubella infection is suspected in the first trimester of pregnancy, the currently available accurate serologic means of diagnosis are mandatory to confirm the diagnosis prior to any consideration of pregnancy interruption. The serologic assessment of rubella immunity should become an integral part of the female premarital examination and is required by law in many states, since attenuated rubella vaccine offers a means of establishing immunity in susceptible women prior to pregnancy.

Finally, within the framework of any program responsible for training physicians, it is important that these physicians become familiar with viral illnesses, institute appropriate control measures when necessary, and provide appropriate prophylaxis for others when this is available. Laboratory documentation of viral illness confirming the clinical impression or in some cases providing an unsuspected diagnosis will improve the judgment of these physicians providing patient care in the years to come.

TREATMENT OF SPECIMENS

A basic understanding of the laboratory procedures is necessary in order to utilize optimally the available virus laboratory facilities. Whenever a specimen is submitted, adequate information must accompany the request. This information is essential, as specimens are often processed in the laboratory according to their source and the physician's provisional diagnosis. Commonly needed information includes patient identification and age, the clinical diagnosis, the referring physician, the source of the specimens, and the date they were obtained. In addition, it is often helpful to know whether the patient has recently received viral vaccines. The responsible laboratory processing these materials should be consulted whenever questions arise, as they will be able to specify the optimal handling of the materials.

Patient materials are cultured for the presence of viruses to detect a specific etiologic agent. Tissue culture systems are usually more readily available and less cumbersome and expensive than embryonated eggs or animals. The cells utilized are dictated in part by the facilities available in the particular laboratory and by the suspected type of infection. It is usually advisable to collect the specimens as early as possible in the course of a clinical illness. Since viral agents are not affected by antimicrobial agents, cultures may be obtained even if therapy for potential bacterial pathogens has previously been initiated. Such materials are obtained from the same site(s) as specimens for bacterial cultures. Respiratory tract secretions,

TABLE 11-1   SUGGESTED SPECIMENS TO BE SUBMITTED FOR VIRAL DIAGNOSIS

	Specimen Source for Culture								
Diagnosis	Nasal secretions throat swab	Stool	Urine	CSF	Skin[1]	Pleural fluid	Pericardial fluid	Acute Serum	Conva-lescent[2] Serum
Respiratory disease[3]	X	X				X (if pres)		X	X
Enteric illness	X	X[9]						X	X
Exanthema	X	X	+		X			X	X
Aseptic Meningitis	X	X	+	X				X	X
Myocarditis-pericarditis	X	X					X (if pres)	X	X
Vesicular eruption[4]	X	+	+		X			X	X
Newborn baby with suspected intrauterine infection	X	X	X	X	+[7]			X[5]	X
Orchitis-parotitis[6]	X		X					X	X
Hepatitis-infectious mononucleosis group	X		X					X[8]	X[8]

X denotes usual sites of culture; + denotes sites which may also be useful

[1]Vesicle fluid, pustules, or skin scrapings.
[2]Convalescent serum 2–4 weeks after onset of illness.
[3]Agents include respiratory syncytial virus, adenoviruses, enteroviruses, influenza, parainfluenza.
[4]Generalized eruption, culture of nasal secretions/throat swab, urine, or stool may be useful.
[5]Blood in newborn period and 6–9 months after birth; mother's serum may also be evaluated.
[6]Usually mumps, occasionally Coxsackie or lymphocytic choriomeningitis.
[7]Mother's cervix can be examined by Papanicolaou stain.
[8]Peripheral leukocytes from heparinized blood may yield EB virus or CMV. HB_sAg and/or anti-HB_sAg determinations may be performed on sera.
[9]Examine stool by EM.

stool, urine, cerebrospinal fluid, pleural fluid, pericardial fluid, blood, bone marrow, vesicle fluid, and skin scrapings may be utilized in attempted viral isolations. Tissue specimens from biopsies or postmortem examinations are suitable for attempted viral isolation if fresh or frozen but not in formalin.

The principles adhered to in the collection of bacterial specimens are applicable to specimens submitted for viral culture. Containers must be clean and sterile and handled accordingly. The more stable viral agents include vaccinia, variola, the enteroviruses, and the adenoviruses. As it is rarely possible to know the particular etiologic agent until the virus is isolated, all specimens should be treated as though the agent were potentially labile. Many viruses tend to be unstable at an acid pH or at temperatures above 4C. All patient specimens should be transported as rapidly as possible to the laboratory. If a delay of even 1 hour occurs, the material should be refrigerated at 4C or transported on ice. Delays of hours or days prior to inoculation usually necessitate freezing of the specimens, preferably at −70C.

In addition to their heat and acid lability, many viruses do not withstand drying. Thus swabs of mucosal surfaces, skin scrapings, and tissues are placed in a transport medium which provides both a protein source and a buffered salt solution essential to preserve infective virus. Several such media are easily made and readily available from any microbiology laboratory. A 1 percent solution of skimmed milk in distilled water, Hanks' balanced salt solution with 10 percent fetal calf serum or 0.5 percent bovine albumin or Minimal Essential Medium and 40 percent sucrose have been used successfully. Each of these media has antibiotics (e.g., penicillin, 50 units/ml; amphotericin, 1 μg/ml; streptomycin, 5μg/ml) added to prevent overgrowth by resident host bacterial flora and to decrease exogenous contamination of the specimen.

IDENTIFICATION OF VIRUS

Isolation of an agent from a patient, especially from a mucosal surface, does not always prove an etiologic relationship to his illness. For this reason, patients being evaluated for viral disease should have both acute and convalescent sera drawn for antibody determinations. The commonly accepted serologic confirmation of acute infection is a four-fold or greater rise in antibody titer. The acute specimen of blood is obtained from the patient as early as possible in his illness, and the convalescent specimen is drawn 2 to 3 weeks later. The serum is separated and kept at refrigerator or freezer temperature. Whole blood should not be frozen, as the erythrocytes lyse and the subsequent assay of hemolyzed serum is difficult and inaccurate. Serologic demonstration of an antibody response may provide evidence of infection in the absence of virus recovery from cultures. Serology may also be used to assay the immune status of an individual following previous infection or vaccination.

The most frequently used techniques to assay viral antibodies include complement fixation (CF), hemagglutination inhibition (HI), and neutralization of the viral cytopathic effect. Complement fixation offers a relatively rapid, efficient way of screening sera for a number of viral antibodies. It is particularly useful in determining that recent infection has occurred. As a general rule, the antibodies thus detected are group-specific and not type-specific. For example, adenovirus antibody can be distinguished from poliovirus antibody, but antibody to type 1 adenovirus cannot be distinguished from antibody to types 7, 8, 12, or any other adenovirus CF antibody. Complement fixation antibodies usually decline more rapidly than specific neutralizing antibodies, so that their absence does not necessarily denote susceptibility.

Hemagglutination-inhibition studies are limited by the fact that not all viruses possess a hemagglutinin. Agents with recognized hemagglutinins are influenza, mumps, measles, the parainfluenzae, rubella, vaccinia, variola, the arboviruses, reoviruses, and some of the adenoviruses and echoviruses. The limitations of the test itself are primarily technical, as it is often difficult to eliminate nonspecific inhibitors of hemagglutination from the serum being tested. The HI antibodies are type-specific, and the method is quite sensitive, correlating well with the antibodies observed in the neutralization test.

Neutralizing antibodies are highly specific for the virus type, appear early in the course of the illness, and persist for a long time thereafter. The determination of neutralizing antibody requires the growth of an agent in a cell culture or an animal system. The serum in question is incubated with the suspected etiologic agent and is then placed into the appropriate culture system. The results are commonly unavailable for at least a week, as the appearance of the cytopathic effects in the cell culture or illness in the animal determines the length of time necessary for the performance of the test.

NEW IMMUNOLOGIC TECHNIQUES

In recent years adaptation of several immunologic techniques has been of great assistance in the viral diagnostic laboratory. Immune adherence hemagglutination (IAHA), agar gel diffusion, countercurrent immunoelectrophoresis (CIE), enzyme-linked immunosorbent assay (ELISA), and radioimmunoassays (RIA) are methods being applied to viral diagnosis (see Chap. 4). Immunoprecipitation has been employed in the case of serum hepatitis to detect viral antigen in the secretions or serum of individuals. CIE increased the speed with which such determinations could be performed, made the test available on a large scale, but has now been superceded by RIA, which is employed in many hospital blood banks to screen donor blood prior to transfusion. The IAHA assay—performed by incubating viral antigen, specific antibody, and complement together—results in the visible agglutination of red blood cells. If antibody or antigen is absent, no agglutination occurs. This assay as applied to varicella serology is a practical tool for assessment of susceptibility to varicella virus. This constitutes a useful epidemiologic parameter when varicella is introduced into the hospital setting. ELISAs are designed with a specific enzyme linked to antibody. The enzyme (e.g., alkaline phosphatase) can produce a color change when provided with the right substrate (e.g., p-nitrophenyl phosphate). The visible color change is appreciated with the eye and can be easily quantitated by a spectophotometer. In practice, antigen is absorbed to a surface (microtiter plate, metal beads, etc.) and the test serum added. Any specific antibody will attach to antigen. Enzyme-labeled antiglobulin is then added and attaches to the bound antigen–antibody complex. After exhaustive washing, addition of the substrate results in the visible color change. If no specific antibody is present, the enzyme labeled antiglobulin will not adhere to the plate and no hydrolysis of substrate can occur. All of these techniques are being applied to the measurement of a variety of viral antibodies, as well as to detection of viral antigens.

Immunofluorescence (see Chap. 4) has been utilized for rapid viral diagnosis of respiratory tract infections, vesicular exanthems, and examination of tissues. Respiratory epithelial cells obtained from nasal secretions may be examined by indirect immunofluorescence for the presence of many of the common respiratory viruses. A large successful experience with diagnosis of respiratory syncytial virus, parainfluenza, influenza, and adenovirus infections has been accumulated in European laboratories. Meticulous preparation of materials is required, and antisera, which are not generally available, must be both sensitive and specific. The basic techniques can be standardized, but still require an experienced person to interpret the results. The method offers possible etiologic diagnosis within one working day, which is a distinct advantage to the clinician. Standardization of reagents and increased availability will render immunofluorescence a more broadly available method.

TISSUE PATHOLOGY

Any laboratory where tissue pathology is done can examine tissues by microscopy, which may facilitate a viral diagnosis. Histologic examination of pertinent clinical specimens with hematoxylin and eosin or Giemsa stains may disclose pathognomonic features of viral infection such as inclusion bodies of multinucleate giant cells. If specific antisera are available, immunofluorescence may also be used.

Measles infection, either that of natural disease or of attenuated virus, may produce multinucleate giant cells in the urine, as well as intranuclear and intracytoplasmic inclusions. Such findings have also been described in the sputum or throat scrapings of patients with giant cell pneumonia or in malnourished patients with measles. Rubella, mumps, herpangina, and variola have been reported to produce eosinophilic cytoplasmic inclusions in the cells of the urine sediment. Agents such as cytomegalovirus and herpes simplex have been observed to produce intranuclear inclusions in cells in the urine. Such positive findings in the urine, sputum, or tissues will point toward the probability of viral infection, but without laboratory culture of the virus cannot identify the specific agent involved. The absence of such findings does not exclude viral disease, and it is very important to recognize that many agents (for example, the enteroviruses) do not cause specific hallmarks of viral infection.

Cutaneous lesions such as vesicles, macules, or pustules can also be examined for the presence of multinucleate giant cells and inclusion-bearing cells. Materials obtained from such lesions can be examined easily and quickly. The vesicle is sponged with alcohol (if no viral cultures are being taken from the lesion) and the roof of the vesicle is reflected. The fluid is blotted, the base of the lesion is scraped, avoiding gross bleeding when possible, and the cellular material is spread on a glass slide and air dried. It can be fixed with methyl alcohol and stained with Wright's or Giemsa stains. Giant cells are readily apparent, and inclusions may be seen with carefully made preparations. The presence of such giant cells will exclude vaccinia and variola from the differential diagnosis but will not distinguish between herpes simplex and varicella-zoster. Similarly, the diagnosis of herpes simplex infection of the female genital tract can be approached by obtaining a swab of the cervix and examining by Papanicolaou stain for giant cells and/or intranuclear inclusions. Again, immunofluorescence may provide specific etiologic information in some of these situations.

ELECTRON MICROSCOPIC EXAMINATION

The electron microscope (EM) has been utilized recently to assist in viral diagnosis. Examination of materials from vesicular lesions by experienced personnel can readily distinguish a viral agent of the herpes group from one of the poxvirus group and identification of the etiologic agent of hepatitis B was accomplished by EM visualization of the virus. Another example of the use of the EM has been the detection of previously unrecognizable viral particles in stool specimens. Ten percent suspensions of stool examined directly will reveal the presence of virus at a concentration of 10^5 particles/g of stool. These observations have led to the identification of several agents (rotaviruses, parvoviruses, caliciviruses) now considered causative in acute gastrointestinal disease. The human rotaviruses have not yet been grown in any cell culture system, so electron microscopy is the most practical diagnostic method at the present time. Other refinements of methodology have broadened the diagnostic applications of the EM. For example, the use of specific antiserum causes adherence of virions to antibody and results in clumps of virus particles. This clumping permits detection of small numbers of virus particles. Such immune electron microscopy provides direct visualization of the antigen–antibody reaction. Although such work is primarily investigational at the present time, new insights into gastrointestinal disease have come from work with the EM. The future may see EM diagnosis supplanted by more practical methods.

The information provided in this brief discussion can be applied in the diagnosis of the viral infections discussed in the subsequent clinical virology chapters.

FURTHER READING

Bradstreet CMP, Pereira MS, Pollock TM: The organization of a national virological diagnostic service. Prog Med Virol 16:242, 1973

Gardner PS, McQuillin J: Rapid virus diagnosis. Application of Immunofluorescence. London, Butterworth, 1974

Herrmann EC Jr: New concepts and developments in applied diagnostic virology. Prog Med Virol 17:222, 1974

Hsiung GD: Diagnostic virology. New Haven, Yale University Press, 1973

Hsiung GD: Laboratory diagnosis of viral infections: General principles and recent developments. Mt Sinai J Med 44:1, 1977

Sever JL, Madden DL (ed): Enzyme-linked immunosorbent assay (ELISA) for infectious agents. J Infect Dis Suppl, October 1977, p S257

CHAPTER 12
Poxviruses

The poxviruses are a group of large, complex DNA viruses which can be divided into six genera that comprise more than 27 members. Although several of the group can initiate human infections, only variola (smallpox) has been of worldwide epidemic significance, while molluscum contagiosum consistently causes isolated illness in man. The remainder of the agents are of importance in their various species, including monkeys, buffalo, swine, camels, sheep, rabbits, mice, goats, cows, pigeons, turkeys, canaries, and other birds. With the eminent success of the World Health Organization (WHO) smallpox eradication program, it has become increasingly important to determine (1) whether any nonhuman reservoir for variola virus permits persistence of the agent and (2) whether any of the poxviruses of other species can produce epidemic disease in man. To date a negative answer can be given to both these questions. However, careful continued epidemiologic and clinical surveillance, bolstered by laboratory studies of poxvirus DNAs by restriction endonuclease analysis, will be required in the coming years after the anticipated eradication of naturally occurring smallpox.

Smallpox (variola)

Smallpox has been recognized as a major health problem for at least three thousand years. The facial pocks of the mummy of the Pharaoh Ramses V, who died about 1160 BC, suggest that he was a victim of the disease. Early epidemics may have been confined to the Orient, Middle East, and Africa until spread into Europe by the Arab expeditions of the sixth century. Increased explorations, military expeditions, commerce, and urbanization all contributed to the disease's continued spread throughout the world. It is probable that early sixteenth century Spanish conquistadors introduced smallpox to the Western Hemisphere with transmission to Aztec natives. Susceptible Indian populations were decimated by the infection in the succeeding centuries as it spread throughout the Americas. Early American colonists were familiar with the illness, recording outbreaks in nearly all principal towns and cit-

ies along the Atlantic coast. A noteworthy epidemic in 1721 involved 5,889 of Boston's 12,000 inhabitants. The Reverend Cotton Mather, of Salem witchcraft trial fame, had urged Boston's physicians to practice variolation as a means of prophylaxis, but only Dr. Zabdiel Boylston responded to Mather's pleas. The publication in 1798 by Edward Jenner of his work with cowpox heralded the beginning of a new era in the prevention of smallpox. Despite widespread acceptance of his vaccination procedure, and its use with increasing frequency throughout Europe and the United States, another century passed before major epidemics were eliminated in these nations. By 1967 there were still 33 countries, mainly in Asia and Africa, where smallpox remained endemic, claiming millions of victims annually. The WHO program for eradication was initiated that year and culminated in success, with the last naturally occurring case detected in Somalia in October 1977.

EPIDEMIOLOGY

Man is the only known natural host of smallpox virus, but nonhuman primates have been successfully infected in the laboratory setting to serve as experimental models. Transmission of virus occurs from person to person, and there are no intermediate vectors or reservoirs. The two major sources of infectious material are respiratory tract secretions with droplet spread and fragments of scabs from skin lesions which become airborne as they are desquamated. The virus seems able to persist for lengthy periods in the protein and cellular debris of the sloughed skin lesions. Although it is highly contagious, smallpox does not compare with measles or chickenpox in its attack rate of susceptible contacts. Members of a patient's household are the most likely to become infected. Another group with a high rate of spread has been hospital personnel caring for a patient with the illness. Importations of smallpox by air travellers from Asia to Western Europe in the 1950s and 1960s frequently produced clusters of cases among nurses, physicians, hospital workers, and other hospitalized patients. These were noted in England, Yugoslavia, Sweden, West Germany, and other nations free of endemic disease. Although respiratory tract shedding of virus may precede the appearance of the exan-

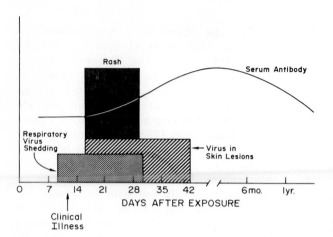

FIG. 12-1. The course of infection with smallpox virus.

them (see Fig. 12-1), the patient is most highly infectious with the onset of rash and remains so for as long as 4 weeks. Those susceptibles in intimate contact with the patient are most likely to become infected and develop disease after about 2 weeks' incubation. However, there have been occasional documented examples of spread to nonimmunes who never came in direct contact with the patient, but were exposed to aerosolized infectious material carried by air currents to more remote areas of a hospital or laboratory.

As noted above, the widespread use of smallpox vaccine has been responsible for striking alterations in the geographic patterns of the disease. In 1945 there were still 91 countries reporting smallpox with millions of cases annually. The last recorded case in the United States was an importation in 1949, but there had been 48,000 patients in 1930. The WHO eradication program began in 1967, when 62 nations still harbored the disease. By 1974 only eight countries reported cases, and it was endemic in only four (India, Pakistan, Bangladesh, and Ethiopia). The last known naturally acquired case was recorded in the town of Merka, Somalia, on 26 October 1977. Because there is no compelling evidence that smallpox virus can persist in any nonhuman reservoir, it is likely that laboratory deep-freezes now hold the only variola virus in the world.

Clinical Illness

Classic smallpox (variola major) is a severe, biphasic, febrile illness which may carry a mortality rate of 10 to 50 percent. An initial prodro-

mal phase is characterized by the abrupt onset of high fever, chills, headache, backache, and malaise lasting from 2 to 4 days. The patient is toxic and may be obtunded or delirious. This is followed by a sudden, dramatic improvement as temperature falls and respiratory symptoms develop, including sore throat and cough. At this time rash begins, often starting on the face and arms, but progressing to involve the trunk and lower extremities. Simultaneously, an enanthem develops with painful ulcers of the buccal, pharyngeal, and bronchial mucosa. The rash starts as macules which within a few hours have become papules. Over the next several days the lesions vesiculate and then between the 7th and 9th day evolve into frank pustules. Fever rises again, and the patient becomes severely ill with constitutional symptoms as well as intensely painful local lesions. By the 10th day the pustules start to rupture and then to crust with sloughing of the scabs over the next 5 to 10 days, leaving scars in the areas of previous involvement.

There is considerable individual variation in the clinical picture, and the rash may range from discrete to confluent lesions. A small number of patients undergo severe hemorrhagic smallpox with a rapid, fulminant course, including hemorrhages in the skin and mucous membranes in the absence of the usual eruption. This form of the disease is almost always fatal. In contrast, a very mild, modified illness may occur in partially immune individuals. The rash is sparse and evolves over a shorter period of time. Because the patient does not feel ill, he may be a source of virus spread throughout the community, since he remains ambulatory and pursues his usual activities.

A less virulent strain of smallpox virus was responsible for outbreaks of a relatively benign form of illness known as variola minor or alastrim. It had a prolonged incubation period of 15 to 20 days, a mild prodromal phase, a discrete rapidly evolving eruption, moderate constitutional symptoms, and a mortality rate of less than 1 percent.

In general, the smallpox rash had a number of reliable characteristics.

1. The lesions had a centrifugal or peripheral distribution, with a greater density over the face, hands, feet, legs, and forearms. On the trunk, thighs and upper arms rash was more sparse and discrete.
2. On any one area of involved skin, the lesions were at a single stage of development (monomorphic).
3. The lesions themselves were round with smooth edges, umbilicated, extended deep into the dermis and felt shotty on palpation.

Complications included pneumonia, keratitis with corneal damage, encephalitis, and peripheral neuritis. Most common were secondary bacterial infections of the skin and respiratory tract with cellulitis, furunculosis, impetigo, and pneumonia. Bacteremia resulted in metastatic foci, including pyogenic arthritis and osteomyelitis. No specific effective antiviral chemotherapy was available, but the oral administration of methisazone (methylisatin-beta-thiosemicarbazone) early in the incubation period, promptly after a known exposure, did diminish the attack rate in susceptible contacts. In a similar pattern of use, vaccinia immune globulin showed some prophylactic efficacy. Supportive therapy with fluids, electrolyte replacement, nutritional supplementation, and antibiotic treatment of secondary bacterial complications were all of importance.

Pathology and Pathogenesis

After its entry via the upper respiratory tract, smallpox virus spread to regional lymphatics and nodes from which a primary viremia then emanated. Dissemination was widespread, with involvement of all tissues, especially epithelial cells. This coincided with the prodromal clinical phase of illness. Lesions began in the areas surrounding infected endothelial cells of small

vessels. The skin lesions evolved as foci of mononuclear cell infiltration and edema, progressing to swollen epithelial cells within which acidophilic inclusion bodies (Guarnieri bodies) appeared in the cytoplasm. Epithelial cell necrosis and accumulation of interstitial fluid resulted in vesicle formation. A similar evolution took place in mucous membranes where rapid unroofing of the vesicle surface initiated ulcerative lesions. Focal necrotic lesions also developed in spleen, testis, liver, lung, and other organs. Virus released from these multiple sites of replication produced a secondary viremia coinciding with the eruptive phase of disease.

As noted above, the severity of smallpox is quite variable and probably correlates with the extent of virus replication. This quantitative aspect depends on the virulence of the strain of infecting virus, the host's status, the actual inoculum, and the age of the host. Those patients at the extremes of life, the young infant and the aged, have more severe disease and are more likely to die as a result. Prognosis correlates well with the severity of the prodromal phase and the type of rash. Striking toxemia and prostration in the first days of illness herald a poor outcome. The hemorrhagic rash may presage an 80 percent or higher mortality; confluent exanthem, 50 percent; discrete lesions, less than 10 percent.

Immunity

There is no known natural immunity to infection with smallpox virus. An attack of the disease generally confers active immunity of lifetime duration. Second cases are said to have occurred, but with great rarity. Prior to the discovery of vaccination, variolation was widely practiced as a means of inducing active immunity. Attributed originally to the Chinese, this consisted either of inhalation of a powder prepared from dried scabs shed by a smallpox patient or the direct inoculation into one's skin of material harvested from the pustules of a patient. A mild form of illness ensued, which in most instances produced immunity of the same degree as that induced by the full-blown disease. However, occasional recipients developed severe smallpox from variolation, and there were deaths in as many as 1 to 2 percent. Lady Mary Montagu, a renowned London beauty who had been earlier disfigured by an attack of smallpox and the wife of the English ambassador to the Turkish court in Constantinople, is credited with the enthusiastic popularization of variolation in Great Britain in the early 18th century after she observed its success in Turkey.

The introduction of vaccination as a method of active immunization against smallpox is one of the fascinating sagas of medical history. On May 14, 1796 Edward Jenner inoculated 8-year-old James Phipps with material obtained from a cowpox lesion on the hand of Sarah Nelmes, a Gloucestershire milkmaid. Six weeks later, James was fully resistant to smallpox when challenged. Over the subsequent years, vaccination achieved a widespread role as the principal prophylactic method for smallpox prevention. It was not until the WHO program began in 1967 that the final stages in disease eradication provided the culmination in 1977 of Jenner's observations 181 years earlier.

Following natural infection or immunization, virus-specific antibodies appear in the patient's serum. These will fix complement, inhibit hemagglutination, neutralize viral infectivity, and/or inhibit agar gel precipitation (AGP). A fourfold rise in paired sera may rarely assist in diagnosis of an acute illness. More often, the presence of antibody in a single specimen is helpful in documentation of immunity due to past infection or vaccination.

Diagnosis

When smallpox was a common disease, diagnosis was often established on the basis of (1) a clinically compatible illness, (2) a known recent exposure in an endemic area, (3) the absence of a history of vaccination or a visible vaccination scar. Now that the infection has become a rarity, laboratory confirmation of any suspected case is absolutely essential because of the public health implications and possible consequences. A number of available techniques combine to offer rapid and reliable laboratory diagnosis.* Table 12-1 lists the methods of laboratory diagnosis. In actual practice those laboratories most familiar with smallpox virus will employ four tests: electron microscopy (EM), AGP, and viral propagation in cell cultures and on the chorioallantoic membrane (CAM). The materials to be

---

*Any suspicious patient or specimen material should be discussed promptly with the Center for Disease Control, Atlanta, Georgia, telephone number (404) 633-3311; attention: Biohazards Control Officer.

TABLE 12-1   LABORATORY METHODS FOR SMALLPOX DIAGNOSIS

A. Direct visualization of poxvirus
   1. Electron microscopy (EM)

B. Visualization of elementary bodies
   1. Light microscopy of stained smears

C. Identification of viral antigens
   1. Agar gel precipitation (AGP)
   2. Fluorescent antibody (FA)
   3. Complement fixation (CF)

D. Isolation of virus
   1. In ovo on chorioallantoic membrane (CAM)
   2. In cell cultures
      a. Primary rhesus kidney cells
      b. Infant foreskin fibroblasts (FS-32)
      c. Vero cells

E. Serologic confirmation
   1. Specific antibody rise (4-fold) in paired sera
      a. Complement fixation (CF)
      b. Virus neutralization (VN)
      c. Agar gel precipitation (AGP)
      d. Hemagglutination inhibition (HI)

tested include aspirated vesicular fluid, smears of lesions, suspensions of ground scabs, and anticoagulated blood. Virus can be visualized with any of the first three types of specimens. From fatal cases, autopsy samples of lung, liver, spleen, and kidneys can be utilized for virus isolation. Electron microscopic identification of a poxvirus may be accomplished within an hour; AGP requires only 1 to 4 hours. Rapid diagnosis is therefore possible with the skill of those working regularly with various poxviruses by these methods.

Because chickenpox (varicella), disseminated herpes hominis, vaccinia, and some noninfectious exanthemata may mimic smallpox, it is not surprising that many "false alarms" have been avoided or aborted by use of the technics listed. Monkeypox, although indistinguishable by EM and AGP, produces distinct CAM lesions which can be identified by pock morphology. Vaccinia and variola are similarly distinguished by varying pock characteristics and growth in cell culture systems. In a 2-year period shortly before the successful eradication in 1977, the Center for Disease Control's laboratory examined specimens from 422 suspect smallpox patients. Using a battery of four tests (EM, AGP, and isolation of virus in ovo and in cell cultures), they successfully excluded the infection in 200 cases, which proved to be herpes hominis, or a nonvariola poxvirus. Among the 222 positive for smallpox, 92 percent were EM positive in about 2 hours; 85 percent were identified within 4 hours by AGP; 95 percent could be recognized in 3 days by CAM inoculation.

Prevention

The appropriate use of vaccination has been the most effective means of preventing smallpox. Vaccinia virus confers protective immunity against variola for a lengthy period of time, albeit less enduring than the active immunization produced by the disease itself. Past field experience with studies of vaccine prophylaxis under conditions of natural exposure demonstrated a 99 percent efficacy for 1 year, 95 percent for 3 years, 85 percent for 10 years, and 50 percent for 20 years. It was also noted that previously vaccinated individuals, whose immunity had waned sufficiently so that they developed smallpox, tended to have a much milder clinical illness and a mortality rate far lower than that of unvaccinated patients. This ameliorating effect persisted as long as 20 years.

If a patient with smallpox is identified, a number of steps must be taken promptly to limit and prevent spread to contacts. The patient should be strictly isolated until scabs have disappeared. All persons in the same household should be vaccinated as soon as possible and kept under careful medical surveillance for 17 days after their last contact with the patient. Careful measures for disinfection of the patient's bedding, clothes and room are recommended. The Public Health Service, notified immediately through the Center for Disease Control, will initiate an investigation of the source of infection and of all possible contacts. Since the last naturally acquired case in Somalia in 1977, the only known patients with smallpox have been individuals whose exposure was traced to laboratories where the virus was under study.

VACCINIA

As mentioned above, Jenner introduced vaccination to the public with a paper published in 1798 based on his studies begun in 1796. It is not possible to establish today what virus he actually employed, or what has happened to it over the ensuing nearly two centuries of multiple human and animal passages. Despite the common belief that vaccinia virus is cowpox or a closely related derivative, its true identity remains a mystery. Jenner believed his strain was horsepox ("grease") which had been passed in cows before transfer to man. Laboratory markers show that current vaccinia strains differ from cowpox as well as from other known members of the poxvirus group. Some investigators suggest that it is a recombinant of cowpox and smallpox.

Vaccine is prepared from dermal vaccinia lesions of infected calves. The harvested material is processed to prepare a lyophilized vaccine that is resistant to heat inactivation and therefore stable in many climatic settings. The WHO standard requires 10^8 CAM pock-forming units per ml of reconstituted vaccine. It may be administered by a number of methods, including scratch, multiple pressure and multiple puncture technics, and intradermal jet injection. Successful vaccination produces a local vesicle which progresses to pustular scabbing over a 14-to-21-day period in a susceptible recipient. Revaccination may induce a more rapid and locally constricted lesion which resolves in 7 to 14 days.

Until 1971 it was recommended that all infants in the United States receive smallpox vaccination as a routine component of the childhood immunization schedule. The absence of endemic smallpox, the absence of disease importations since 1949, and the recognition of certain common and other rare complications of vaccination (Table 12-2) led public health and pediatric authorities to discontinue this practice after 1971–72. Until that time nearly four million primary vaccinations and eight

TABLE 12-2 COMPLICATIONS OF SMALLPOX VACCINATION

A. Common, ordinarily benign
   1. Autoinoculation or accidental implantation
   2. Toxic eruptions ("erythema multiforme")
   3. Secondary bacterial local infection

B. Rare, serious
   1. Generalized vaccinia
   2. Postvaccinial encephalitis
   3. Eczema vaccination
   4. Vaccinia necrosum (progressive vaccinia or vaccinia gangrenosa)

million revaccinations had been performed annually. With careful surveillance of vaccine complications, it was possible to exclude certain candidates who were at increased risk of these problems. Patients with exfoliative dermatitis (especially eczema) were likely to develop diffuse vaccinial lesions over the involved skin. This followed accidental vaccination of the patient himself or exposure to a household or other close contact who had an active vaccinial lesion. Individuals with immunodeficiency or immune-suppressive therapy, particularly those with compromised T-cell function, were more apt to develop progressive vaccinia. It was not possible to predict the rare patient who might develop encephalitis in the 10 to 14 days after inoculation, but there seemed to be an increased risk correlated with increasing age at the time of primary vaccination. Autoinoculation was relatively common among infants who scratched the site of their primary lesion and then introduced the virus elsewhere on their uninvolved skin. This was not a source of any serious problems unless the secondary site was the eye, where keratitis was a possible outcome. Local bacterial infections with pyogenic skin flora, especially staphylococci or streptococci, occasionally precipitated a cellulitis at the lesion site.

MOLLUSCUM CONTAGIOSUM

This is a specifically human disease spread by direct contact and inoculation of the virus into minute abrasions in the skin. Two to eight weeks after inoculation a small papillomatous lesion appears at the site of inoculation. These lesions are easily distinguished from other papillomata by their umbilicated appearance produced by central degeneration and by the formation of satellite nodules at the periphery of the parent lesion. On pathologic examination, the epidermis shows ballooning degeneration, acanthosis, hyperplasia, and the presence of large acidophilic inclusions (molluscum bodies) in superficial epithelial cells. Ordinarily, the lesions, which are white in color and painless, disappear after a few months and require no therapy. The incubation period varies widely from 14 to 50 days. The disease is worldwide, but much more common in some localities than in others. Swimming pools may be a source of infection.

ORF

This old Saxon term refers to the infection of man with the virus of contagious pustular dermatitis (scabby mouth) of sheep, which is a parapoxvirus. In lambs the disease is characterized by the development of watery papillomatous lesions on the cornea, lips, and mouth, which usually resolve in 4 to 6 weeks without consequence. In man, vesicles usually develop at the site of abrasions on the hand or face, which may evolve into hyperplastic nodular masses accompanied by regional lymphadenopathy. Ordinarily, lesions require no treatment. Since parapoxviruses and orthopoxviruses are only very distantly related immunologically, vaccination provides no protection. Orf is an occupational disease associated with handling of sheep, particularly in shearers.

MILKER'S NODULES (PSEUDOCOWPOX)

This condition is also caused by a parapoxvirus. It is a cutaneous disease of cattle that may be transmitted to man by contact. Lesions appear 1 to 2 weeks after contact and are similar to those produced by contagious pustular dermatitis. Recovery is complete in 4 to 8 weeks. Reinfection may occur, indicating that immunity does not last long.

COWPOX

Cowpox produces a vesicular eruption on the udders of cattle from which it may be transmitted to man. Usually the illness in man is a limited vesicular eruption of the skin, though rarely the disease develops into a widespread eruption with systemic symptoms. The disease occurs occasionally in Europe but appears to be virtually unknown in the United States.

FURTHER READING

Bauer DJ, St. Vincent L, Kempe CH, et al: Prophylaxis of smallpox with methisazone. Am J Epidemiol 90:130, 1970

Baxby D: The origins of vaccinia virus. J Infect Dis 136:453, 1977

Baxby D: Identification and interrelationships of the

variola/vaccinia subgroup of poxviruses. Prog Med Virol 19:216, 1975

Cho CT, Wenner HA: Monkeypox virus. Bacterial Rev 37:1, 1973

Esposito JJ, Obijeski JF, Nakano JH: Orthopoxvirus DNA: Strain differentiation by electrophoresis of restriction endonuclease fragmented virion DNA. Virology 89:53, 1978

Fenner F: The eradication of smallpox. Prog Med Virol 23:1, 1977

Goldstein JA, Neff JM, Lane JM, Koplan JP: Smallpox vaccination reactions, prophylaxis, and therapy of complications. Pediatrics 55:342, 1975

Kempe Ch: Smallpox: Recent developments in selective vaccination and eradication. Am J Trop Med Hyg 23:775, 1974

Moss B: Reproduction of poxvirus. In Fraenkel-Conrat H, Wagner RR (eds): Comprehensive Virology, Vol 3. New York, Plenum, 1974, pp 405–474.

Ricketts TF, Byles JB: The diagnosis of smallpox, Vols. I & II (Reprinted from the 1908 London edition), US Dept of HEW, Bureau of Disease, Prevention and Environmental Control, Division of Foreign Quarantine, 1966

CHAPTER 13

Herpesviruses

The herpesviruses of man include herpes simplex virus, the varicella-zoster virus, cytomegalovirus (CMV), and the Epstein-Barr virus (EBV). Herpesviruses in general (though not EBV) have a tendency to infect derivatives of the ectoderm; it is not surprising, therefore, to find that their infections manifest skin or nervous system involvement. Some herpesviruses possess oncogenic potential; at least two of the human herpesviruses are strongly suspected of causing cancer in man. The other fascinating characteristic of herpesviruses is that they may establish latent infections, with reactivation possible after variable periods of quiescence.

Herpes simplex virus

The word "herpes" derives from the Greek, *herpos,* to creep. Hippocrates probably knew and described herpetic infections, Galen may have used the name herpes to describe zoster, and Richard Morton, who gave us the name "chickenpox," also wrote a good account of herpes simplex infections.

Herpes simplex virus infections commonly involve the skin, mucous membranes, eyes, and central nervous system. The initial or primary infection is generally acquired through the mouth, mucous membranes, or broken skin. Infection may also be acquired by venereal contact. Most primary infections are unrecognized or subclinical. In many (and perhaps all) cases the virus becomes latent, and the patient develops antibody. Subsequently, in response to a variety of stimuli, the latent virus may be reactivated. Reactivation is not usually accompanied by a significant antibody change and may occur with or without recognizable accompanying lesions. The spread of recrudescent herpes simplex virus very likely occurs cell to cell, since otherwise it is difficult to imagine how the virus would avoid the neutralizing effects of circulating antibody.

Recurrent herpes infections may be stimulated by fever, menstruation, exposure to sunlight (ultraviolet irradiation), emotional upsets, or intercurrent infections. Not all febrile diseases are equally efficient in stimulating reactivation of these recurrences. Herpetic lesions are very common in malaria and in pneumococcal pneumonia, less frequent in brucellosis, and rare in typhoid fever.

Primary Herpes Simplex Virus Infections

Gingivostomatitis This condition affects children between the ages of 1 and 6 years, occurs without any seasonal distribution, and is accompanied by fever and a sore mouth. The gums become painful and swollen and then vesiculated. Ultimately the vesicles ulcerate, and the gingival surface bleeds readily. Vesicles also appear on the buccal mucosa, tongue, and lips. The lesions less commonly involve the tonsillar pillars and pharynx. The child remains febrile and ill for about a week, though the ulcers may require longer to heal completely.

It is apparent that the most heavily involved tissues are those of ectodermal derivation. In contrast, herpangina, a reflection of enterovirus (Coxsackie A) infection, is accompanied by painful vesicular lesions, involving primarily the posterior oropharynx (tissues of endodermal origin). This distinction is not absolute, however, and herpangina and herpetic gingivostomatitis are commonly confused.

The major problems encountered by children with herpetic gingivostomatitis relate to the extreme discomfort, as well as the difficulty in maintaining fluid balance and adequate nutrition during the acute phase of disease.

Vulvovaginitis In this case the primary lesions involve the mucous membranes and skin of the labia and lower vagina. The ulcers are accompanied by fever and regional lymphadenopathy. Primary vulvovaginitis is recognized more frequently than is its counterpart in males, herpes progenitalis.

Meningoencephalitis Many cases of herpetic meningoencephalitis follow a primary infection of the central nervous system. However, in some instances this illness follows viral recurrence in individuals with preexisting antibodies. These illnesses are severe and may be quite damaging. They are accompanied by significant swelling, necrosis, and destruction of the involved brain. The mortality may be as high as 70 to 80 percent, and survivors frequently manifest residual brain damage. Encephalitis frequently involves the temporal lobe and may present as a mass lesion. Indeed, radiologic studies may suggest the presence of a brain tumor.

Keratoconjunctivitis The eye represents an important site of serious herpetic infection. Corneal ulcerations induced by herpesvirus may be quite deep and can result in blindness. Primary herpetic involvement of the eye can be severe and should be treated promptly.

Eczema Herpeticum (Formerly called Kaposi's varicelliform eruption. This confusing eponym, which was also applied to eczema vaccinatum, should not be used.) This condition is a complication of eczema or severe atopic dermatitis. The abraded weeping and denuded skin is inoculated with virus, which spreads widely in the absence of the protective cornified epithelium. The onset of disease is heralded by fever and nonspecific symptoms, followed by the appearance of vesicles which evolve and crust in a fashion similar to those of varicella. A Tzanck preparation (pp. 239 and 241) may be used to identify herpetic infections. Lesions may appear for more than a week. The course is variable, ranging from a mild illness to a severe fulminant and fatal disease.

Involvement of Abraded and Injured Skin Herpetic infections may occur in the skin of burned patients, in wrestlers (herpes gladiatorum), and at the site of injured cuticles on the fingertip (herpetic whitlow). The latter condition is an occupational hazard if personnel do not wear gloves when providing mouth care to debilitated patients.

Disseminated, Visceral, and Congenital Herpes Simplex Virus Infections When a primary herpesvirus infection occurs in a pregnant woman, the fetus is at risk just as in the case of maternal rubella or CMV infection. Unprotected by maternal antibodies, the fetus may become infected in utero and develop stigmata of severe disseminated congenital viral infection. In the case of herpes simplex virus, the major involvement is of the liver and the central nervous system.

If the maternal infections involve the cervix or the labia, the infant may be infected during passage through the birth canal. In this case, the manifestations of infection may be delayed several days postnatally. It has been suggested that if an ongoing significant maternal genital infection is recognized prepartum, it is an indication for caesarean section, although in some instances even operative removal of the fetus has failed to prevent evolution of the disease, suggesting that transmission occurred earlier by an ascending route or due to a maternal viremia.

Disseminated herpes simplex virus infections in the newborn carry a grave prognosis. Although most of these babies succumb, a few survive, making difficult the evaluation of the few (uncontrolled) trials of systemic therapy.

Individuals other than the fetus and neonate subject to disseminated herpes simplex virus infections include those patients severely debilitated and/or malnourished, individuals receiving immunosuppressive therapy or having a severe underlying immunologic deficit, and occasional patients undergoing certain intercurrent infections (such as measles) which abrogate normal cellular immune mechanisms.

Recurrent Herpes Simplex Virus Infections

Individuals who experience recurrent herpes simplex virus infections always have preexisting serum antibody to the virus. The manifestation of recurrence is most frequently a skin eruption—grouped vesicles surrounded by a halo of erythema. Prior to the appearance of the vesicles the patient may experience a burning or itching sensation at the site. Most often these eruptions occur on the face at mucocutaneous junctions, such as the nostrils or, more frequently, the lips (herpes labialis). The vesicles evolve, crust, and fade in approximately one week.

Recurrent herpetic eruptions may also occur at the site of traumatic virus inoculation of the skin. In addition, the vesicles may appear in other areas of mucocutaneous junction, including the penis (herpes progenitalis), the urethral orifice, and the vulva.

Although eruptions recur in the same site with little variation, latent virus does not reside in the skin. Areas of affected skin have been excised between recurrences, and fresh unaffected skin has been transplanted. After this procedure, the recurrent disease may return at the original site. It is likely that virus is latent in the ganglia of sensory nerves innervating the affected area of skin. In support of this concept, recurrent herpetic lesions of the skin occasionally follow the distribution of sensory nerves, imitating zoster in appearance (zosteriform herpes) and herpes simplex virus has been recovered from explanted neural ganglia in man and experimental animal hosts. It has been proposed that the virus remains latent in sensory nerve ganglion cells and, under certain circumstances, travels down the associated neural axons to involve the skin and associated organs of the innervated dermatome.

Corneal herpes may also recur, with resultant repetitive attacks of keratoconjunctivitis. These repeated episodes of herpetic activation may result in progressive corneal injury and, finally, blindness.

Epidemiology

Herpes simplex virus infections are virtually universal, and there are no appreciable seasonal patterns. Most people are infected by the age of 6 or 7 years. Only about 1 percent of those infected experience a recognizable associated illness. A much larger percentage of the general population possessing antibodies will develop recurrent herpetic eruptions at some time. Herpetic infection is more common in lower socioeconomic groups. Spread appears to be primarily by direct contact.

Two subtypes of herpes simplex virus have been distinguished: herpes simplex virus type 1, which usually infects the mouth, throat, respiratory tract, eyes, and central nervous system; and herpes simplex virus type 2, which involves primarily the genitalia. These types are commonly termed "oral" and "genital" strains. Although either strain may involve the oral or genital mucosa, there is a striking tendency for oral strains to produce infections above the waist, while herpetic infection below the waist is usually caused by genital strains. These subtypes may be distinguished on the basis of serologic and biologic studies. These variants can also be clearly distinguished biochemically.

For a discussion of the role of herpes simplex virus in the etiology of cancer, see Chapter 9.

Diagnosis

Diagnosis of herpesvirus infections may usually be made on clinical grounds. However, where uncertainty or therapeutic considerations arise, virologic and serologic techniques are available. Because of the rapid replication of *H. simplex* in vitro, a more prompt diagnostic study may be performed in this instance than in the case of most viruses.

A Tzanck smear (fixed stained scrapings from the vesicle base) will reveal multinucleated giant cells and intranuclear inclusions in the case of herpesvirus infections. One cannot by this means, however, distinguish between varicella-zoster virus and herpes simplex virus infections.

Rapid (hours) identification of herpesviruses may be made employing examination of fixed negatively stained samples, though these methods cannot distinguish the various herpesviruses one from another. Specific, accurate, and rapid identification of herpes simplex antigens may be achieved by employing immunofluorescent (FA) techniques. Diagnoses based on FA and EM usually require the presence of a substantial concentration of virus and—if the specimen contains lesser quantities of virus (such as may occur when a brain biopsy taken for the identification of herpes encephalitis does not include the areas of heaviest virus concentration)—supplementation of FA and EM techniques with tissue culture inoculation is desirable.

Treatment and Prevention

In 1962 Kaufman demonstrated that herpetic keratitis could be ameliorated with the topical use of 5-iodo-2'-deoxyuridine (IDU). This halogenated pyrimidine competes with thymidine for incorporation into DNA, thus yielding a biologically inactive nucleic acid (Chap. 8). These studies gave vast impetus to the field of antiviral chemotherapy. However, although treatment with IDU did have a significant effect on the superficial keratitis, it did not actually cure the disease or heal deep lesions. Recurrences were not prevented, and resistant mutants appeared after prolonged therapy. Topical treatment with IDU of recurrent herpetic skin eruptions and systemic IDU therapy have both been disappointing, and the results of controlled studies have been negative or at best equivocal.

Whereas steroids are helpful in certain viral infections or postviral immunologic diseases, these substances are contraindicated during most herpes virus simplex infections. Whether applied topically or administered systemically, they appear to enhance virus spread and should be avoided, although steroids have been administered to patients with herpetic encephalitis to reduce brain swelling and edema.

Attempts to treat recurrent herpes simplex infections with topically applied chemicals have been unrewarding. The use of photodynamically activated dyes, though reported by some workers to reduce the effects of recurrent herpes infections, has been abandoned because of the enhanced oncogenic potential of the affected virus.

For the management of life-threatening encephalitis caused by herpes simplex type 1, systemic adenine arabinoside (ara-A) has been found to have a good therapeutic index and is currently approved for use by the FDA. A brain biopsy is recommended in order to make an accurate diagnosis, and in order to achieve maximum benefit, it is recommended that the drug be given as early as possible—preferably before the onset of coma.

Experimental studies indicate the dramatic effectiveness in arresting herpes simplex infections of 9-(2-hydroxyethoxymethyl) guanine—or acycloguanosine—a nucleoside analog. This compound, of low toxicity for uninfected cells, is phosphorylated and thus activated only by the herpes-specified thymidine kinase. The potent and selective antiherpes activity of acycloguanosine coupled with its apparent lack of toxicity have raised expectations that an effective means of antiherpes therapy may be at hand. Plans for human trials of acycloguanosine have been approved, and preliminary investigations in man are currently being initiated.

Vaccines to interrupt and prevent herpes simplex infections have been considered for some time. Crude preparations (Lupidon G and H) have been available and in use in Germany for several years. A double-blind study of the Lupidon vaccines is currently underway. A purified noninfectious herpes simplex subunit vaccine is currently being developed in the United States. Consideration of vaccines for herpesviruses raises many difficult questions, some of which are listed in Table 13-1.

TABLE 13-1 HERPESVIRUS VACCINE? SOME QUESTIONS TO BE CONSIDERED*

1. Is disease serious enough to warrant a vaccine?
2. Do high-risk groups exist?
3. How shall the effectiveness of a vaccine be evaluated?
4. Does immunization prevent the establishment of latent infections?
5. What happens if titers wane after immunization? Will this change the clinical course?
6. Is it necessary to evaluate cell-mediated as well as serum immunity?
7. What is the degree of cross-reactivity among herpesvirus strains?
8. What type of vaccine should be developed and what are the associated attendant risks?
9. How might further information on establishment of latency and reactivation influence the approach to prevention and therapy?
10. Is there any reason to vaccinate a person with a latent infection?
11. What type of treatment can be offered to persons with primary, latent, or recurrent herpesvirus infections?

*Modified from a list of questions posed by Dr. Abner Notkins at The Experimental Herpesvirus Vaccine Workshop held in Bethesda, Maryland, February 8–9, 1979.

Varicella (chickenpox)

Clinical Course

In nosocomial outbreaks the incubation period is between 15 and 18 days. The existence and severity of the prodrome is variable, ranging from nothing recognizable in the pre-eruptive phase to a moderate period of fever, headache, anorexia, and malaise. In most children systemic symptoms and the exanthem appear simultaneously. The course of the disease is illustrated in Fig. 13-1.

The characteristic rash of varicella may be preceded by a brief scarlatiniform eruption. Thereafter the exanthem appears abruptly, usually first on the trunk, then spreading to involve the extremities. Macules evolve quickly into papules and then into superficial clear watery vesicles surrounded by an erythematous area. These water dropletlike vesicles are in

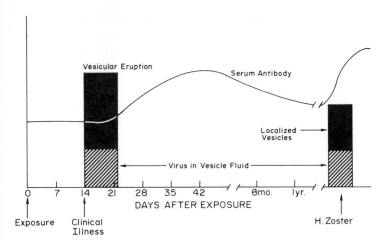

FIG. 13-1. The course of varicella virus infection.

contrast to the lesions of variola and vaccinia, which are more firm and deeply seated. Lesions frequently appear on the scalp as well as on the palms and soles. Where creases of the skin are involved the vesicles may become oval and elongated. The vesicles are usually not umbilicated and do not feel hard except occasionally over areas of heavy cornification such as the palms and soles. The vesicular fluid becomes cloudy (bacteria and inflammatory cells) for two to three days and subsequently dries as the vesicles crust (after approximately one week). The lesions evolve rapidly and continue to spread, with new vesicles making their appearance for up to one week. For this reason, the varicella patient characteristically appears with crops of lesions in a variety of stages.

The concentration of lesions is generally heavier on the trunk and the central portion of the extremities. In contrast to variola and vaccinia, the vesicles and pustules of chickenpox tend to involve the axillae and generally favor covered, protected areas of the trunk. Varicella lesions concentrate in body hollows rather than on prominences. The palate and tonsillar areas may also be involved.

Pathologically, varicella skin lesions consist of ballooned injured and degenerating cells in the deeper layers of epidermis, accompanied by a massive outpouring of extracellular fluid. In the base of the lesion, giant cells may be found, some multinucleated, with intranuclear inclusions. In contrast, scrapings from the vesicles of vaccinia yield cells with cytoplasmic inclusions only. These observations form the basis of the Tzanck test, in which scrapings from the base

of fresh vesicles are prepared with Giemsa stain and examined microscopically.

When varicella occurs in the adult, systemic symptoms may be severe, the rash very profuse, and the entire disease much more intense than in childhood. The rash may become hemorrhagic, and occasionally bullous lesions appear. Varicella pneumonitis, rare in children, is much more frequent in older individuals. In (rare) fatal cases of varicella pneumonia, an associated hepatitis may be found at autopsy. Varicella pneumonia may be followed by the appearance of scattered calcified nodules visible on chest x-ray. Significant secondary bacterial infection of varicella lesions may occur. These are usually staphylococcal or streptococcal infections. In the latter instance, complications of the streptococcal invasion may occur, including septicemia and acute glomerulonephritis. Acute rheumatic fever does not occur following streptococcal invasion of varicella skin lesions.

Reye has described the syndrome of acute hepatic failure, encephalopathy, and hypoglycemia which may appear following several infectious entities. The syndrome, which has now been recognized by many additional clinicians, has in a significant number of cases followed varicella, in some instances accompanied by a history of the administration of large doses of salicylates.

Rarely varicella is accompanied by encephalitis, which may appear before, coincident with, or following the eruption. Although most individuals with varicella encephalitis survive and recover without sequelae, there is an incidence of mortality (about 5 percent) and brain damage (7 of 59 in the series of Applebaum reported in 1953).

In recent years, several new problems have been posed for certain special patients exposed to varicella virus. These individuals include those being treated with high doses of corticosteroids, cytotoxic or immunosuppressive drugs, and patients with disseminated malignancies, who, with the benefit of chemotherapy, are maintained alive, although with severely compromised immunologic functions. Patients receiving immunosuppressive therapy may develop herpes zoster, which can in some cases spread to become generalized. Generalized zoster is, essentially, the same condition as varicella. Very elderly and/or debilitated patients may develop varicella even in the presence of a definite past history of childhood varicella.

Epidemiology

Varicella is primarily a disease of children. There is a peak incidence between the ages of 2 and 8 years. Cases do occur in adults who have escaped infection during childhood and occasionally in individuals with compromised cellular immunity on the basis of disseminated malignancies, extreme age and disability, or immunosuppressive drug therapy. If a primary infection occurs during pregnancy, the virus is capable of crossing the placenta and infecting the fetus. Congenital infections may be associated with the appearance at birth or shortly thereafter of varicella. Neonatal varicella may be severe, and it carries a significant mortality. There is some suggestive evidence, however, that if the eruption is present at birth or shortly thereafter, the illness is more often mild, possibly reflecting the transmission of some maternal antibody. It has been recognized recently that primary maternal varicella occurring early in gestation may result in congenital infection manifested at birth by cicatrizing lesions of the extremities as well as neuromuscular disorders.

The disease is worldwide in distribution and occurs throughout all seasons, with a somewhat higher incidence in winter and spring in temperate climates. Large fluctuations in case occurrence and significant epidemics do not occur. The illness occurs sporadically and endemically in most urban centers in the

United States. Microepidemics occur especially within families and schools, reflecting the lability of this agent and the need for relatively close direct contact to effect transmission. With direct contact varicella is a highly transmissible infectious disease. Although there exists some evidence for transmission of varicella by the respiratory route, the virus has been very difficult to recover from the nasopharynx, possibly because the presence of virus in the mouth and pharynx occurs very early during the incubation period.

Diagnosis

The diagnosis of chickenpox is usually made on clinical grounds. The features of the rash already described will usually serve to distinguish this exanthem. The differentiation of varicella and variola is no longer a problem, since naturally occurring smallpox has been eradicated.

The viruses of varicella and zoster are almost certainly one and the same, and the agent is frequently called V-Z virus or VZV. The identity of these viruses is based on their behavior in tissue cultures, the interchangeable results of cross-immunologic studies, and inoculation studies. In the latter, susceptible children inoculated in the skin with vesicle fluid from zoster and varicella patients develop typical chickenpox. In the presence of a past history of chickenpox or zoster, inoculation with either fluid is without effect.

Treatment, Protection, and Prevention

Treatment of varicella is usually symptomatic and revolves in most cases around control of fever and itching. Other than this, the therapy of varicella is directed at the complications of this infection. Steroids have been advised in the treatment of varicella encephalitis, as well as in hemorrhagic chickenpox. The latter seems paradoxical, since hemorrhagic varicella has occurred as a complication of long-term steroid therapy. There exists no good evidence for the efficacy of steroids in treating these complications, although steroid therapy does not appear hazardous. The key factors here are apparently the dose and duration of steroid therapy. Intercurrent varicella in a patient on long-term steroid medication is frequently severe. Short-term therapy instituted after the onset of disease is apparently, in contrast, without significant danger.

Exposed individuals are usually not protected by administration of pooled gamma globulin. Brunell and associates prepared zoster immune globulin (ZIG) from the plasma of patients recovering from herpes zoster and found that a 2-ml dose of ZIG given within 72 hours after the initial exposure prevented varicella. Infection was apparently prevented rather than modified (failure of antibody rise). It has been shown that ZIG or VZIG (prepared from varicella-convalescent plasma) is protective when administered to varicella-exposed susceptible adults, as well as to patients with cancers, those under therapy with corticosteroids or antimetabolites, and exposed newborns whose mother contracted chickenpox at or near term.

Prolonged isolation of infected children is generally not indicated, although in order to protect the highly susceptible patients who populate modern hospitals, individuals with varicella should only be admitted to a general medical ward when absolutely necessary and then under strict isolation precautions.

After the prodromal symptoms develop or the eruption of varicella appears, preventive measures have no effect. In the face of life-threatening varicella infections the use of experimental drugs is justified. Adenine arabinoside (ara-A) has been employed for the therapy of complicated VZV infections and those occurring in immunologically compromised individuals. In immunosuppressed patients with *H. zoster* ara-A therapy accelerated clearance of vesicle virus and reduced both the appearance and time to resolution of new lesions.

A live attenuated VZ vaccine has been developed and tested in Japan. A similar vaccine is presently under study in the United States. The Japanese strain (Oka) has been tested with minimal side effects in normal children and in a highly susceptible group with malignant disease. Seroconversion has been satisfactory, though at this time the duration of antibody persistence is unknown, as is the tendency or lack of same for the establishment of latency and reactivation. Though there has been considerable debate, the current consensus seems to be that some form of VZ vaccine will be useful in certain high-risk populations.

Herpes zoster (shingles)

Clinical Course and Epidemiology

It is thought that herpes zoster in most cases probably represents a reactivation of latent virus (acquired during an earlier bout of varicella). The length of the incubation period is thus uncertain. In some patients, after chickenpox, the V-Z virus remains latent in host nerve root cells, in most cases cells of the posterior root ganglia or cranial nerve roots. The factors that reactivate the virus are ill defined. Cold and trauma have been cited. Since zoster may appear following exposure to varicella, some authors propose that the condition may also follow invasion of a partially immune host by virus newly acquired. That altered immunity plays a role in any case seems likely, since zoster frequently appears in patients with diminished immunologic competence. Zoster may appear in a newborn or young child after maternal chickenpox.

Once activated, the virus apparently moves down the sensory nerves to the innervated skin. The most common site of sensory nerve involvement is the skin innervated by the thoracic ganglia. The most frequently affected cranial nerve is the fifth (trigeminal).

The outstanding prodromal symptom of zoster is pain. The eruption appears following three or four days of pain. The skin lesions resemble those of varicella, but their distribution is strictly along the course of the affected sensory or cranial nerve(s). The evolution of vesicles is identical to that described for varicella. Regional nodes may be enlarged and painful.

The distribution of lesions is usually unilateral. When the ophthalmic branch of the trigeminal nerve is involved, the uvea may become inflamed. Involvement of the mandibular branch may lead to the appearance of vesicles on the mouth and tongue, while maxillary branch zoster may yield vesicles on the uvula and tonsil. When the geniculate ganglion is involved, the Ramsay-Hunt syndrome appears (ear pain, facial paralysis, and vesicles in the external auditory canal). It is of interest that an identical syndrome may be produced by infection with herpes simplex virus.

Pathologically, the involved skin is indistinguishable from that seen in the case of varicella. Infected sensory nerves may show axonal degeneration and demyelination.

Diagnosis

In most cases the diagnosis is readily made on the basis of the clinical findings. It should be noted that herpes simplex virus may cause identical (zosteriform) eruptions. Giemsa stain of vesicular scrapings cannot distinguish V-Z infection from that of herpes simplex virus, but V-Z virus may be grown in tissue culture and can be serologically differentiated from herpes simplex virus. Serologic studies may be helpful, since zoster patients develop high levels of antibody very rapidly after the appearance of lesions (immunologic recall).

Treatment

During the acute stages of zoster, treatment is directed primarily at control of pain. Occasionally, this pain may persist, leading to a severe painful neuralgia. Extreme measures have been attempted to control this postinfectious neuralgia. These include injection of the involved ganglion with alcohol, leukotomy, or application of cold to the affected area. The number and variety of therapeutic maneuvers employed is a reflection of the lack of consistent success resulting from the use of any one.

Cytomegalovirus

At the turn of the century, several pathologists reported the presence of large swollen cells in the liver, kidney, and lungs of infants dying with presumed congenital syphillis. Goodpasture and Talbot in 1921 and Lipschütz, in the same year, recognized that these cells resembled the swollen, injured inclusion-bearing cells seen in some virus infections, and they coined the term "cytomegaly."

Farber and Wolbach recognized salivary gland cytomegaly in an astonishing 12 percent of infants dying of all causes. Infants with overwhelming CMV infection were said to have cytomegalic inclusion disease.

A broad spectrum of clinical illnesses has been associated with CMV infections. The picture emerging from epidemiologic surveys is that of a ubiquitous virus which may be shed persistently by infected hosts. In the face of persistence and widespread occurrence, it has been difficult to assign to CMV etiologic responsibility even when the virus is recovered in simultaneous association with a clinical entity. Certain conditions have, however, been definitely or tentatively associated with CMV infections.

Congenital Infections

When a primary CMV infection occurs during gestation, the virus may invade the placenta. The infection then spreads to involve the fetal villi, followed by a viremia in the developing fetus. The results of fetal CMV infection may be dependent on the timing and extent of the virus invasion.

Many babies survive to term though infected and damaged. Typically the baby manifesting congenital CMV infection is small-for-dates and has hepatosplenomegaly and associated abnormalities of liver function. Petechial rash is common (Fig. 13-2), associated with a moderate thrombocytopenia and hemolytic anemia. The most common and devastating injuries and abnormalities are those of the central nervous system including chorioretinitis, infantile seizures, intracerebral calcifications, and, occasionally, massive necrosis of the brain. Infected infants may be microcephalic or (less commonly) hydrocephalic. Deafness has been reported, as have microphthalmia, cataracts, and glaucoma. Although these latter abnormalities occur less frequently than they do in association with congenital rubella, their occurrence emphasizes the difficulties which may be encountered in distinguishing these conditions.

In addition to arrested and/or abnormal development, some organs of affected infants manifest evidence of active or burned-out inflammation and injury. Thus one may find bile stasis associated with periportal inflammatory infiltrates, giant-cell hepatitis, focal pancreatic fibrosis, and pulmonic and renal infiltrates.

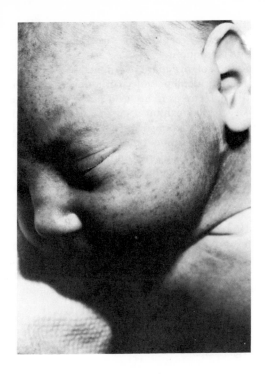

FIG. 13-2. Fine macular and petechial rash in a neonate with congenital cytomegalovirus infection.

vated and/or persistent CMV can be transmitted transplacentally. Still unresolved is the critical issue of damage: though maternal antibody may not prevent prenatal CMV infection, transplacental transfer of antibody may influence and restrict the extent of virus-induced injury. These issues and questions must be resolved before a decision can be made concerning the application of specific immunoprophylaxis for the modification of congenital CMV infection and the prevention of injury.

Studies have shown gestational and congenital CMV infections to be associated with impairment of some virus-specific cell-mediated responses. Whether these observations reflect cause or effect is unresolved. The pathogenesis of these important gestational, prenatal, and perinatal infections remains to be elucidated.

Hepatitis

Acquired CMV infections are frequently associated with hepatitis. The manifestations of CMV hepatic disease are usually milder than those associated with hepatitis A or B.

Mononucleosis

In 1965 Klemola and associates reported the association with CMV of a heterophil-negative, mononucleosislike syndrome. Affected patients may exhibit high spiking fevers and chills associated with a mild hepatitis, enlarging spleen, and large numbers of atypical lymphocytes. A similar syndrome occurring in approximately 3 to 5 percent of patients following cardiopulmonary bypass perfusion during open-heart surgery was also found to be associated with CMV infection. Blood transfusions seem to be incriminated, with the implication that CMV may be carried in the blood of asymptomatic blood donors, though pretransfusion identification of CMV carriers (in blood) has not been satisfactorily achieved.

Opportunistic Infections

Patients with disseminated malignancies, those with immunologic deficiencies, especially of the cellular variety, and individuals receiving steroids and/or immunosuppressive therapy experience clinically significant CMV infections

Recognition of congenital CMV infections was originally based on pathologic studies of fatal cases, which created the impression that CMV infection was rare and usually fatal. More recent studies, using virologic as well as exfoliative techniques, have shown that cytomegaloviruria can be detected in 1 to 2 percent of neonates. Between 15 and 20 percent of these infants eventually exhibit evidence of central nervous system damage.

Based on the experience with and understanding of congenital rubella, it had been believed that transplacental passage of CMV occurred only following primary maternal infection during gestation. However, several investigators have reported the birth of more than one infant infected prenatally with CMV. That this may occur even with identical strains of virus has been shown by the studies of Stagno and co-workers, and investigations in the United States and the Ivory Coast have shown that primary infection before pregnancy does not protect the infant from in utero CMV infection. Indeed, prenatal infection has been found to occur more frequently in an immune cohort than in a corresponding unselected population, and there is now no doubt that reacti-

very frequently. As many as 90 percent of recipients of organ transplants exhibit evidence of CMV infection during life or at autopsy. Although some of these infections are apparently recently acquired and primary, there is evidence that many represent reactivation of latent virus. Some patients develop interstitial pneumonia terminally, and at autopsy evidence may be found of extensive CMV involvement of the liver, kidneys, and central nervous system.

CMV infections have been associated with rejection and failure of organ and tissue grafts and are temporally related to many illnesses and deaths among renal and marrow graft recipients. Acquired and reactivated CMV infections have been found to cause significant chorioretinitis in these and other patients who are immunologically compromised.

It has been suggested that CMV infection, acquired with the transplanted organ or transfusions, may induce allograft rejection. It has been proposed alternatively that the rejection process reactivates latent virus, whether carried in recipient or donor cells, and experimental data has been derived to support both theses.

Epidemiology

Anywhere from 20 to 80 percent of a population possess CF antibody by adulthood. Cytomegalovirus spreads slowly and apparently requires close contact for transmission. It is presumed that CMV is transmitted in salivary secretions, and it has been suggested that venereal infections also occur. Virus may be recovered from the saliva, the urine, and the uterine cervix. Lang and Kummer have demonstrated the prolonged presence of CMV in semen in the absence of urologic symptoms or significant alterations in the composition of semen. Virus was recovered from semen when simultaneous urine specimens were negative for CMV. In addition, virus may be present in blood and has also been recovered from human milk.

The persistence of viruria following an initial infection has made assessment of the role of CMV in associated illnesses difficult. Viruria has been demonstrated in as many as 40 percent of children living in institutions. Surveys of children from 2 to 8 years of age living at home have revealed as many as 5 to 10 percent excreting CMV in the urine. Beyond the age of 10, however, recovery of virus from urine becomes infrequent.

Studies of pregnant women have shown an increasing incidence of viruria during gestation until, by term, as many as 3 percent of women surveyed may exhibit a viruria. Cultures of the cervix also demonstrate an increase in the number of positive cultures during the final trimester of gestation. Younger women tend to have a higher incidence of viruria, of virus recovered from the uterine cervix, and of symptomatic infected babies.

Congenitally infected infants may excrete CMV in the urine indefinitely, and viruria of four and five years duration has been documented. During early infancy these babies may have very high titers of virus in the urine and can be an efficient source of contagion to contacts.

Diagnosis

Identification of CMV is usually made on the basis of its characteristic cytopathology, inability to replicate in cells derived from other species, and serologic studies, most of which employ the complement-fixation test, which is usually performed with a crude CMV antigen. Indirect fluorescent antibody (FA) techniques have also proved useful in the study of CMV. Based on the observation that the fetus and neonate produce IgM antibodies which do not ordinarily cross the placenta, a CMV-specific IgM FA test has been developed which permits the identification of congenital CMV infection with the analysis of a single serum. The presence of specific CMV IgM antibody in initially asymptomatic CMV-infected infants correlates closely with the recognition later of significant central nervous system damage.

When available, laboratory isolation of CMV is the most sensitive method of identifying the presence of an active infection. Urine should be collected for this purpose in a sterile fashion, maintained cold, and transported promptly to the viral laboratory. If nasopharyngeal secretions are obtained on a cotton or nylon swab, this should be immersed in transport medium. Blood specimens for isolation should be anticoagulated to facilitate removal and inoculation of the leukocyte-rich plasma.

When no virus laboratory is available, urine

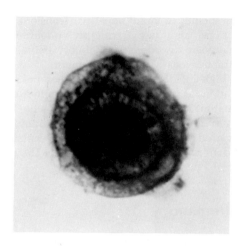

FIG. 13-3. Typical inclusion-bearing cell found in the urine of a CMV-infected individual (Papanicolaou stain). ×400.

specimens may be examined in the cytology laboratory for the presence of characteristic inclusion-bearing cells (Fig. 13-3). Absence of such cells does not rule out CMV infection.

Prevention and Treatment

Systemic administration of interferon and drugs such as halogenated pyrimidines, cytosine arabinoside (ara-C), and ara-A have been tried in some CMV infections. Though associated in some instances with transient depression of virus shedding, no consistent definite clinically significant improvement has been achieved.

Experimental CMV vaccines are under study in the United Kingdom, Europe, and the United States, although some workers feel that this approach is premature. Preliminary results of one vaccine trial in transplant recipients indicate that a CMV specific immune response is induced though acquisition and/or reactivation of CMV is not prevented. The clinical impact of these vaccines remains to be determined.

Oncogenic Potential

Ultraviolet-inactivated CMV can induce malignant transformation of hamster cells, and human cells infected with active CMV transiently acquire some characteristics of transformed cells. CMV may therefore possess oncogenic potential. This possibility is being investigated actively in several laboratories.

Epstein-Barr virus (EBV)

In 1958 Burkitt first described a malignant lymphomatous disease which affected children and occurred with a peculiarly high frequency in Central and West Africa. These tumors were found to constitute about 50 percent of all malignancies seen in African children. The neoplasm, which frequently involved the jaw and abdomen, had a characteristic cytopathology termed by pathologists a "starry sky" appearance, reflecting descriptively the presence of diffuse, primitive, poorly differentiated, large, immature cells of lymphoblastic type. Epidemiologic features suggested that a transmissible agent might be involved. In 1964 Epstein, Achong, and Barr noted that a proportion of cultured cells derived from a Burkitt lymphoma contained a herpeslike virus visible on electron microscopy (Chap. 2). This virus was named the Epstein-Barr virus (EBV), and antibody tests were derived enabling seroepidemiologic studies which indicated that infection with EBV was common and distributed worldwide. An association between EBV and infectious mononucleosis (IM) was established and extensively confirmed. Antibody to EBV capsid antigens was usually present by the time of appearance of clinical symptoms, and the etiologic association of EBV and IM was initially questioned. Subsequently, in vitro transmission of EBV to cord blood lymphoid cells was accomplished by the demonstration of morphologic transformation, and this technique enabled the recovery of EBV from pharyngeal secretions of patients with IM, as well as the specific neutralization of EBV by antibody. Additional serologic tests which have been developed for early viral antigens and structural components of EBV have made firm the association of EBV and IM. The timing of the EBV infection in relation to the appearance of IM as well as the nature of the clinical findings suggest that this illness probably reflects an immunopathologically mediated response to EBV infection.

The significance of EBV as the etiologic agent of Burkitt's lymphoma, nasopharyngeal carcinoma, and IM is discussed in Chapter 9.

Infectious Mononucleosis

In 1889 Pfeiffer described a glandular fever, "Drüsenfieber," which probably encompassed a melange of febrile disorders associated with lymphoid proliferation, including IM. In the 1920s Sprunt and Evans coined the name "infectious mononucleosis," described the clinical features more exactly, and identified the presence of the associated atypical lymphocytes. In 1932 Paul and Bunnell described the presence of IM patients of agglutinins for sheep red blood cells, heterophil antibodies. The specificity of the heterophil test was markedly improved by Howard and Davidsohn, who noted that the heterophil antibodies of IM were absorbed by beef cells but not by an extract of guinea pig kidney. Those heterophil antibodies, occasionally found in normal sera, were found to be absorbed by guinea pig kidney but not by bovine erythrocytes, while the heterophil antibody encountered in serum sickness was absorbed by both the kidney preparation and the beef cells.

Clinical Course

Since the exact mode of transmission is uncertain, the time of infection is usually unknown, and so the incubation period is indeterminate. Guesses range from 4 to 50 days.

The initial symptoms are usually headache, fatigue, fever, chills, anorexia, and general malaise, followed by lymphoid proliferation and sore throat. Various of these features may predominate, leading some authors to subclassify IM into glandular, febrile, and anginose (sore throat) types.

The illness may be brief, or it may persist for weeks. The sore throat can be quite severe and is occasionally accompanied by a striking membranous exudate. Lymph nodes are frequently enlarged and tender, particularly in the cervical chain, and splenic enlargement is almost always present.

The most common additional clinical fea-

tures include a mild hepatitis, transient skin eruptions, and, less frequently, pneumonitis, hematologic disorders (thrombocytopenia), and central nervous system involvement.

A high incidence of skin rash has been observed in IM patients, including prominent erythematous eruptions among those who have been treated with ampicillin. The ampicillin-associated IM rash has not yet been explained, but does not in most instances reflect the presence of a penicillin sensitivity.

Epidemiology

Infectious mononucleosis is thought of as a disease of young adults of higher socioeconomic strata, thus college students. Cases do occur among children, although the course in younger patients is usually less severe. Infection early in life is probably frequently inapparent and may occur more often among lower socioeconomic groups.

The age-associated attack rates (per 100,000) for recognized IM in the United States are approximately as follows:

45 in the nation as a whole
66 between the ages of 10 and 14
343 between 15 and 19 years
123 for individuals between 20 and 24

No definite large epidemics have been noted, and close contact is probably required for transmission.

Diagnosis

The diagnosis may be suspected on clinical grounds alone. The occurrence of fever, malaise, sore throat, enlarged lymph nodes, and splenomegaly in an otherwise healthy adolescent or young adult should arouse the suspicion of IM. The finding of a lymphocytosis and the presence of significant numbers (10 to 50 percent) of atypical lymphocytes further support the diagnosis. Occasionally IM patients manifest a leukopenia with diminished granulocytes. Thrombocytopenia is sometimes present, although in IM the diminished platelet count is rarely associated with clinically significant bleeding. The presence of heterophil antibodies absorbed by bovine cells and not by guinea pig kidney completes the identification of heterophil-positive IM. The heterophil test has recently been modified by the substitution of horse erythrocytes (formalinized), which permits a sensitive, rapid spot test performed on a very small sample of serum.

Although most patients have developed peak titers of EBV anti-capsid antibody at the time of recognition of IM, about 15 to 20 percent of patients may show a significant rise in this antibody (EBV-VCA) when acute and convalescent samples are compared. In comparison to VCA antibodies, antibody to EBV nuclear antigens (EBNA) may be more delayed in appearance and—if obtainable—this test may be helpful in making a specific diagnosis. Anti-EBV macroglobulins have been shown to appear during acute disease and fade by about 6 months after the illness. When available, these tests also may be diagnostically useful.

FURTHER READING

Herpes simplex virus

BOOKS AND REVIEWS

Baringer JR: Herpes simplex virus infection of nervous tissue in animals and man. Prog Med Virol 20:1, 1975

Juel-Jensen BE, MacCallum FO: Herpes simplex, varicella and zoster: Clinical manifestations and treatment. Philadelphia, Lippincott, 1972

Kaplan AS (ed): The Herpesviruses. New York, Academic, 1973

Nahmias AJ, Roizman B: Infection with herpessimplex viruses 1 and 2. N Engl J Med 289:667, 719, 781, 1973

Rapp F: Herpesviruses and cancer. Adv Cancer Res 19:265, 1974

SELECTED PAPERS

Baringer JR: Recovery of herpes simplex virus from human sacral ganglions. N Engl J Med 291:838, 1974

Bastian FO, Rabson AS, Yee CL, Tralka TS: Herpesvirus hominis: Isolation from human trigeminal ganglion. Science 178:306, 1972

Corey L, Reeves WC, Holmes KK: Cellular immune response in genital herpes simplex virus infection. N Eng J Med 299:986, 1978

Hanshaw JB, Dudgeon JA: Herpes simplex infection of the fetus and newborn. In Major Problems in Clinical Pediatrics, Vol. 17, Viral Diseases of the Fetus and Newborn. Philadelphia, Saunders, 1978

Kaufman HE: Clinical cure of herpes simplex keratitis by 5-iodo-2'-deoxyuridine. Proc Soc Exp Biol Med 109:251, 1962

Kaufman HE, Brown DC, Ellison EM: Recurrent herpes in the rabbit and man. Science 156:1628, 1967

Lehner T, Wilton JMA, Shillitoe EJ: Immunological basis for latency, recurrences, and putative oncogenicity of herpes simplex virus. Lancet 2:60, 1975

Nahmias AJ, Naid ZN, Josey WE: Herpesvirus hominis Type II infection—Associated with cervical cancer and perinatal disease. Perspect Virol 7:73, 1971

Varicella-Zoster

BOOKS AND REVIEWS

Burgoon CF Jr, Burgoon JS, Baldridge GD: Natural history of herpes zoster. JAMA 164:265, 1975

Downie AW: Chicken pox and zoster, Br Med Bull 15:197, 1959

Gordon JE: Chickenpox: An epidemiologic review. Am J Med Sci 244:362, 1962

Juel-Jensen BE, MacCallum FO: Herpes simplex, varicella and zoster: Clinical manifestations and treatment. Philadelphia, Lippincott, 1972

Taylor-Robinson D, Caunt AE: Varicella virus, Virology Monograph No. 12. New York, Springer, 1972

SELECTED PAPERS

Baba K, Yabuuchi H, Okuni H, et al: Studies with live varicella vaccine and inactivated skin test antigen. Pediatrics 61:550, 1978

Dodion-Fransen J, Dekegel D, Thiry L: Congenital varicella-zoster infection related to maternal disease in early pregnancy. Scand J Infect Dis 5:149, 1973

Feldman S, Hughes WT, Daniel CB: Varicella in children with cancer. 77 cases. Pediatrics 56:388, 1975

Gershon AA, Steinberg S, Brunnell, PA: Zoster immune globulin, a further assessment. N Engl J Med 290:243, 1974

Kempe CH, Gershon AA: Varicella vaccine at the crossroads. Pediatrics 60:930, 1977

Merigan TC, Rand KH, Pollard RB, et al: Human leukocyte interferon for the treatment of herpes zoster in patients with cancer. N Engl J Med 298:981, 1978

Takahashi M, Otsuka T, Okuno Y, et al: Live vaccine used to prevent the spread of varicella in children in hospital. Lancet 2:1288, 1974

Weller TH: Serial propagation in vitro of agents producing inclusion bodies derived from varicella and herpes zoster. Proc Soc Exp Biol Med 83:340, 1953

Whitley RJ, Chien LT, Dolin R, et al: Adenine arabinoside therapy of herpes zoster in the immunosuppressed. N Engl J Med 294:1193, 1976

Cytomegalovirus

BOOKS AND REVIEWS

Krech U, Jung M, Jung F: Cytomegalovirus infections of man. Basel, Karger, 1971

Plummer G: Cytomegaloviruses of man and animals. Prog Med Virol 15:92, 1973

Symposium—The risk of cytomegalovirus infection transmitted by blood transfusion. Yale J Biol Med 49:1, 1976

Weller TH: The cytomegaloviruses. N Engl J Med 203:267, 1971

SELECTED PAPERS

Hanshaw JB, Scheiner AP, Moxley AW, et al: School failure and deafness after "silent" congenital cytomegalovirus infection. N Engl J Med 295:468, 1976

Lang DJ: Cytomegalovirus infections in organ transplantation and posttransfusion: An hypothesis. Arch Gesamte Virusforsch 37:365, 1972

Lang DJ, Kummer JF: Cytomegalovirus in semen: Observations in selected populations. J Infect Dis 132:472, 1975

Luby JP, Johnson MT, Jones SR: Antiviral chemotherapy. Annu Rev Med 25:251, 1974

Neff BJ, Weibel RE, Bunyak EB, et al: Clinical and laboratory studies of live cytomegalovirus vaccine Ad-169. Proc Soc Exp Biol Med 160:32, 1979

Niederman JC, Miller G, Pearson HA, et al: Infectious mononucleosis: Epstein-Barr virus shedding in saliva and the oropharynx. N Engl J Med 294:1355, 1976

Pagano JS: Infections with cytomegalovirus in bone marrow transplantation: Report of a workshop. J Infect Dis 132:114, 1975

Phillips CF: Congenital cytomegalovirus disease. Is prevention possible? Prog Med Virol 23:62, 1977

Rapp CE Jr, Hewetson JF: Infectious mononucleosis and the Epstein-Barr virus. Am J Dis Child 132:78, 1978

Rola-Pleszczynski M, Frenkel LD, Fucillo DA, et al: Specific impairment of cell-mediated immunity in mothers of infants with congenital infection due to cytomegalovirus. J Infect Dis 135:386, 1977

Sawyer RN, Evans AS, Niederman JC, McCollum RW: Prospective studies of a group of Yale University freshmen. I. Occurrence of infectious mononucleosis. J Infect Dis 123:263, 1971

Schopfer K, Lauber E, Krech U: Congenital cytomegalovirus infection in newborn infants of mothers infected before pregnancy. Arch Dis Child 53:536, 1978

Simmons RL, Lopez C, Balfour H Jr, et al: Cytomegalovirus: clinical virological correlations in renal transplant recipients. Ann Surg 180:623, 1974

Steel CM, Ling NR: Immunopathology of infectious mononucleosis. Lancet 2:861, 1973

CHAPTER 14

Adenoviruses and Adenovirus-associated Viruses

Adenoviruses

The adenoviruses were named by Enders and colleagues in 1956 to reflect the initial demonstration of these viruses in explanted adenoid tissue. Rowe and colleagues had observed degeneration secondary to virus cytopathology which occurred spontaneously in surgically removed adenoid tissues, and in which, by implication, this virus resided latent. The initial recovery of adenovirus thus highlights the predilection of adenoviruses to infect lymphoid cells and to establish latent or persistent infections.

Soon after the demonstration of these viruses in adenoidal tissues, related viruses were recovered from military recruits with acute respiratory disease, and it was subsequently shown that adenoviruses are widespread and constitute important respiratory pathogens. Adenoviruses have also been recovered from gastrointestinal tract, the conjunctivae, urine, and the central nervous system. Clinically significant adenovirus infections may involve each one or more of these organs, cells, and body fluids. Viruses having similar physicochemical and morphologic properties have been recovered from many animal species. All except the chicken adenovirus have a similar, but not identical, group antigen, which can be detected in complement-fixation tests with appropriate antisera. Adenoviruses are relatively species-specific in their host range, though certain serotypes of human adenoviruses produce tumors in newborn hamsters. Only human strains, comprising 33 different serotypes, will be considered here. These strains have been classified into three subgroups on the basis of hemagglutination patterns. Group I viruses, which agglutinate rhesus monkey red blood cells, include types 3,7,11,14,16,20,21,25, and 28. Group II viruses agglutinate rat erythrocytes and include types 8,9,10,13,15,17,19,22,23,24,26,27,29, and 30. Group III viruses, which partially agglutinate rat cells, include numbers 1,2,4,5,6,12,18, and 31, adenovirus types which are frequently recovered from humans.

Clinical Associations

Adenoviruses show a predilection for infection of conjunctival, respiratory, and intestinal epithelium, and regional lymphoid tissue. Incubation periods for disease have been 1 to 2 weeks where discernible. Latent tissue infections, healthy persons shedding virus, and prolonged virus shedding (particularly intestinal) following illness have all been described.

Adenovirus types 1, 2, and 5 have been recovered from surgically removed tonsils and adenoids, and have also been associated with sporadic mild respiratory illness of infants and children. Types 3 and 7 also cause similar illness sporadically or sometimes in epidemics. Characteristically, the patient is febrile with pharyngitis, cervical adenitis, and conjunctivitis. Coryza and cough are frequently present. Atypical pneumonia may occur, and adenoviruses have been recovered from the lungs of fatal cases. A whooping cough syndrome has also been associated with adenovirus infection. Gastrointestinal symptoms are occasionally prominent, and occasionally adenovirus-induced lymphoid hyperplasia in the gastrointestinal tract can serve as a focus promoting intussusception.

Adenoviruses have been found to be responsible for a significant proportion of serious, acute, lower respiratory tract infections in children. Since adenoviruses do not have a recognizable pattern of seasonal occurrence, and since prolonged excretion of virus (pharynx and feces) may follow infection, the association of adenovirus infection and disease in civilians has been established by patient-control studies. Longitudinal "virus-watch" programs are particularly important in this regard, since analysis of lower respiratory tract infections in hospitalized patients provides an incomplete and distorted picture of the responsibility of specific viruses for clinical disease. On the basis of differences in morbidity in infected children compared with that in controls, the respiratory pathogenicity of adenoviruses has been established in many studies. Types 1,2,5, and 6 have been recovered frequently in association with respiratory illnesses in children. Investigations in Sweden showed that between 11 and 19 percent of febrile upper respiratory illnesses in infants were due to adenoviruses. Though most adenovirus infections in civilian pediatric practice are associated with undifferentiated upper respiratory disease, acute laryngitis, and influenza-like syndromes, some severe and even fatal cases of adenovirus bronchitis, bronchiolitis, and pneumonia are seen.

Studies of previously well children recovered from adenovirus pneumonia or bronchitis have shown that a proportion develop evidence of chronic pulmonary disease thereafter. It has been suggested that adenovirus infections may injure the elastic fibers of the bronchiolar walls, leading to the development of bronchiectasis.

Adenovirus types 3, 4, 7, 14, and 21 have been associated with epidemics of febrile respiratory illness in institutional populations, particularly among military recruits. Syndromes include an influenza-like illness, febrile pharyngitis and atypical pneumonia. The conjunctival mucosa is frequently involved, and there are occasionally gastrointestinal symptoms.

Epidemic keratoconjunctivitis is most commonly caused by adenovirus type 8, although other types may sometimes be responsible. The disease is characterized by the acute onset of tearing, erythema, suffusion, lymphoid follicles beneath the conjunctiva, and preauricular adenopathy. As the conjunctivitis begins to subside after 1 to 2 weeks, discrete corneal infiltrates appear. The latter may persist for 1 or 2 years.

Several other syndromes and illnesses have been associated with adenovirus infection. Among them are myocarditis, hepatitis, renal disease, gastroenteritis, acute hemorrhagic cystitis, acute and subacute meningoencephalitis, and generalized exanthems. Many of the higher adenovirus serotypes have been recovered from persons with inapparent infections, and their significance in disease is uncertain.

Epidemiology

Human adenoviruses spread from person to person with no other known reservoir. Serologic evidence of infection with one or more low-numbered serotypes is very common by age 5. Approximately 5 to 10 percent of civilian respiratory disease appears to be due to adenovirus infection. Occasional outbreaks of infection have been associated with swimming pools, but most have no such association.

The method of acquisition of adenovirus appears to be important in determining disease. Attempts to artificially induce conjunctivitis in volunteers are unsuccessful unless the conjunctival surface is mildly irritated by swabbing. Dusty environments, swimming pool water, and optical instruments such as tonometers can all provide the necessary conjunctival irritation,

and in some cases transmit the virus as well.

Nasopharyngeal inoculation of volunteers with adenovirus and ingestion of the virus usually results in asymptomatic infection or mild, afebrile illness. However, inhalation of adenovirus aerosols into the lower respiratory tract has resulted in the full range of clinical syndromes. It is thus suggested that the epidemic adenovirus disease seen in army recruits is the result of airborne spread facilitated by close contact of a large group of susceptibles.

Diagnosis

The differential diagnosis of the clinical syndromes caused by adenoviruses includes infections with several bacteria, chlamydia species, *Mycoplasma pneumoniae*, and several viruses. A frequent differential diagnosis that is important for therapeutic reasons is between streptococcal and adenovirus pharyngitis. Characteristics more common for adenovirus pharyngitis include exudate on the pharyngeal wall (as opposed to tonsils), a granular appearance to the mucosa, only moderate tenderness of the enlarged cervical nodes, and a normal or minimally elevated white blood cell count; however, a culture for streptococci should be obtained whenever possible.

The atypical pneumonia caused by adenoviruses cannot be clinically distinguished from that caused by mycoplasma. As many as one in five persons with adenovirus atypical pneumonia have modest elevations in cold agglutinins. One must depend on epidemiologic data to guide therapy (see Mycoplasmas) and on cultural and serologic data for retrospective diagnosis.

Adenoviruses can be recovered from respiratory secretions, throat swabs, conjunctival swabs, and feces of infected persons. The specimen is inoculated into a continuous human cell line (HeLa, HEP II, KB) and/or human embryonic kidney tissue culture. Virus growth may be slow, necessitating passage into fresh tissue culture tubes so that 3 to 4 weeks of observation can be achieved. Typical adenovirus cytopathic effect consists of grapelike clusters of rounded, refractile cells. Identification of the agent as an adenovirus can be accomplished using hyperimmune antiserum to detect the group antigen in tissue culture fluid by a complement fixation test. Adenoviruses can be subgrouped according to their hemagglutination reaction with rhesus and rat blood cells. Serotype identity is accomplished using specific antisera in hemagglutination inhibition and/or neutralization tests.

Because adenoviruses can be recovered from healthy persons, increases in antibody titer between acute and convalescent sera should be sought to document recent acute infection. This may be done using the complement fixation test and any adenovirus serotype. Neutralization or hemagglutination inhibition tests employing type-specific antiserums may be employed to identify the specific adenovirus serotype causing the infection. Sera to be used for the hemagglutination inhibition test must be pretreated to remove nonspecific inhibitors.

Treatment

Treatment of adenovirus infections is symptomatic and supportive. Secondary bacterial infection is not common.

Prevention

Immunity following adenovirus infection is serotype specific and appears to be long-lasting. Because of the problem that adenovirus infections pose in military recruit camps, vaccines have been developed and tested in these populations. Experience with subcutaneous inoculation of an inactivated polyvalent vaccine containing types 3, 4, and 7 indicated that such vaccines could control infection. Use of this vaccine was discontinued when adenovirus types 3 and 7 were shown to be capable of inducing tumors in hamsters.

More recently, infectious adenovirus vaccines of types 4 and 7 have been developed for oral administration. The virus is encased in an enteric-coated capsule and is released in the intestine, where it causes an asymptomatic, nontransmissible infection. Good protection has been provided by these vaccines, particularly when both are used simultaneously, so that the vaccine-suppressed serotype is less likely to be replaced by a nonsuppressed serotype. Currently, efforts are being directed toward the exploration of additional forms of immunoprophylaxis, including new modes of vaccine administration and the development of purified subunit vaccines.

Evidence for Adenovirus Oncogenicity

In 1962 it was shown that certain adenoviruses of man could induce tumors in newborn hamsters. Subsequently, human adenoviruses were classified on the basis of their oncogenic potential into (a) a highly oncogenic group consisting of types 12, 18, and 31; (b) less oncogenic types (longer latency and fewer neoplasms), including 3, 7, 14, 16, and 21, and (c) a group including types 1, 2, 5, and 6 which do not cause tumors in inoculated hamsters but can transform cells in vitro (which cells can, in turn, cause tumors).

In spite of these observations made in animals and tissue cultures, the search for evidence of adenovirus oncogenicity in man has been entirely negative. Nonetheless, these findings have led to a reduction in the enthusiasm for development and widespread use of attenuated adenovirus vaccines and to an increased interest in purified subunit vaccines.

Adenovirus-associated viruses

The adenovirus-associated viruses (AAV) replicate only in cells coincidentally infected with adenoviruses. The name adeno-satellite virus (ASV) has also been suggested for these small viruses originally found during electron-microscopic examination of adenoviruses.

Four antigenic types of AAVs have been described. Serologic studies (complement fixation, hemagglutination inhibition, neutralization) have indicated that AAV 1 may originate from rhesus monkeys and that AAV 4 probably originates from green monkeys. Both cultural and serologic evidence indicates the occurrence of human infection with AAV 2 and 3. The relationship of AAV infection to disease is obscured by the concomitant occurrence of adenovirus infection and also by inadvertent immunization with AAVs contaminating various vaccines grown in monkey kidney. Possible associations of AAV infection with pneumonia and with exanthems (hypersensitivity ?) have been suggested.

FURTHER READING

Adenovirus

BOOKS AND REVIEWS

Chanock RM: Impact of adenoviruses in human disease. Prev Med 3:466, 1974

Ginsberg HS: Adenoviruses. Am J Clin Pathol 57:771, 1972

Jackson GG, Muldoon RL: Viruses causing common respiratory infection in man. IV. Reoviruses and adenoviruses. J Infect Dis 128:811, 1973

Knight V, Kasel JA: Adenoviruses. In Viral and mycoplasmal infections of the respiratory tract. Philadelphia, Lea & Febiger, 1973

Norrby E: The structural and functional diversity of adenovirus capsid components. J Gen Virol 5:221, 1969

Rose HM, Lamson TH, Buescher EL: Adenoviral infection in military recruits. Arch Environ Health 21:356, 1970

Rowe WP, Hartley JW: A general review of the adenoviruses. Ann Acad Sc 101:466, 1962

Van Der Veen J: The role of adenoviruses in respiratory disease. Am Rev Resp Dis (Suppl) 88:167, 1963

SELECTED PAPERS

Belshe RB, Mufson MA: Identification by immunofluorescence of adenoviral antigen in exfoliated bladder epithelial cells from patients with acute hemorrhagic cystitis. Proc Soc Exp Biol Med 146:754, 1974

Brandt CD, Kim HW, Vargosko AJ, et al: Infections in 18,000 infants and children in a controlled study of respiratory tract disease. I. Adenovirus pathogenicity in relation to serologic type and illness syndrome. Am J Epidemiol 90:484, 1969

Chany C, Lepine P, LeLong M, et al: Severe fatal pneumonia in infants and young children associated with adenovirus infections. Am J Hyg 67:367, 1958

Dudding BA, Top FH Jr, Winter PE, et al: Acute respiratory disease in military trainees. The adenovirus surveillance program, 1966–1971. Am J Epidemiol 97:187, 1973

Fox JP, Brandt CD, Wassermann FE, et al: The Virus Watch Program: A continuing surveillance of viral infections in metropolitan New York families. VI. Observations of adenovirus infections: virus excretion patterns, antibody response, efficiency of surveillance, patterns of infection, and relation to illness. Am J Epidemiol 89:24, 1969

Glezen WP, Denny FW: Epidemiology of acute lower respiratory disease in children. N Engl J Med 288:498, 1973

Kelsey DS: Adenovirus meningoencephalitis. Pediatrics 61:291, 1978

Lang WR, Howden CW, Laws J, Burton JF: Bronchopneumonia with serious sequelae in children with evidence of adenovirus type 21 infection. Br Med J 1:73, 1969

McAllister RM, Gilden RV, Green M: Adenoviruses in human cancer. Lancet 1:831, 1972

Nelson KE, Gavitt F, Batt MD, et al: The role of adenoviruses in the pertussis syndrome. J Pediatr 86:335, 1975

Numazaki Y, Kumasaka T, Yano N, et al: Further study on the hemorrhagic cystitis due to adenovirus type II. N Engl J Med 289:344, 1973

Rowe WP, Baum SG: Evidence for a possible genetic hybrid between adenovirus type 7 and SV40 viruses. Proc Natl Acad Sci USA 52:1340, 1969

Sprague JB, Hierholzer JC, Currier RW, Hattwick MAW, Smith MD: Epidemic keratoconjunctivitis. A severe industrial outbreak due to adenovirus type 8. N Engl J Med 289:1341, 1973

Steen-Johnson J, Orstavik I, Attramadal A: Severe illness due to adenovirus type 7 in children. Acta Paediatr Scand 58:157, 1969

Top FH Jr, Dudding BA, Russell PK, et al: Control of respiratory disease in recruits with types 4 and 7 adenovirus vaccines. Am J Epidemiol 94:142, 1971

Adenovirus-Associated Virus

BOOKS AND REVIEWS

Hoggan DM: Adenovirus associated viruses. Prog Med Virol 12:211, 1970

Mayor HD: Satellite viruses. In Busch H (ed): Methods in Cancer Research. New York, Academic, 1973, Vol 8, p 203

SELECTED PAPERS

Atchison RW, Casto BC, Hammon WM: Adenovirus-associated defective virus particles. Science 149:754, 1965

Blacklow NR, Hoggan MD, Kapikian AZ, et al: Epidemiology of adenovirus-associated virus infection in a nursery population. Am J Epidemiol 88:368, 1968

Boucher DW, Melnick JL, Mayor HD: Nonencapsidated infectious DNA of adeno-satellite virus in cells coinfected with herpesvirus. Science 173:1243, 1971

Hoggan MD, Blacklow NR, Rowe WP: Studies of small DNA viruses found in various adenovirus preparations: physical, biological, and immunological characteristics. Proc Natl Acad Sci USA 55:1467, 1966

Mayor HD, Ito M: Distribution of antibodies to type 4 adeno-associated satellite virus in simian and human sera. Proc Soc Exp Biol Med 126:723, 1967

Mayor HD, Kurstak E: Viruses with separately encapsidated complementary DNA strands. In Kurstak E, Maramorosch K (eds): Viruses, Evolution and Cancer. New York, Academic, 1974, p 55

Parks WP, Boucher DW, Melnick JL, Taber LH, Yow MD: Seroepidemiological and ecological studies of the adenovirus-associated satellite viruses. Infect Immunol 2:716, 1970

Melnick JL, Rongey R, Mayor HD: Physical assay and growth cycle studies of a defective adenosatellite virus. J Virol 1:171, 1967

Rosenbaum MJ, Edwards EA, Pierce WE, et al: Serologic surveillance for adeno-associated satellite virus antibody in military recruits. J Immunol 106:711, 1971

CHAPTER 15

Human Papovaviruses

BACKGROUND

Awareness and concern about papovavirus infections of man dates to the recognition in 1960 that SV40 virus, its presence unrecognized in rhesus monkey kidney cells used for poliovirus culture, had been inoculated into many of the initial recipients of polio vaccine virus. This initial anxiety was heightened when it was found that SV40 and polyoma viruses produce tumors when inoculated into varieties of newborn rodents and were capable of transforming cells in vitro. In spite of intense concern and study, to date no recognizable disease, illness, or increased incidence of neoplasia has been associated with these documented instances of the introduction of SV40 into humans. Nevertheless, since the prototype papovaviruses have been associated with the presence and/or the induction of neoplasms, the recognition and characterization of papovaviruses of man has continued to stir considerable interest.

PAPILLOMAVIRUSES

Papillomaviruses can be visualized by electronmicroscopy in several varieties of human warts. Though these virus particles are more easily visualized in cutaneous than in genital or laryngeal warts, the appearance of all of these viruses is similar.

The papillomaviruses of man do not replicate in cultured cells. However, the ready availability of excised virus-containing warty material permits the extraction of sufficient quantities of viral antigens and DNA to allow the preparation of reagents for serologic tests, as well as (more recently) the performance of some biochemical analyses. By employing restriction endonucleases and nucleic acid hybridization, it has been possible to define distinct subtypes of human papillomavirus. Type-specific associations of papillomaviruses have been described with plantar and hand warts (verrucae plantares and vulgares), genital warts (condyloma acuminata), laryngeal papillomas, and the red spots and flat warty lesions associated with a rare heredofamilial disease (epidermodysplasia verruciformis). Additional classification and mapping will undoubtedly continue, though really significant progress in papillomavirus research awaits the in vitro cultivation of these viruses (see Chapter 9).

POLYOMAVIRUSES

The two known human papovaviruses which are members of the polyomavirus genus (BK and JC viruses) have both been replicated in vitro, and thus much more virologic work has been published relevant to the human polyomaviruses than to the papillomaviruses of man.

JC Virus

In 1965 electronmicroscopic examination of brain tissue from patients with progressive multifocal leukoencephalopathy (PML) revealed aggregates of small round particles in distended oligodendroglial nuclei (Fig. 15-1). After unsuccessful attempts by several investigators, Padgett and co-workers recovered a papovavirus by inoculating cultures of human fetal astrocytes with dispersed brain cells taken from lesions of a patient with PML. The virus (named JC for the patient), which did not induce grossly recognizable cytopathology, grew very slowly, and its presence was revealed by staining and microscopic study.

Initially, immunofluorescent techniques were employed to show that JC was distinct

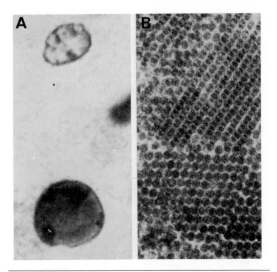

Fig. 15-1. **A.** Abnormal oligodendroglial nucleus (bottom) with hetrochromatic patches suggestive of inclusions. Astrocytic nucleus on top. **B.** Paracrystalline array of virus particles in glial nucleus. (From zuRhein: Acta Neuropathol 8:57, 1967)

from polyomavirus, SV40, and human wart viruses. JC virus was found to agglutinate human, chicken, and guinea pig erythrocytes, and these observations led to the development of hemagglutination inhibition tests which facilitated serologic studies.

Extensive biochemical, antigenic, and even biologic studies of JCV have been hampered by its slow rate of replication in vitro. Restriction endonuclease cleavage maps of JCV DNA are now available.

Newborn hamsters inoculated with JCV develop diverse brain tumors (after intracerebral inoculation) and visceral neoplasms (following subcutaneous or intraperitoneal infection). The oncogenic potential of JCV for other animals is largely unexplored (see also Chapter 9).

BK Virus

In 1971 Gardner reported the recovery of BK virus from the urine of a renal allograft recipient. It is related to, but distinct from, SV40. Many isolations of BKV have been reported subsequently from patients with chronic diseases and altered immunologic competence. It has been demonstrated in the urine of as many as 15 percent of immunosuppressed transplant patients, from some children with hereditary immunodeficiences, and from certain individuals with diffuse malignancies—especially those requiring therapy with cytotoxic or immunodepressive medications. Although BKV has been recovered from patients with reduced immunocompetence, no definite relationship has been established between infection with it and any specific disease entity. BKV agglutinates erythrocytes from humans, guinea pigs, and young chickens, and after the removal of nonspecific inhibitors of these reactions, serum may be tested for antibody by the hemagglutination inhibition test.

SV40

SV40 is a virus of simian primates, specifically of Asiatic macaques. It is thought not to be associated with human conditions or illnesses. Nevertheless, a virus very similar to, if not identical with it has been recovered from a few patients with PML. These isolates were designated "SV40-PML." Isolates of SV40 have also been reported by one investigator from malignant melanoma, and the presence of SV40-specific antigens has been described in cells derived from several meningiomas. Thus although SV40 is primarily a simian papovavirus and may exist only as a "passenger" in human tissues, these reports indicate the possibility that this virus may be a potential pathogen for man in rare cases.

EPIDEMIOLOGY AND CLINICAL ASSOCIATIONS

Papillomaviruses

The papillomaviruses are responsible for a variety of skin growths, or warts. Warts reflect the effects of an infectious disease of skin which results in the production of epidermal hypertrophy. These tumors are largely benign, although some may undergo malignant degeneration. The clinical description of warts is generally related to their location and/or morphology. Thus warts may be classified as plantar, digitate, venereal, laryngeal, flat, filiform, etc.

The incidence of warts increases at the time of puberty, which may reflect the opportunity for increased interpersonal contacts. Virtually all individuals have been infected by some wart viruses by the age of 20 years. Endocrine factors may play a role in the occurrence and spread of some warts. Cellular immune function is an important feature of papillomavirus control. Immunosuppressed renal allograft recipients may experience a diffuse spread of warts which can be very difficult to control. In one study, 18 of 49 pediatric renal transplant patients developed warts. In only 5 of these 18 instances were the warts successfully eradicated. Pregnancy may be accompanied by considerable proliferation of genital warts. The appearance of laryngeal papillomas in infancy is frequently associated with the occurrence of maternal genital warts during gestation and probably reflects the transmission of papilloma viruses at the time of delivery.

Though transmission to dogs of human laryngeal papilloma has been reported, the human papillomaviruses are largely species-specific and most interspecies inoculation attempts have failed. Nevertheless, some serologic cross-reactivity exists between human and other animal papillomaviruses. Investigators in Finland have reported that 25 percent of dogs surveyed possessed antibody to human papillomaviruses. Conversely, 18 percent of adults tested have detectable antibody to bovine papilloma antigens.

BK Virus

Seroepidemiologic studies of human papovaviruses has shown that BK virus infections are widespread in many parts of the world. Age-related prevalence data indicate that infection with BKV occurs early in childhood, and in the United States and Western Europe as many as 70 to 90 percent of adults possess antibody to this virus. In another study, 100 percent of children were BK antibody positive by 10 to 11 years of age. No animal reservoir for BKV has been detected, and all indications are that this virus is species-specific for man.

JC Virus

JC virus seems also to be limited to human populations, although the epidemiology of JC and BK viruses seems to be entirely independent of one another. Serologic surveys for antibody to JC virus have shown it to be distributed worldwide, and indications are that primary infections are common in childhood. For example, Padgett and Walker found that 65 percent of children in Wisconsin were JC seropositive by 10 to 14 years of age. In another study, fifty percent of children had antibody to JC virus by the age of 3 years. It is more difficult to determine the incidence of naturally occurring infection with SV40-PML, but it seems to be much more rare. In 80 randomly sampled sera, antibodies to SV40 were found in 1.3 percent and antibody to JC virus in 58.8 percent.

PROGRESSIVE MULTIFOCAL LEUKOENCEPHALOPATHY

Progressive multifocal leukoencephalopathy has been observed only in adults, most commonly in the fifth to seventh decades of life. The course is relentlessly progressive, usually leading to death within 3 to 6 months from the onset of symptoms. The clinical signs are variable, depending on the location and size of the lesions. The cerebral hemispheres, both gray and white matter, are most frequently involved, but any part of the neuraxis can be affected, including cerebellum, brainstem, and spinal cord. The patients usually develop visual disturbances, progressive mental deterioration, and at times focal motor weakness, cranial nerve dysfunction, ataxia, and aphasias. There is no associated fever or spinal fluid abnormality. Occasionally the course fluctuates with periods of apparent remission, but inevitably death ensues. Whether this is primarily the result of the underlying systemic disease or a sequel to the brain involvement is often difficult to determine. The disease that occurs in patients with JCV infection differs in no appreciable way from that seen in patients from which SV40-PML was isolated.

Pathologic lesions occur in both the gray and white matter. These are classically small, discrete areas of demyelination which may become confluent. Cytologically, the outstanding features are the unusual astrocytes with bizarre chromatin patterns and the atypical oligodendroglia with enlarged, ill-defined inclusions. As a result of the altered function of these oligodendroglia, demyelination ensues, making this the first reported instance of a virus-induced demyelinating disease in man.

The pathogenesis of the nervous system infection with JCV is unknown. The leading hypotheses propose either that it is a result of the activation of virus latent in the host since early life or that it results from a recently acquired infection in an immunocompromised host not previously exposed to the virus. Nothing is known at this time of the mode of transmission of the virus to human beings, the sites of extraneural replication, or the clinical expression of primary disease.

SV40-PML, JC, and BK viruses are antigenically distinguishable from each other and share common antigens with SV40. With the use of immune electron microscopy, the antigenic nature of the virions in suspect cases of PML can be identified prior to virus isolation in cell culture. This is done by cross-reacting type-specific antiserum prepared with PML isolates with virions extracted directly from brain. Not only has this permitted rapid identification of virus type, but it has now been shown conclusively that the SV40-PML agent was a brain isolate and did not result from recombination with latent simian agents in cell cultures, as was originally suspected.

No proven specific therapy is yet available for the treatment of PML.

As already noted, the shedding of BK and JC viruses has been detected in association with immunologic compromise. In addition, the proliferation of papillomata in connection with immune suppression suggests that papillomaviruses are similarly activated in hosts with diminished cell-mediated immunity. These features of papovavirus epidemiology are reminiscent of those associated with herpesvirus infection, latency, and reactivation. The lack of recognition of clinical syndromes in association with papovavirus infection in normal hosts may reflect the results of longstanding host–parasite mutual adaptation. The development of simple techniques for papovavirus recovery may lead to the recognition of clinical associations presently not perceived. For example, the presence of papovaviruses has been reported in acute hemorrhagic cystitis of childhood, unassociated with host immune compromise.

RELATIONSHIP OF HUMAN PAPOVAVIRUSES TO THE ETIOLOGY OF NEOPLASIA IN MAN

In experimental circumstances many papovaviruses, including those recovered from man, induce neoplastic cellular transformation. The detection of papovaviruses in association with some tumors of man has provided impetus to the investigation of the role of these viruses in the induction of some cancers. Thus far it seems that human papovaviruses of the polyomavirus genus play no role in the etiology of human malignancies.

Human papillomaviruses are also logical candidates for study in the search for oncogenic viruses. Papillomaviruses are widespread in all populations and are tropic for epithelium (most human cancers are carcinomas). Papillomaviruses influence infected skin epithelium to induce abortive transformation of germinal cells. These precancerous lesions may become malignant depending on variable host and environmental factors.

Epidermodysplasia verucciformis (EV) is a rare genetically determined disease characterized by the occurrence of disseminated skin lesions and warts. Transformation of some of these lesions into intraepidermal carcinomas occurs in about 25 percent of patients with EV. Papillomavirus can be detected in the original EV lesions but cannot be found in the resulting malignancies. This malignant transformation of benign virus-associated papillomas of man is reminiscent of the malignant transformation of some (Shope) papillomas of rabbits. Further exploration of the role of papillomaviruses in human cancer will be accelerated by, and may await, the development of convenient reliable systems for the in vitro replication of these viruses.

FURTHER READING

Books and Reviews

Fareed GC, Davoli D: Molecular biology of papovaviruses. Ann Rev Biochem 46:471, 1977

Orth G, Breitburd F, Favre M, et al: Papillomaviruses: possible role in human cancer. In Origins of Human Cancer. Cold Spring Harbor, New York, Cold Spring Harbor Laboratory, 1977

Padgett BL, Walker DL: New human papovaviruses. Prog Med Virol 22:1, 1976

Shah KV, Nathanson N: Human exposure to SV40: review and comment. Am J Epidemiol 103:1 1, 1976

Takemoto KK: Human papovaviruses. Int Rev Exp Pathol 18:281, 1978

Selected Papers

Brown P, Tsai T, Gajdusek DC: Seroepidemiology of human papovaviruses: discovery of virgin populations and some unusual patterns of antibody prevalence among remote peoples of the world. Am J epidemiol 102:331, 1975

Gissmann L, Pfister H, zur Hausen H: Human papilloma viruses (HPV): characterization of four different isolates. Virology 76:569, 1977

Gissmann L, zur Hausen H: Human papilloma virus DNA: physical mapping and genetic heterogeneity. Proc Natl Acad Sci USA 73:1310, 1976

Hashida Y, Gaffney PC, Yunis EJ: Acute hemorrhagic cystitis of childhood and papovavirus-like particles. J Pediatr 89:85, 1976

Ingelfinger JR, Grupe WE, Topor M, et al: Warts in a pediatric renal transplant population. Dermatologica 155:7, 1977

Mäntyjärvi RA, Meurman OH, Vihma L, et al: A human papovavirus (B.K.), biological properties and seroepidemiology. Ann Clin Res 5:283, 1973

Martin JD, Frisque RJ, Padgett BL, et al: Restriction endonuclease cleavage map of the DNA of JC virus. J Virol 29:846, 1979

Orth G, Jablonska S, Favre M, et al: Characterization of two types of human papillomaviruses in lesions of epidermodysplasia verruciformis. Proc Natl Acad Sci USA 75:1537, 1978

Padgett BL, Walker DL: Prevalence of antibodies in human sera against JC virus, an isolate from a case of progressive multifocal leukoencephalopathy. J Infect Dis 127:467, 1973

Padgett BL, Walker DL, zuRhein GM, Eckroade RJ: Cultivation of papova-like virus from human brain with progressive multifocal leukoencephalopathy. Lancet 1:257, 1971

Portolani M, Barbanti-Brodano G, LaPlaca M: Malignant transformation of hamster kidney cells by BK virus. J Virol 15:420, 1975

Reid TMS, Fraser NG, Kemohan IR: Generalized warts and immune deficiency. Br J Dermatol 95:559, 1976

Shah KV, Daniel RW, Warszawski RM: High prevalence of antibodies to BK virus, an SV40 related papovavirus in residents of Maryland. J Infect Dis 128:784, 1973

Takemoto KK, Rabson AS, Mullarkey MF, et al: Isolation of papovavirus from brain tumor and urine of a patient with Wiskott-Aldrich syndrome. J Natl Cancer Inst 53:1205, 1974

Walker DL, Padgett BL, zuRhein GM, Albert AE: Human papovavirus (JC): Induction of brain tumors in hamsters. Science 181:674, 1973

Weiner LP, Herndon RM, Narayan O, et al: Isolation of virus related to SV40 from patient with progressive multifocal leukoencephalopathy. N Engl J Med 286:385, 1972

CHAPTER 16

Enteroviruses

The human enteroviruses comprise three subgroups, namely, the polioviruses, the Coxsackie viruses, and the echoviruses. These three were originally grouped together because their common portal of entry includes the gastrointestinal tract, wherein they replicate and initiate many different forms of disease. There are three distinct poliovirus serotypes, 33 echovirus serotypes, 23 Coxsackie A serotypes, and 6 Coxsackie B serotypes.

HISTORY

Sporadic cases of paralytic disease are as old as recorded history. A famous Egyptian stele from the period 1580 to 1350 BC depicts a priest with a flail atrophic lower limb typical of paralytic polio. Sir Walter Scott underwent an illness in 1772 which was one of the earliest and certainly the most renowned case of polio described in the British Isles. Acute paralytic illness in children was first described by Underwood in his textbook in 1789, while the name anterior poliomyelitis is attributed to Kussmaul in the late nineteenth century. The term poliomyelitis was derived from the Greek for gray marrow of the spinal cord and the Latin (*itis*) for inflammation; the location of the involved cells in the anterior horns of the spinal cord contributed the designation anterior. Outbreaks of illness were regularly identified in the nineteenth and twentieth centuries by the observation of paralysis among young children. The first isolations of poliovirus were achieved in 1908 by the inoculation of central nervous system tissue into susceptible monkeys via the intracerebral route. In 1949 Enders, Robbins, and Weller reported their classic experiments on the cultivation of poliovirus in tissue cultures of nonneural human cells. These techniques paved the way for accelerated development of the fields of animal and human virology.

In contrast to the centuries' old descriptions of poliovirus infections, the histories of the other enteroviruses are relatively recent. In 1948 Dalldorf and Sickles isolated a filterable agent from the stool of a patient from Coxsackie, New York, with paralytic illness. They utilized suckling mice to test for the presence of this virus. Subsequently, a large group of antigenically related agents has been characterized and designated the Coxsackie viruses. The division into subgroups A and B was based on differing pathology in the infected suckling mice. Members of the A group produced paralysis with extensive degeneration of skeletal muscle; the B viruses produced focal myositis, inflammatory lesions in many viscera, and fat necrosis. The echoviruses were also isolated initially from human fecal specimens, frequently from patients without overt disease, and their name represents an acronym (*e*nteric *c*ytopathogenic *h*uman *o*rphan). With increasing clinical study, they have been associated with a wide variety of illness, so that most no longer are "orphans." They replicate in monkey kidney cell cultures but are not pathogenic for mice or monkeys. As distinct antigenic serotypes were identified, they were assigned sequential numbers. With further refinement of our knowledge of the physical and biochemical features of viruses, these properties have now been utilized for more accurate assignment of viruses to different groups. As a result, several of the agents originally listed as echoviruses have been reallocated to other groups.

EPIDEMIOLOGY

The epidemiology of all human enteroviruses is quite similar. The pattern is most clearly defined for the polioviruses because paralytic disease has been so readily identifiable. As early as 1916 the clinical epidemiologic features were defined on the basis of an outbreak which occurred that year in New York City. These principles were (1) that poliomyelitis is, in nature, exclusively a human infection, transmitted from person to person without the necessary intervention of a lower animal or insect host. (2) The infection is far more prevalent than is apparent from the incidence of clinically recognized cases, since a large majority of persons infected become "carriers" without clinical manifestations. It is probable that during an epidemic such as that in New York City, a very considerable proportion of the population became infected, adults as well as children. (3) The most important agencies in disseminating infection are the unrecognized carriers and perhaps mild abortive cases ordinarily escaping recognition. It is fairly certain that frank paralytic cases are a relatively minor factor in the spread of infection. (4) An epidemic of 1 to 3

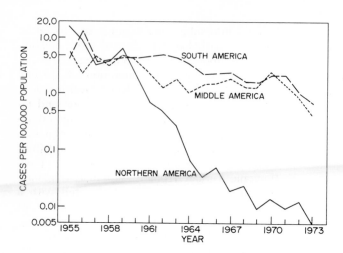

FIG. 16-1. Reported cases of polio (per 100,000 population) in the Western Hemisphere by regions, 1955–1973. (Adapted from the Bulletin of the World Health Organization)

recognized cases per 1000, or even less, immunizes the general population to such an extent that the epidemic declines spontaneously, due to the exhaustion or thinning out of infectable (i.e., susceptible) material. Apparently an epidemic incidence relatively small in comparison to that prevailing in an epidemic may produce a population immunity sufficient to definitely limit the incidence rate in a subsequent epidemic.

The utilization since the mid-1950s of poliovirus vaccines has dramatically reduced the occurrence of paralytic disease in this country and in other nations. The existence of only three distinct serotypes of poliovirus made vaccine development feasible, in contrast to the large numbers of serotypes of the other enteroviruses. Certain tropical countries, especially in Central and South America, Africa, and parts of Asia, have had problems with the

utilization and distribution of vaccines. These have prevented such a dramatic decrease. The annual case rates for the Western Hemisphere are shown on Figure 16-1.

In temperate climates the recognized, reported numbers of patients with enterovirus infections are greatest during the summer and early autumn (Fig. 16-2). In tropical zones they may occur year round. Because the usual route of transmission appears to be a fecal-oral one, improvements in sanitation and hygiene alter the epidemiology. Epidemics of polio originally occurred almost exclusively in children less than 6 years of age. In the early twentieth century, outbreaks were reported which involved teenagers and adults. Improvement in sanitation and living conditions was felt to have been responsible for this age shift, as childhood exposure with resultant infection and immunity was less likely to occur. In the presence of crowding

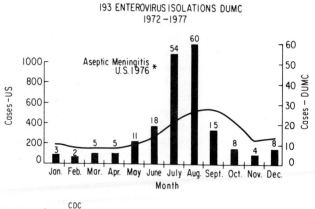

FIG. 16-2. Seasonal occurrence of enterovirus infections in the USA. Reported aseptic meningitis in the United States for 1976 (solid line). Enterovirus isolations at Duke University Medical Center, 1972–1977 (bars).

and less optimal hygiene, infants and young children were infected early in life with the enteroviruses. With improved sanitation, exposures to the agents were postponed, so that children, in the absence of the immunizing experience of early life infection, became susceptible adults. Paralytic disease more frequently resulted from poliovirus infection of an adult than a child.

Many outbreaks with nonpolio enteroviruses have been studied over the past several decades. Coxsackie B epidemics tend to recur every 2 to 5 years. Young infants constitute the largest susceptible population and are the most commonly affected. Older children and adults are more likely to be immune to a given agent if it has circulated in the community within the past 5 or more years. The characteristic features of an outbreak of a specific agent will vary considerably. With virologic studies it is possible to show a range of clinical manifestations from occult infection to overt disease with varying clinical symptoms. (Table 16-1). Transmission of virus occurs most readily under intimate contact situations. These include household settings, day care centers, nurseries, children's institutions, and other similar environments. Early in the infection both pharyngeal secretions and stools will be positive for virus. The pharynx after a few days will be free of infectious material, but the virus will persist in stools for periods as long as 6 to 8 weeks. Studies of families indicate nearly a 100% in-

fection rate among children and only slightly lower rates among the adult members. Monitoring of community sewage has been used as an epidemiologic tool to study community spread of the various enteroviruses.

CLINICAL ILLNESS

The broad spectrum of clinical disease produced by the enteroviruses overlaps among groups. A listing of various syndromes is included in Table 16-1. The more common manifestations associated with infection are discussed briefly.

Febrile Illness The great majority of infections with enteroviruses produce no specific clinical hallmarks. In young children, undifferentiated febrile illness, nonspecific malaise, and myalgias are frequently associated with enterovirus infections. There is nothing unique about this type of clinical presentation. Some infants and young children have had vomiting and diarrhea with echovirus infections, but it is dubious whether these agents play a major role in gastroenteritis.

Respiratory Disease Mild upper respiratory tract illness has been associated with several of the Coxsackies and echoviruses. A very few cases of pneumonia have been attributed to Coxsackie infection.

TABLE 16-1 CLINICAL MANIFESTATIONS OF ENTEROVIRUS INFECTIONS

Clinical Syndrome	Viruses Implicated*			
		Coxsackie		
	Polio	A	B	Echo
Asymptomatic infection	X	X	X	X
Non-specific febrile illness	X	X	X	X
Respiratory disease		X	X	X
Exanthems		X	X	X
Enanthems		X		
Pleurodynia			X	
Orchitis			X	
Myocarditis			X	
Pericarditis			X	X
Aseptic meningitis and meningoencephalitis	X	X	X	X
Disseminated neonatal infection			X	
Transitory muscle paresis	X		X	X
Paralytic disease	X			

*X, Involvement of multiple serotypes.

Exanthems and Enanthems Various enteroviruses (particularly echos 9 and 16, Coxsackies A9 and 16, Coxsackie B5) have been associated with large outbreaks of febrile rash disease. Younger children are more likely to develop exanthem, and they vary widely in their characteristics. Small vesicular lesions have been described on the hands and feet in association with ulcers of the buccal mucosa ("hand-foot-mouth disease"), particularly with Coxsackie A16. Macular and maculopapular eruptions indistinguishable from rubella have been observed with a number of the Coxsackie and echoviruses. Petechiae have accompanied some rashes, especially with echovirus 9. The presence of virus has been demonstrated in the skin lesion themselves. Some of the syndromes include an associated enanthem. Herpangina, one of these, is most commonly associated with Coxsackie A infections. Ulcerative lesions of the mucosa appear on the posterior pharynx and soft palate. The associated discomfort, fever, and sore throat are prominent. Because many different serotypes of the echo and Coxsackie groups can cause identical clinical pictures, virus isolation is necessary to identify the specific etiologic agents.

Pleurodynia (Bornholm Disease) This is a febrile illness with extreme myalgia, pleuritic chest pain, and headache. Most often it has been associated with Coxsackie B viruses, but other agents have occasionally been recovered from patients.

Orchitis Although viral orchitis most often is due to mumps, it has accompanied infections with the Coxsackie B viruses.

Myocarditis, Pericarditis Newborn infants may develop severe myocarditis as a part of generalized Coxsackie B virus infection. Isolated myocarditis and/or pericarditis in older children and adults has been shown to result from Coxsackie B or echovirus infections. The spectrum has ranged from benign, self-limited pericarditis to severe, chronic, fatal myocardial disease.

Viral Meningitis and Meningoencephalitis Many of the enteroviruses have been isolated from the cerebrospinal fluid (CSF) of patients suffering from aseptic meningitis and/or encephalitis. These illnesses are characterized by varying degrees of fever, headache, nuchal rigidity, malaise, and altered central nervous system function. The most severely affected patients may be obtunded or comatose. The least involved patients may complain only of mild headache and/or stiff neck. A CSF pleocytosis is detected, with usually less than 500 cells per microliter. Early in the disease these cells are likely to be predominantly polymorphonuclear leukocytes, but after 24 to 48 hours of illness, mononuclear cells are the majority. The CSF protein is moderately elevated, but the glucose is normal. The prognosis for such patients is far better than that for similar pyogenic infections of the central nervous system and meninges. Even the youngest infants appear to recover completely, but there are very few longitudinal studies of such patients to determine their eventual intellectual, motor, and emotional function.

Congenital and Neonatal Infections Transplacental and neonatal transmission have been demonstrated with Coxsackie B viruses, resulting in a serious disseminated disease which may include hepatitis, myocarditis, meningoencephalitis, and adrenal cortical involvement. Myocarditis as an isolated phenomenon has also been observed with Coxsackie B virus infection of the newborn. Several clusters of such cases have been noted in nurseries, suggesting postnatal nosocomial transmission. Many of these infections have been fatal. When poliovirus infections were common, examples of transmission from infected mother to fetus were apparent in the birth of a paralyzed infant. Several echoviruses have been associated with neonatal hepatic necrosis, usually fatal.

Paralytic Disease Poliovirus infection, especially with types 1 and 3, was responsible for almost all the paralytic disease associated with the enteroviruses. Occasional cases of transient paralysis and muscle weakness have been noted with other members, particularly the Coxsackie B agents. However, these are very few in number. With classic paralytic polio, there is a 2 to 6 day incubation period with an initial nonspecific febrile illness. This probably coincides with early replication of virus in the pharynx and gastrointestinal tract. With the subsequent hematogenous spread of virus, central nervous system involvement may result in meningitis and anterior horn cell infection. From 1 to 4

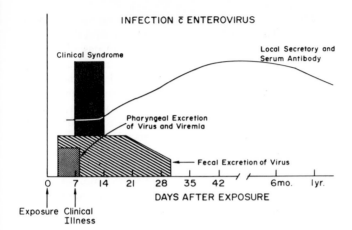

FIG. 16-3. The course of typical enterovirus infection.

percent of susceptible patients infected with polioviruses develop central nervous system involvement. The spectrum of paralytic disease is enormously variable and may involve only an isolated muscle group or extensive paralysis of all extremities. Characteristically, the picture is one of asymmetrical distribution, the lower extremities more frequently involved than the upper. Large muscle groups are more often affected rather than the small muscles of the hands and feet. Involvement of cervical and thoracic segments of the spinal cord may result in paralysis of the muscles of respiration. Infection of cells in the medulla and the cranial nerve nuclei results in bulbar polio with compromise of the respiratory and vasomotor centers. With the return of the patient's temperature to normal, the progress of paralysis ceases, and the subsequent weeks and months reveal a varying spectrum of recovery ranging from full return of function to significant residual paralysis. Recovery may be exceedingly slow, and its full extent cannot be judged for 6 to 18 months. Atrophy of involved muscles becomes apparent more rapidly, after 4 to 8 weeks.

PATHOLOGY

Because most Coxsackie and echovirus infections are short-lived and nonfatal, little is known of their histology. The pathologic changes of poliovirus in the central nervous system are most prominent in the spinal cord, medulla, pons, and midbrain. After initial cytoplasmic alterations in the Nissl substance of the motor neurons, nuclear changes develop next

and pericellular infiltrates of polymorphonuclear and mononuclear cells accumulate. The final stage is destruction with neuronophagocytosis and "dropping out" of the necrotic cells.

IMMUNITY

Enterovirus replication within the gastrointestinal tract results in the production of local secretory immunoglobulin A (IgA) at the site of contact of virus with lymphoid cells. The polioviruses have been extensively investigated, and the localized nature of the antibody response carefully documented. Serum antibody specific to the virus is found shortly after the clinical syndrome as shown in Figure 16-3. Development of type-specific antibody provides lifelong protection against clinical illness due to the same agent. Local reinfection of the gastrointestinal tract may recur, but this is accompanied by only an abbreviated period of viral replication without clinical illness. Serum antibodies to the enteroviruses can be demonstrated in three immunoglobulin classes, IgM, IgG, and IgA. Usually the earliest response is in the IgM fraction, overlapping with the later development of IgG antibodies. Complement fixation, virus neutralization, immunoprecipitation, and, in a few instances, hemagglutination inhibition are the most widely used techniques for assaying enterovirus antibodies. Neutralizing antibodies are type-specific, whereas complement fixation demonstrates group-reactive antibodies. In the course of his lifetime, humans sustain multiple infections, occult or overt, with a variety of enteroviruses. A specific infection elicits the production of antibody specific to that virus type but also may prompt an anamnestic response demonstrated by an increase in group-reactive antibody and by parallel rises in antibodies to serotypes of some of the other enteroviruses previously encountered. The concomitant serological rises in antibody titer create some problems with serologic surveys, rendering the complement-fixation test inadequate to define a specific infection. Measurement of virus-neutralizing antibody is a far more specific technique.

The aspects of cellular immunity for enterovirus infection are not well defined. Circulating peripheral white blood cells have been a source of virus isolation during acute illness. The infected cells are probably lymphocytes or macrophages, but this has not been determined with certainty. There are also several in vitro assays of white blood cell response to enterovirus infection, but no extensive evaluation of host cell-mediated immunity to these agents has been reported. The abnormal response of some children with immunodeficiency disease to infection with attenuated polioviruses may offer further insight to the immune processes normally stimulated by enterovirus infection. These are discussed further in the section on prevention (see below).

DIAGNOSIS

The clinical illness in some instances may permit a presumptive diagnosis of enterovirus infection. As shown in Table 16-1 the spectrum of illness is wide. The time of year may also be helpful, with a predilection for summer and early autumn circulation of enteroviruses in communities in the temperate zones. Specimens for virus isolation should be obtained early in the course of illness from various sites. The CSF of patients with aseptic meningitis and/or meningoencephalitis has been a rich source of enteroviruses, except for the three polio types. Materials such as pleural and pericardial fluid should also be cultured when available and appropriate to the clinical illness. Various simian and human cell culture systems will support replication of most of the enteroviruses with cytopathic effects revealing their presence. However, some of the Coxsackie A viruses are more fastidious, and their recovery

may require the inoculation of suckling mice. The specimens submitted most often for attempted virus isolations are nasopharyngeal swabs and stool specimens. As noted, an enterovirus may be excreted in fecal material for several weeks after a clinical illness. Recovery of an enterovirus from throat or stool of a patient does not in itself establish this as the etiologic agent of the illness observed. The temporal association of illness, virus recovery, and an antibody rise specific to that agent provide firmer evidence of a causative relationship.

Acute and convalescent serum samples obtained from 7 to 21 days apart will help to define quantitative changes in antibody titers. The isolation of a specific virus provides the opportunity for assessment of the patient's antibody against his own viral agent. In the absence of the recovery of a virus, one faces the problem of seeking specific antibody rises against the whole family of enteroviruses which might have been responsible for a clinical syndrome.

PREVENTION

Because there currently is no specific treatment for enterovirus infections, efforts have focused on means of prevention. The multiple antigenic types, and the usually benign, self-limited course of most echovirus and Coxsackie virus infections have resulted in little stimulus for the development of vaccines. The story of the poliovirus vaccines, however, has been one of the most exciting and rewarding sagas in microbiologic history. Prior to the work of Enders and colleagues with successful tissue culture techniques for growth of the polioviruses, there had been several ill-fated vaccines prepared from emulsions of spinal cord removed from monkeys infected with wild type poliovirus. These preparations were treated with formalin or other inactivating agents. Trials of such "vaccines" in 1935 proved unsuccessful and unacceptable.

Enders' tissue culture techniques lent themselves to the propagation in vitro of sufficient amounts of relatively pure poliovirus, so that controlled formaldehyde inactivation could be used to produce noninfectious virus which retained its antigenicity. Salk and his colleagues pursued this line of research and by 1954 were able to embark on a field trial which established the efficacy of an inactivated poliovirus vaccine in the prevention of paralytic disease. This was a trivalent preparation incorporating the three poliovirus types. After an initial series of two or three injections spaced several weeks to months apart, followed by a booster 6 to 12 months later, there was demonstrable serum antibody to all three polio serotypes. The vaccine was widely used in the United States during the five years from 1956 through 1960. The results were dramatic. Previous years had seen from 10 to 20 thousand cases of paralytic disease reported annually. With the widespread use of Salk vaccine, this rapidly dropped to 2 or 3 thousand cases annually, as increasingly large numbers of susceptible individuals were immunized.

By the early 1960s, a second vaccine was available. Strains of poliovirus which Sabin had selected and studied in his laboratory were proven attenuated for monkey and man. Their ingestion resulted in intestinal infection and virus excretion, so that humoral and gastrointestinal tract immunity developed without any illness. Because these could be administered more readily, by the oral route, and because their multiplication in the gastrointestinal tract more closely mimicked natural infection, they offered certain selected advantages, which led to their replacing the injectable Salk vaccine. Over the first 5 years of the 1960s, more than 400 million doses of oral vaccine were distributed in the United States. At the same time, trials also were successfully conducted in European nations, Japan, and other countries. The use of oral vaccine in this country was accompanied by a further decrease in the annual reported polio cases, so that beginning in 1966 fewer than 100 have occurred each year. By 1978 with continued use of the oral vaccines, only 8 cases of paralytic disease were reported in the United States. In less than 20 years, a disease which had claimed thousands of victims annually, and which had been the source of indescribable community anxiety, was reduced to a rarity!

In the complex processes of vaccine development, commercial production, and widespread utilization, a number of unexpected events transpired which merit consideration. After the highly successful field trials of 1954, commercial manufacture of the Salk type vaccine was licensed. Within a few weeks of its use, paralytic disease was observed in April–June 1955 among children in California and Idaho who had received some of the first lots of commercial vaccine manufactured by the Cutter Laboratories. By the time this had been fully investigated and resolved, it was learned that there were 204 cases of vaccine-associated disease. Seventy-nine were among children who had received the vaccine, 105 were among their family contacts, 20 were in community contacts. Nearly three quarters of the cases were paralytic, and there were 11 deaths. The agent isolated from these patients was type 1 poliovirus. Laboratory tests on vaccine distributed by Cutter Laboratories revealed viable virulent type 1 poliovirus in 7 of 17 lots. Revisions of the federal regulations governing the steps in vaccine manufacture were promptly promulgated and implemented to prevent recurrence of such a tragic episode.

Manufacturers faced further difficulties in maintaining the fine balance between the complete elimination of the infectious live virus from the production process and the retention of effective antigenicity of the inactivated components. A number of lots of vaccine subsequently proved to be poorly antigenic for type 3 poliovirus. As a result, when community polio outbreaks occurred among well-immunized groups, there were "breakthroughs" with paralytic disease, especially due to type 3 virus, in previously immunized subjects. Such an outbreak was studied in 1959 in Massachusetts, where an analysis of polio cases revealed that 47 percent (62 of 137) of the patients had previously received three or more inoculations of inactivated vaccine.

With the availability of attenuated oral vaccine, enthusiasm for its use was enhanced by the aforementioned unfortunate episodes involving inactivated vaccine. A final unanticipated discovery in 1960 was the detection of a simian virus, SV-40, as a contaminant of the monkey kidney cell cultures utilized in the preparation of both inactivated and oral attenuated vaccines. Once again, a revision of standards for preparing and testing the tissue cultures was necessary to exclude this previously unrecognized agent. In the United States, inactivated vaccine has been used sparingly in the past 15 years. Almost all immunization has been conducted with the oral attenuated product. A number of European countries, especially those in Scandinavia, have adhered to the use of inactivated vaccine. With successful production of fully potent antigens, their record of achievement in the control of polio has been parallel to that of this country.

The marked decrease in paralytic disease due to "wild" polioviruses has disclosed a small but significant number of cases of oral vaccine recipients who have developed paralytic illness in temporal association to the ingestion of vaccine. In addition to these few cases among recipients of vaccines, there have also been paralytic episodes reported among susceptible family or community contacts of the vaccine recipients. These have been few in number and are difficult to characterize with complete clarity. A small portion of these patients have been found to be immunodeficient children, particularly those with congenital hypogammaglobulinemia. One concern has always been that the attenuated strains of virus might prove genetically unstable in human intestinal passage, so that increased neurovirulence might result from widespread dissemination. This has not been demonstrated. Monovalent oral polio vaccine (MOPV) was used extensively until 1964, when it was supplanted by trivalent vaccine (TOPV). With MOPV the risk of vaccine-associated illness in recipients was estimated to have been 0.19 per million doses distributed. With TOPV an overall figure of risk to recipients and their contacts has been calculated at 0.28 per million or 1 case per 3.6 million doses.

The achievements with poliovirus vaccination have been impressive in the United States, Canada, most of Europe, Australia, and some Asian and African nations. Polio remains an endemic disease in many tropical lands. It is premature therefore to relax the use of poliovirus vaccination in those parts of the world where disease has been nearly eradicated. The possibility of the inadvertent introduction of virulent virus is omnipresent. A number of such episodes have occurred already on the Texas–

Mexico border. Although polio immunization is now confined principally to infants and children, it is recommended that adult Americans travelling abroad to endemic areas receive polio vaccine prior to departure. Armed Forces recruits are routinely administered oral poliovirus vaccine. Its major use remains that for infants and children in the first 18 months of life.

FURTHER READING

Enders JF, Weller TH, Robbins FC: Cultivation of the Lansing strain of poliomyelitis virus in cultures of various human embryonic tissues. Science 109:85, 1949

Gear JHS, Measroch V: Coxsackievirus infections of the newborn. Prog Med Virol 15:42, 1973

Horstmann DM: Viral exanthems and enanthems. Pediatrics 41:867, 1968

Kibrick S: Current status of Coxsackie and Echo viruses in human disease. Prog Med Virol 6:27, 1964

Melnick JL: Advantages and disadvantages of killed and live poliomyelitis vaccines. Bull WHO 56(1):21, 1978

Morens DM: Enteroviral disease in early infancy. J Pediatr 92:374, 1978

Nathanson N, Langmuir AD: The Cutter incident, I, II, III. Am J Hyg 78:16–81, 1963

Nightingale EO: Recommendations for a national policy on poliomyelitis vaccination. NEJM 297:249 1977

Ogra PL: Effect of tonsillectomy and adenoidectomy on nasopharyngeal antibody response to poliovirus. N Engl J Med 284:59, 1971

Paul JR: A history of poliomyelitis. New Haven, Conn, Yale University Press, 1971

Sabin AB: Poliomyelitis vaccination. Am Soc Clin Pathol 70:24, 1978

Schonberger LB, McGowan JE Jr, Gregg MB: Vaccine-associated poliomyelitis in the United States, 1961–72. Am J Epidemiol 104:202, 1976

Wilfert CM, Lauer BA, Cohen M, Costenbader ML, Meyers E: An epidemic of Echovirus 18 meningitis J Infect Dis 131:75, 1975

CHAPTER 17

Orthomyxoviruses and Paramyxoviruses

Influenza viruses

Members of this group of ubiquitous myxoviruses produce epidemic illness so regularly that all readers of this chapter will have experienced several infections by them. They have been man's inconstant companion for at least as long as recorded history, and the broad age range of those afflicted, as well as the broad spectrum of illness observed, was well described in English texts as early as the sixteenth century, when it was called the "newe acquaintance." Because of the seasonality of influenza epidemics, it was thought initially to be a disease affected by celestial movement ("malattia influenza per le stelle"), and this belief permanently baptized both the viruses and the clinical illnesses which they produce. Although several major epidemics occurred during the early part of the twentieth century, the viral etiology of this disease was not demonstrated until the 1930s, and an understanding of the antigenic variability of these viruses and thus the periodic occurrence of epidemic illness had to await the more advanced virologic techniques of the 1940s and 1950s.

Clinical Features

As with most viral diseases, influenza represents a very broad spectrum of severity of clinical illness, from the occasional asymptomatic infection to the more common, bothersome and irritating "grippe," to the often fatal primary influenza pneumonia. The reasons for this wide variation are poorly understood and may be partly, but not exclusively, related to the age, health, and antigenic background of the patient, the presence of preexisting abnormalities of the pulmonary parenchyma or vasculature, and possibly to the virulence of the virus itself.

The significant histologic abnormalities induced by the virus are relatively uniform, beginning with virus replication in all superficial respiratory cells, destruction of the ciliated columnar epithelium, and subsequent denudation of the tracheobronchial tree. These anatomic changes invariably lead to impairment in respiratory function and physiology even in patients without obvious clinical evidence of pulmonary involvement. Defects include diminished defusing capacity, diminished tracheobronchial clearance, increased sensitivity to irritation, and variable degrees of small airway obstruction. The pharyngitis and tracheitis thus induced explain the severe throat and substernal pain, cough, and shortness of breath often seen during acute illness, as well as the paroxysms of cough that may be exacerbated by sudden exposure either to cold or environmental pollutants for weeks after the subsidence of clinical illness. Some of the functional pulmonary defects may require weeks or occasionally months to disappear. The tracheobronchial tree which has been denuded of its usual defense mechanisms is also more susceptible to invasion by bacteria. This occurs most commonly in patients with chronic bronchitis (where large numbers of bacteria may already be present in the lower respiratory tract), in aged individuals and in others with poor cough reflexes and poor efficiency of expectoration, or in any patient where pulmonary mechanics or host defense mechanisms are already abnormal.

The usual clinical disease begins rather suddenly within 1 or 2 days after exposure to infected aerosol droplets (Fig. 17-1), most often in epidemic circumstances and primarily during the colder months of the year. A sudden rise in temperature, sometimes to 102F, though rarely higher in adults, is accompanied by rigors and myalgias, and is then followed by the variable appearance of sore throat, nasal congestion and dryness, conjunctivitis, a nonproductive cough, and headache. Marked lassitude and moderate anorexia accompany these symptoms and are frequently so severe that even the most stalwart and ambitious individual must spend the first day or two of illness in bed. Physical examination will reveal generalized vasodilation of the dermal and mucosal surfaces of the head and neck, as well as edema and little to moderate discharge from the upper respiratory tract. On auscultation, the lungs are generally clear in individuals with otherwise healthy pulmonary function. Young children, however, are prone to develop laryngotracheobronchitis, bronchiolitis, croup, gastrointestinal upset, and the common sequela of otitis media. Laboratory studies will generally show a moderate lymphocytopenia, and blood gas analysis

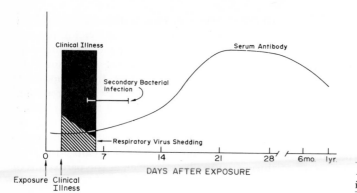

FIG. 17-1. The course of influenza virus infection.

will generally indicate a surprising level of hypoxia even in individuals without physical or roentgenologic evidence of pulmonary involvement. Patchy diffuse infiltrates of small size can occasionally be noted on chest x-ray. However, these are usually evanescent and of little consequence, but must be differentiated from the more severe forms of pneumonia seen during influenza epidemics and described below. Acute bronchitis, possibly bacterial, and exacerbations of chronic bronchitis are seen more commonly. The symptoms generally subside within several days, and the patient is able to return to his normal daily functions shortly thereafter. However, a severe postviral aesthenia, as well as a depressed affect, may be quite striking and persist for several weeks or more after resolution of the illness. The etiology of this weakness and easy fatigability is unknown, and its perceived severity is often proportionate to the expectations of the patient, as is often the case with infectious mononucleosis occurring during adolescence.

Pneumonias are the most common serious complications of influenza and can be divided into two generally distinct types. The more common is that produced by secondary bacterial invasion. It occurs particularly in patients with preexisting lung disease and in elderly or debilitated individuals. It generally follows a brief quiescent period after the initial viral illness, at a time when the patient feels he is beginning to recover. The most common infecting organisms are pneumococci, *Haemophilus influenzae* and staphylococci, although any species of bacteria can cause such an infection. The pneumonias themselves differ in no way from those caused by the same organisms under any other circumstance, and mortality is

generally low (less than 20%). As with all bacterial pneumonias, nevertheless, the precise prognosis of the patient depends on the virulence and sensitivity to antibiotics of the infecting organism, the age and general health of the patient, the expertise of his physician, and the zeal and skill with which supportive care is given.

A second type of infection caused solely by the virus itself is much less common, but considerably more lethal. The individual's clinical illness gives him no respite from the explosive onset of the "grippe" and continues inexorably through coughing, often in paroxysms, and marked shortness of breath, finally requiring admission to hospital. The sputum is scant with little mucus and is occasionally tinged with blood. Hypoxia is marked and chest x-rays often reveal more than does physical examination. Initially there are small patches and streaks of hazy infiltration, usually extending from the hilum to involve one or several lobes. These infiltrates increase in size, distribution, and density each day in spite of increasing levels of respiratory support and the desperate use of antibiotics until near total opacification of both lung fields occurs, and the patient expires from extreme hypoxia. This sequence may evolve so rapidly that occasionally a patient may die suddenly after only a few days' or hours' symptoms, whereas patients with severely depressed cell-mediated immunity may rarely manifest a much more indolent course. On pathologic examination, the lungs appear firm and red and the cut surfaces give the appearance of raw beef. Microscopic examination reveals no bacteria, but rather severe interstitial edema and hemorrhage, as well as occlusion of alveoli with proteinaceous débris and blood. While this disease is seen most commonly in patients with mitral stenosis and, in some epidemics, in women during the last trimester of pregnancy, it may occasionally be seen in otherwise healthy individuals.

It should be emphasized that many more patients die as a result of influenza than can be accounted for solely by those succumbing from pneumonia. Total "excess mortality" (the difference between the observed and expected number of deaths in a given time period) is generally two to three times greater during influenza epidemics than mortality from pneumonias. This can be easily understood if one accepts that influenza infection causes major derangements in many of the body's homeostatic mechanisms, and a large number of patients with marginal cardiac, metabolic, neurologic, and renal function may succumb rapidly from their primary disease when overcome by the virus-induced stress of fever and hypoxia.

A number of other less common complications or sequelae of influenza have been observed at variable frequencies, and these include myositis, myocarditis, pericarditis, and aseptic meningitis, as well as a postinfectious neuritis impossible to distinquish from the Guillain-Barré syndrome.

Epidemiology

The interpersonal spread of influenza viruses is caused by the formation of virus-laden aerosol droplets when an infected patient sneezes or coughs. Individuals in the immediate vicinity inhale these particles, the larger ones being deposited in the nares and upper nasopharynx, the smaller ones reaching the lower tracheobroncheal tree. If the virus is not neutralized, as for example by local IgA-specific antihemagglutinin antibody, then initial replication occurs. Further replication, and thus clinical illness, is dependent on a host of factors, including local IgA, serum IgG, and possibly various effectors of cell-mediated immunity. The efficiency of aerosol spread is exceedingly high and is in direct proportion to the number of particles generated, the density of susceptibles, and their physical and temporal contiguity to the aerosols, as well as the degree of stagnation of the ambient atmosphere. It is not surprising therefore to observe high attack rates

(at times approaching 90 percent) in nursing homes, classrooms, barracks, ships, and aircraft, where these conditions prevail. There is little evidence that other modes of contagion play a significant part in the spread of influenza, although the role of certain fomites such as nasal spray applicators and drinking glasses remains to be investigated.

Young children of preschool or gradeschool age appear to be the most effective disseminators of this disease, probably as a direct result of their limited prior experience with influenza (and thus high attack rates), as well as their primitive concepts of hygiene when they are actively excreting virus. An explanation of the appearance of epidemic influenza in the winter in temperate zones has been assiduously sought, but few positive conclusions have been reached. It does not appear to be directly related to temperature, relative humidity, or the duration of sunlight exposure. Likewise, wetting or chilling of the susceptible, so prominent a factor in avian respiratory infections, also does not appear to influence the susceptibility to or virulence of human influenza infections. It must be remembered that influenza epidemics may occur at any time in the tropics, and there have been epidemics recorded in every month of the year in the United States. Our largest recorded epidemic occurred in the late summer and early fall of 1918.

An understanding of the periodicity of influenza epidemics and pandemics requires a basic knowledge of the ultrastructure of the virus as delineated in Chapter 2, for the surface antigens of the virus and the antibodies they induce within a population determine the epidemiology of influenza. The lipid envelope of the virus is covered by a number of glycoprotein spikes made up primarily of hemagglutinin with a smaller number of separate neuraminidase spikes. Antibody to hemagglutinin prevents infection, either in tissue culture or in people, whereas antibody to neuraminidase limits virus spread in tissue culture and, if antihemagglutinin antibody is absent, lessens the severity of clinical influenza. Changes in the amino acid sequences of these glycoproteins lead to changes in their antigenic identity. The mechanisms which engender these changes have been a matter of great speculation for many years.

Generally, when a new virus with a different

hemagglutinin type appears there is little antibody against it in the population at large. The first wave of the epidemic infects a substantial proportion (25 to 60 percent) of the population. The virus also has the opportunity to undergo spontaneous, albeit limited, mutation ("antigenic drift"), so that a separate wave occurring 1 to 5 years later infects many of the remaining antibody-negative individuals, as well as some of the individuals who were previously infected, but whose antihemagglutinin antibody response was either initially low or had waned over the intervening years. This sequence continues until the vast majority of the population has developed protective levels of antibody. At that point, a new hemagglutinin type, which shares no obvious antigenic cross-reactivity with the first, then begins the cycle once again.

The grouping of these major antigenic types is based on the ability to observe serologic cross-reactivity in hemagglutination- or neuraminidase-inhibition (HI or NI) tests. To date, there have been five major types of human influenza hemagglutinin identified (H swine, H0, H1, H2, and H3) and two major types of neuraminidase (N1 and N2). Antibody to one will not be protective against infection with the other. That a greater hierarchy occurs than can be detected by serologic HI assays, however, is demonstrated by the fact that one can note antigenic "priming" between members of the H swine, H0, and H1 group, and similar "priming" between the H2 and H3 groups. Such "priming" is observed when prior infection with one member of the group causes an accelerated antibody response when subsequent infection or immunization occurs with another member of the group.

Serologic analysis of individuals who were alive during the epidemics occurring in the latter part of the nineteenth century and the early part of the twentieth have clarified and greatly expanded our understanding of the periodic recurrences of influenza epidemics. Epidemics of viruses bearing the H2 hemagglutinin occurred in 1889 to 1890, and these viruses were later replaced by viruses probably bearing the H3 antigen around 1900. The H swine antigen first appeared in 1918 and gave rise to one of the greatest epidemics in recorded history. Next to appear in the 1930s was the H0 antigen, and viruses carrying the H1 antigen became epidemic in 1946. The H2 antigen then reappeared in 1957 to be supplanted by

the H3 antigen in 1968, a sequence similar to that noted in 1889 and 1900. After the abortive reappearance of viruses bearing the H swine antigen in 1976, the H1 antigen reappeared in Southeast China in mid-1977 and caused widespread epidemics with apparently low mortality throughout the world in 1977 and 1978.

Many recent pandemics (1957, 1968, 1977) have begun in China. The reasons for this are not clear, although it has been hypothesized that the density of human and swine populations in certain areas of that country provide an opportunity for dual infections with viruses indigenous to each species. Subsequent recombination may produce new viruses (see Chap. 7), some of which may be novel or more virulent, or against which there is little protective antibody in the population. Although spread from China to the rest of the world may take several months or more, the inexplicable simultaneous appearance of the virus in several widely separated parts of the world is well documented and was observed even before the advent of rapid intercontinental travel. Epidemics tend to sweep rapidly through individual communities and rarely last more than a month from beginning to end. Mortality rates are difficult to quantitate, but are generally quite low, often being quoted as 0.01 to 0.1 percent for the population at large, although the case fatality rate is considerably higher in the very young and very old. No valid explanation has yet been brought forth to account for the high fatality rate, particularly in patients in their third and fourth decades of life, observed in the 1918 H swine epidemic.

Influenza B viruses also undergo major and minor antigenic changes, though at lower frequencies and longer intervals. Epidemics caused by influenza B viruses have rarely been associated with excess mortality, but the significance of major antigenic changes, as occurred for example in 1973, is more obvious in younger rather than in older age groups. Children are more readily infected with this virus, are more obviously symptomatic once infected and are more likely to suffer one of the most serious sequelae of infection by this virus, namely, Reye's syndrome. The association of this disease, characterized by hepatocerebral fatty infiltration and degeneration, with influenza B and a few other viral infections is extremely strong and over 300 cases were noted in the United States during the brief influenza B epidemic of 1973–74. The pathomechanisms of this disease remain to be elucidated.

Diagnosis

While the influenza viruses can produce a wide variety of clinical illnesses, it must also be remembered that a broad spectrum of infectious and noninfectious agents can produce clinical syndromes which are similar to those of "influenza." Any of the multitudinous diseases which cause fever and myalgias are described in many medical texts as producing "influenzal symptoms," and patients infected by other common respiratory pathogens (RSV, adenovirus, coronavirus, rhinovirus, parainfluenza viruses, etc.) will often state that they have "influenza." Although influenza is more easily diagnosed during epidemics, a pathognomonic complex of signs and symptoms does not exist, and there is little to distinguish it on clinical grounds from other illnesses which may also occur during these periods. This difficulty in precise clinical diagnosis becomes particularly prominent when several epidemics occur simultaneously, as can happen in temperate zones when influenza and RSV epidemics occur in close temporal proximity. Recent experience in the Caribbean has shown that epidemic dengue is difficult to distinguish from influenza unless the patient has prominent respiratory signs and symptoms.

Rapid confirmation of the clinical impression of influenza can be obtained by examination of material obtained by swabbing the posterior pharynx. Viral antigen can be detected within nasopharyngeal cells by direct or indirect immunofluorescent stains directed at any of the viral antigens. Such tests, however, are not generally available, require excellent reagents and appropriate positive and negative controls, as well as experienced observers.

The virus may also be grown from nasopharyngeal swabs after inoculation into any of several substrates. One of the more convenient and effective are continuous Madin-Darby canine kidney (MDCK) cells. Other effective substrates are primary monkey kidney cells and the allantoic or amniotic cavity of 10-day old embryonated hen's eggs. In most instances, little viral cytopathic effect is induced, and the presence of virus must be determined by immunofluorescence, hemadsorption, or hemag-glutination. Plaques can be formed in MDCK cells if the overlay contains trypsin, but this is of little use in primary isolation. Viral culture of the blood is of no use, as viremia is exceedingly rare in influenza.

It should be remembered that if the initial swab cannot be immediately inoculated into the appropriate growth substrate, the virus may remain viable for several days without loss of titer if the swab is placed in appropriate media such as veal heart infusion broth and placed in the refrigerator at 4C. A common mistake is to freeze the sample for preservation until culture can be arranged within several days. Such a single freeze-thaw cycle will generally reduce the viral titer significantly, and in those samples which have marginal or low titers, this reduction may be sufficient to render the specimen negative for growth.

It is generally not practical, necessary, or even desirable to attempt virus isolation in most clinical situations. Nonetheless, programs of surveillance that include virologic culture of select populations, are highly necessary for the general well-being. Individuals involved in such programs serve as sentinels and permit the early awareness of epidemic influenza. These surveillance systems operate best when centralized, as through the Center for Disease Control or the World Health Organization, and when they disseminate their information rapidly and effectively. Public health authorities may then make plans for immuno- or chemo-prophylaxis.

Serologic confirmation of influenza infection is rarely required. If such a confirmation is deemed necessary for scientific investigational purposes, however, several facts should be borne in mind. Complement-fixation (CF) tests depend on the interaction between antibody and internal (nucleocapsid) proteins. Since these proteins are antigenically (if not chemically) identical for all viruses within a group (A, B, or C), positive CF tests can signify infection by a member of that group but cannot more specifically identify the virus. CF titers generally rise relatively rapidly (within 1 to 2 weeks), but their decline is somewhat variable, often in terms of several weeks to several months. In some individuals, however, they may persist for longer periods of time, though usually at a rather low titer. A single postillness serum sample, therefore, cannot be considered absolutely diagnostic, but does become more useful if the titer is elevated ($\geq 1{:}32$). Studies of postinfection CF titers are generally of more epidemiologic than clinical value.

A more precise method of determining acute influenza infection is the quantitation of HI antibodies simultaneously on samples obtained before or at the beginning of illness and a second sample obtained 3 or more weeks later. A fourfold rise in antibody titer is considered diagnostic of influenza if the patient has not been recently vaccinated. Additionally, because this antibody is directed against the specific surface glycoprotein, the precise identity of the originally infecting organism can generally be determined with some precision.

Since most influenza infections are self-limited, the major goal in diagnosis is really to determine whether other diseases causing "influenzal" syndromes are present or whether bacterial superinfection pneumonia has occurred. The criteria for establishing the latter diagnosis include sustained pyrexia, leukocytosis, evidence of pulmonary consolidation on physical or x-ray examination, and, most importantly, the presence of substantial numbers of granulocytes and bacteria in the sputum. It cannot be overemphasized that the ability to perform a proper Gram's stain of the sputum and examine it with accuracy is the *sine qua non* of a physician competent to treat all pneumonias, including those following influenza. The proper treatment and final clinical outcome of the patient will often be dependent on the physician's ability to distinguish granulocytes from epithelial cells (and thus his ability to distinguish a sputum from spit) and his recognition of properly stained bacteria.

Treatment

General supportive measures have been prescribed for centuries to alleviate the bothersome symptoms of common influenza. There is little evidence that any of these measures shorten the duration of illness or diminish the frequency of complications. Antipyretics for fever, fluids for thirst, analgesics for muscle pains, and bed rest for lassitude all support the comfort of the patient. Topical and systemic decongestants relieve nasal obstruction, and antihistamines promote restful sleep.

Evidence is accumulating that amantadine, a symmetrical primary amine, is effective in the treatment of clinical disease caused by influenza A viruses. This effect must be carefully differentiated from its prophylactic effect, which will be discussed below. The interpretation and performance of studies on the therapeutic effectiveness of this synthetic inhibitor of viral uncoating is difficult because influenza is a fairly rapidly evolving and resolving illness, and only a few days of clinical disease are ordinarily observable after the patient has consulted with his physician and sought therapy. Earlier studies during the 1960s demonstrated that amantadine definitely lessened the number of days of pyrexia, but this effect appeared little different from that of aspirin. More recent studies, however, have shown that amantadine often diminishes the severity of most symptoms to a significant extent and hastens clinical recovery by a day or more. Although this improvement may appear to be of marginal benefit, any improvement will be welcomed by most patients. Amantadine also reduces the titer and duration of virus excretion, which, while of little direct benefit to the patient himself, may be very significant in terms of limiting the spread of disease. Most importantly, it has been demonstrated that the duration of altered pulmonary function (such as diminished frequency-dependent compliance), which may last for several weeks after clinical influenza, can be markedly shortened following amantadine therapy. The dose of amantadine given is usually 100 mg twice a day. Side effects can occur with variable frequency and are more common in the elderly. The most common effects in young adults appear to be psychologic. Insomnia, excitement, and a poor attention span or difficulty in mental concentration may be seen in up to 10 percent of recipients. In older individuals, the drug can occasionally precipitate congestive heart failure, cause more severe mental changes or be responsible for other less common adverse effects.

In those patients who have developed bacterial pneumonia, appropriate respiratory support and antibiotic therapy are necessary. Since pneumococci, *Haemophilus influenza*, staphylococci, and a variety of gram-negative bacilli can cause this pneumonia, it is imperative, as noted above, that a sputum examination and culture be performed, as there are few if any single antibacterial agents which are highly active against all of these organisms. If the patient will not spontaneously produce sputum, then consideration should be given to obtaining lower tracheobronchial samples by transtra-cheal aspiration, provided that the patient is cooperative, has normal bleeding parameters, and is not excessively hypoxic (pO$_2$ less than 60). Should this not be possible, then empirical therapy may be initiated with a semi-synthetic penicillin (or a cephalosporin) and possibly an aminoglycoside. This coverage will generally be rather poor against *Haemophilus influenzae* unless the cephalosporin used is one that has reliable activity against it.

Prevention

Because the therapy of influenza is rather lim-ited, major attempts have been made to pre-vent this disease. That these efforts have not been successful in the past is confirmed by the large number of people who die during epi-demics and even during interpandemic years. The etiology of this lack of success is primarily logistical and attitudinal (as was the case for smallpox) and not because effective prophylac-tic measures are not available. Early attempts at preventing the disease involved growing in-fluenza viruses in the allantoic cavity of embry-onated hens' eggs and inactivating the virus in this slurry with formaldehyde. These vaccines generally induced high levels of HI antibody which were closely correlated with protection from infection. Serum HI antibodies of $\geq 1{:}32$ to 40 are generally associated with solid protec-tion from infection, whereas lesser levels of antibody (1:8 and 1:16) are associated with ei-ther moderate protection from infection or at-tentuation of clinical disease. These early vac-cines required large volumes of allantoic fluid for each vaccine dose and were associated with "endotoxinlike" reactions consisting of fever, chills, lassitude, and myalgias, which the recipi-ent often considered to be similar to (or worse than) the prodromata of clinical influenza. The etiology of these reactions was not clear. It was thought that the large amount of nonviral (egg) protein present in the vaccines caused these reactions, although precise clinical investiga-tions both in the 1940s and the 1970s did not support this hypothesis, and vaccines contain-ing larger amounts of egg protein (such as the 17D yellow fever vaccine) do not induce such reactions. An equally valid hypothesis could be that the large amount of virus with its lipopro-tein coat may cause these reactions. Vaccines

produced in the late 1960s and early 1970s by density gradient centrifugation (see Chap. 2), which eliminated most of the nonviral protein, were associated with fewer adverse reactions and equal immunogenicity, although very young individuals continued to experience at times severe and debilitating adverse reactions, including seizures. Disruption of the virus by a variety of detergents and solvents yielded vac-cines which produced fewer adverse reactions and were equally immunogenic in "primed" populations. Single doses of such vaccines, how-ever, are less immunogenic in individuals who have little or no prior antigenic experience with influenza. A second dose given a month or more later, however, ordinarily induces protec-tive antibodies in over 85 percent of these re-cipients.

Inactivated vaccines, nonetheless, are little used for a variety of reasons, the first being the memory of both the patients and doctors of older, more highly reactogenic vaccines. Sec-ond, these vaccines are effective only when the antigens in the vaccine are identical or closely related to the antigens of the epidemic virus. Since the preparation of these inactivated vac-cines requires 6 to 9 months of "lead time," a certain amount of guessing must occur each winter in an effort to establish which virus will be epidemic in the succeeding year. These prognostications are sometimes accurate. In addition, the immunity induced is not highly effective for much more than a year, so that annual vaccination must be given. Finally, the economics of vaccine production are difficult and marginal. Manufacturers must deal with an everchanging target which only intermittently captures the attention of patients, public health authorities, and physicians. As a result, they tend to produce only the number of doses of vaccine they feel will be sold in the succeeding year. This has generally been in the range of 15 to 20 million doses annually in the United States. The practical problems of administering all of these doses at the optimal time (generally several weeks to a few months before the ex-pected epidemic) can be easily imagined.

Because of these logistical problems, vaccine is currently recommended only for those who are at high risk of dying should they acquire influenza. They would include any patient with a chronic illness, particularly pulmonary, but also any chronic cardiac, metabolic, and neuro-logic disease and the elderly (those over 65 years of age). Its use in "vital community personnel" such as physicians, nurses, policemen, and firemen has been tempered by the occurrence of the Guillain-Barré syndrome in approximately 1 out of 100,000 recipients of the A/New Jersey (swine) strain influenza vaccine. About 10 percent of these cases were fatal. It is presumed that this polyneuritis can occur following the use of vaccines made from other strains of influenza, as it occurs following natural illness and following the use of other viral and even non-viral vaccines.

Because it is thought optimal by some to induce local (nasal and/or tracheobronchial) immunity to prevent respiratory disease, great enthusiasm has been generated over the past three decades in Eastern Europe and for the past one to two decades in the West for live influenza vaccines which may be applied topically in the nostrils. The viruses in these vaccines have been attenuated by a variety of means, including multiple egg passages, recombination with highly attenuated strains, cold adaptation, selection of temperature-sensitive mutants, or the selection of strains resistant to nonspecific inhibitors. Clinical studies with several of these types have shown that immunity can be induced at a level and duration roughly equivalent to that induced by inactivated vaccines. One advantage of these vaccines is that a much greater number of doses can be produced per embryonated egg. Fertilized eggs are often a limiting economic and supply factor in the production of influenza vaccines. A disadvantage of these vaccines, on the other hand, is that because they contain infectious virus, much more stringent testing in humans must be performed before their widespread general use in tens or hundreds of millions of recipients can be encouraged. When such testing is completed, the original epidemic strain for which the vaccine was designed has often disappeared from the population. Attempts to use nonhuman and in vitro models of immunogenicity and safety are being intensely investigated, but to date none has been sufficiently consistent and convincing to obviate or circumvent extensive and time-consuming clinical trials in humans.

Amantadine has also been shown to be effective in the prevention of influenza. Its efficacy is approximately that of inactivated vaccines. Its advantage is that its prophylactic activity is not as strain-specific as that of vaccines and therefore may be used when the virus epidemic in the population is different from that in the vaccine or when vaccines are not available. Its advantages are that it will induce a substantial number of side effects in patients who are otherwise healthy, and it must be taken regularly during the entire epidemic exposure season.

Parainfluenza viruses

The four serotypes of human parainfluenza viruses were isolated from throat swabs of infected individuals in the years 1955 to 1958. Information which has accumulated as a result of studies over the subsequent two decades has described the illnesses attributable to these agents and the epidemiology of these infections, and has led to initial attempts to prevent natural infection.

Clinical Illness

Parainfluenza virus type 1 has been the predominant agent associated with laryngotracheobronchitis (croup). This agent tends to occur in biennial epidemics which peak in the fall months. Children usually acquire infection between the ages of 6 months and 3 years. The vast majority of young adults have demonstrable serum antibodies against parainfluenza 1.

Parinfluenza type 2 is also associated with acute laryngotracheobronchitis in the fall, but occurs in sporadic outbreaks. Parainfluenza 2 will often alternate years of occurrence with parainfluenza type 1. Although children of the age group 8 to 36 months are at greatest risk of infection, the morbidity from parainfluenza 2 infections is substantially less than that with parainfluenza type 1.

Parainfluenza virus type 3 is associated with bronchiolitis and pneumonia in infants less than 1 year of age, with croup in children 1 to 3 years old, and with tracheobronchitis in older children. This virus is similar to respiratory syncytial virus, with its predilection for causing infection in infants during the first 6 months of life, as well as through 3 years of age. The majority of primary infections in infants and children are associated with febrile illness of approximately 4 days duration, and at least one third of these illnesses are complicated by pneumonia or bronchiolitis. Serologic studies have demonstrated that approximately one-half of children have been exposed to this agent by the end of their first year of life.

Parainfluenza type 4 has been recovered much less frequently from children and adults with mild respiratory illness. The laboratory isolation of this agent is more difficult, which contributes to its less frequent association with illness. The common symptoms of infection with any of the parainfluenzae agents in adults are nasal congestion, sore throat, and malaise. These viruses have been associated with exacerbations of chronic bronchitis, pharyngitis, and tonsillitis in college-age students, and asymptomatic reinfection.

Infection with these agents is associated with production of specific HI, neutralizing, and CF antibodies. Reinfection produces anamnestic responses with heterotypic antibodies to other serotypes of the parainfluenza viruses. Reinfections occur frequently, although the presence of antibody may modify the clinical manifestations of infection. Studies with parainfluenza type 2 have demonstrated that protection against clinical reinfection is best correlated with the presence of detectable antibody in nasal secretions.

Several recent studies have demonstrated excretion of parainfluenza virus type 3 for 2 to 5 months in a population of adult patients with chronic bronchitis and emphysema. Several of these patients had virus-specific local antibody detected in their respiratory tract prior to detection of virus excretion. This persistence of parainfluenza type 3 infection did not produce a significant increase in local respiratory antibody. This could be contrasted to those patients showing symptomatic illness, brief excretion of virus, and an associated increase in titers of serum and respiratory secretory antibody. Additionally, prolonged excretion of parainfluenza type 2 has been documented in human immunocompromised hosts, particularly those with a deficit in their cell-mediated immunity.

Pathology

Examination of pulmonary tissue from infected children demonstrates an increase in the bronchial-associated lymphoid tissue, peribronchial infiltrates, and exudates within the lumen of the bronchus. The infiltrate around the bronchi is made up of small lymphoid cells. The parenchyma of the lung may have interstitial pneumonia with areas of associated atelectasis.

An experimental animal model employing Syrian hamsters has allowed some study of the pathogenesis of this type of respiratory tract infection. Intranasal inoculation results in virus replication within parenchymal tissue. The peak titers are reached within 5 days, and the pathology is very similar to that described in children.

This model of parainfluenza type 3 pneumonia in hamsters has been utilized in studies of possible interaction of antibody and virus. Administration of antibody to the animal prior to infection did not produce any enhancement of the pneumonitis. There was somewhat less pulmonary parenchymal disease, and the titers of virus were slightly lower in animals receiving antibody. These animals with passive homotypic antibody did not develop an active antibody response to the infection, and subsequent reinfection did not produce more severe disease. Thus these experiments fail to demonstrate that passive antibody enhanced the production of pneumonitis either in the presence of passively administered antibody or with reinfection occurring after the passive protection had waned. This is an important observation in light of the hypothetical explanations for severe disease in very young infants at a time when maternal antibody ought to be present.

Virus infection of tracheal ring cultures with normal differentiated respiratory epithelium produces a loss of ciliary motion. The epithelial layer loses its pseudostratified columnar organization, and some of the ciliated cells are lost, whereas others show fusion with formation of multinucleate giant cells. The giant cells may also be seen in the lamina propria and the cartilage. Virus particles budding from the cell membrane are visible by electron microscopy. Tracheal ring organ culture will support replication of virus over a 2-week period of time. Later there is focal destruction of the respiratory epithelium. Direct immunofluorescence applied as early as 24 hours after inoculation reveals cytoplasmic fluorescence within epithelial cells.

Diagnosis

The overlapping spectrum of respiratory disease produced by a number of viral pathogens can be etiologically defined only by specific identification of the infecting virus. Rapid and accurate diagnosis by immunofluorescence of nasal secretions has been accomplished by several laboratories. Careful collection of materials and fluorescent staining of the exfoliated epithelial cells from the upper respiratory tract provide a diagnosis within 1 day of obtaining the specimen. The fluorescence is cytoplasmic and has been observed in the ciliated epithelial cells.

Isolation and identification of these agents in susceptible cells in tissue culture may require up to 2 weeks. The infected tissue culture can be examined by immunofluorescence with successful demonstration of viral antigens during the first several days of culture. More commonly, hemadsorption employing guinea pig red blood cells is utilized in 4 to 7 days to detect the presence of a hemadsorbing agent.

Tissue culture cells that are used for parainfluenza virus isolation include primary human embryonic kidney, human anmion, and primary monkey kidney cells. Unfortunately, the presence of hemadsorbing simian viruses must be taken into consideration when working with these cell cultures.

The parainfluenza viruses do hemagglutinate guinea pig and chicken red blood cells. Thus hemagglutination inhibition can be useful for virus identification and for determining the titers of specific serum or secretory antibody against known virus strains. With primary infection, the response to the infecting agent tends to be type-specific. With each subsequent infection there is an increased heterologous antibody response.

Treatment

The treatment of these respiratory infections is only by the provision of supportive therapy. Hydration, maintainence of an adequate airway, and therapy of bacterial superinfection are important in the management of these patients. The severity of the lower respiratory tract disease is greatest in the youngest age group.

Parainfluenza virus 3 infection has been documented to be a significant nosocomial infection and with increased morbidity to the patients who acquire these infections in the hospital. All of the viruses causing severe respiratory infections in children should be regarded as communicable within the hospital, and appropriate precautions should be taken to prevent spread of the infection within the hospital.

Prevention

As implied in the preceding paragraphs, immunity to parainfluenza viruses is only partial. That is, reinfection may occur naturally, although the severity of the illness may be considerably diminished.

Parainfluenza viruses grown in embryonated eggs or tissue culture have been administered as formalin-inactivated vaccines. Two to three injections of the individual parainfluenza types are necessary before seroconversion occurs in young infants. These materials have been tested as monovalent or trivalent immunogens. The vaccine can produce a 95 to 100 percent seroconversion rate in antibody-negative subjects, but no resulting significant protection against naturally occurring infection has been demonstrated. Parenteral administration of these inactivated products does not produce local secretory antibody.

Significant morbidity results from infection with these agents in the youngest infants. Prevention of infection is a desirable goal, but the available immunogens have not been successful. Further understanding of the pathogenesis of infection is essential before successful immunization will be possible.

Mumps

Clinical Illness

Mumps virus infection is an illness prevalent in the childhood years and was recognized in epidemic form as early as the fifth century BC by Hippocrates. The most frequently discussed and recognized symptom is parotitis, although the illness may have multiple manifestations which are indicative of the generalized nature of the infection. The portal of entry is thought to be the upper respiratory tract as the virus gains direct access to the mucous membranes of the mouth and nose. Virus is transmitted by the saliva of infected persons or with materials recently contaminated with virus.

The time interval which elapses after exposure to virus and prior to the appearance of clinical symptoms is usually 14 to 21 days. After entering the host, the virus replicates and viremia occurs. Viremia, which has been demonstrated with uncomplicated infection, may then secondarily infect one of several organ systems, including the salivary glands. The specific tissues in which primary virus multiplication occurs, and the factors or conditions which favor mumps virus replication within a given tissue, are not known. Such organs as the already mentioned salivary glands (predominantly the parotids), meninges, testes, pancreas, ovaries, thyroid, and heart can show evidence of infection. Virus is also excreted in the urine, and transient abnormalities in renal function have been found in adult males with mump infection.

Figure 17-2 presents the time relationships of the clinical features of mumps virus infection and correlates the time periods of virus excretion, communicability of the illness, and clinical symptoms. Salivary gland infection with pain, edema, and consequent swelling results in the characteristic parotid gland enlargement diagnosed as mumps. This infection may involve the parotids, submandibular, and, less often, the sublingual salivary glands. Salivary gland involvement usually precedes other clinical symptoms, lasts for 2 to 7 days, and may be unilateral or bilateral. Any of the other manifestations of mumps infection such as central nervous system (CNS) disease occasionally precede, coincide with, or occur in the absence of salivary gland involvement.

Involvement of the CNS with any infectious agent is cause for concern for both the physician and the patient. Fortunately with mumps virus infection, the vast majority of such recognized illness is transient aseptic meningitis with few sequelae. The incidence of this type of CNS infection has been estimated to be as high as 65 percent when lumbar punctures have been done on a group of hospitalized patients with clinical mumps. The CNS involvement was manifest in each case by a predominantly lymphocytic pleocytosis of the cerebrospinal fluid and only one half of these individuals evidenced clinical signs of meningitis. Signs of meningeal involvement are most often manifest 2 to 10 days after the onset of parotitis, last for 3 to 4 days, and are self-limited.

A more serious, and fortunately much less common, CNS manifestation of mumps is postinfectious encephalitis or encephalomyelitis. The ratio of mumps encephalitis to mumps cases is reported to be 2.5/1000 cases for each year from 1968 to 1972, and the total number of reported cases has significantly declined with the advent of vaccine. It is important to distinguish between this and the aseptic meningitis to be able to put the illness of the individual patient in the proper perspective for the family and patient. The time of onset is usually later than the transient aseptic meningitis and occurs 10 to 14 days after the clinical salivary gland involvement. The patient appears severely ill, is deeply obtunded, and may succumb to the illness. Although occurring less often than the postinfectious encephalitis associated with measles or varicella, it is clinically and pathologically indistinguishable from them. The typical manifestations of the preceding illness (e.g., rash with measles) provides the clinical diagnosis of the underlying illness.

An observation recorded in 1967 demonstrated the development of hydrocephalus in hamsters after experimental mumps infection. Suckling hamsters were inoculated intracerebrally with mumps virus isolated from the cerebrospinal fluid of a patient and passaged only once in cell culture. Sequential studies demonstrated virus replication and associated inflammatory changes during the first week after inoculation. Virus was localized almost entirely within ependymal cells of the ventricles and choroid plexus. The animals did not show any signs of acute illness, but hydrocephalus became evident in 3 to 6 weeks after infection. These animals had aqueductal stenosis secondary to the resolving inflammation. Thus an inapparent acute inflammation resulted in later sequelae at a time when viral antigen and in-

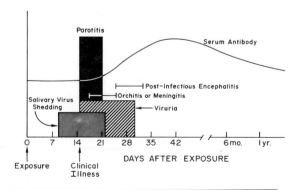

FIG. 17-2. The course of mumps virus infection.

fectious virus were no longer evident. Similar effects have now been described in experimental models employing influenza A_0, parainfluenza 2, and a temperature-sensitive variant of measles virus. The clinical application of such observations remains to be established, although there are now several case reports recording mumps infection in children who subsequently developed aqueductal stenosis. It is not proven that the infection caused the subsequent hydrocephalus.

An occasional sequel of mumps virus infection is deafness, which may occur even in the absence of other evidence of CNS involvement. The loss of hearing may be preceded by tinnitus and a sense of fullness of the ear. Such deafness is relatively uncommon but occurs suddenly during the period of parotid swelling. It is usually unilateral, but an estimated 20 percent of those so affected may have bilateral disease. Once deafness has occurred, the damage is irreversible and is apparently caused by inflammation and subsequent destruction of the organ of Corti.

The complication of mumps infection best known by non-medical persons is the involvement of the testes. Orchitis occurs predominantly in postpubertal males with 20 to 30 percent of this age group manifesting testicular involvement during the course of the mumps infection. Bilateral disease is present in approximately 2 to 6 percent of the total number of patients with orchitis. Inflammation of the testes is extremely painful, but the fear of this affliction stems from its association with sterility. Accurate knowledge of the incidence of sterility following mumps orchitis is not available because the occurrence is sufficiently infrequent that extremely large numbers of cases of mumps must be observed during the original illness to document accurately mumps virus infection as the cause of subsequent sterility. Once it has occurred, therapy of orchitis is symptomatic, without any clearly demonstrated benefits from either steroid therapy or immune globulin.

Pancreatitis with typical symptoms of abdominal pain, fever, and vomiting may occur in association with mumps infection. Recent in vitro work has demonstrated by fluorescence the infection of beta cells and non-beta cells of human pancreas in cell culture. Much less frequently, thyroiditis, mastitis, myocarditis, and oophoritis are seen.

In the normal individual, a single infection with mumps virus confers permanent immunity against clinically evident infection. It is probable that reinfection, defined as an antibody rise after exposure to the virus, may occur, but neither virus shedding nor clinical illness has been demonstrated with such reinfection. Second attacks of parotitis have been observed in the same individual, but it is wise to remember that other causes of parotitis include Coxsackie or lymphocytic choriomeningitis infections, starch ingestion, sarcoidosis, iodine sensitivity, and thiazide therapy. At the time of this writing, there is no documentation by culture or serology of two clinical attacks of mumps virus infection in a single individual.

The host initially responds to mumps virus infection with serum IgM antibodies and subsequently antibodies predominantly in the IgG class. Neutralizing antibodies are formed within the first week of symptoms and ordinarily persist for a lifetime. Hemagglutinating-inhibiting and CF antibodies may appear from the first through the third weeks and usually reach a peak within 3 to 6 weeks after the onset of symptoms. Complement-fixing antigens have been differentiated into two distinct types. The first is the soluble (S) or nucleoprotein antigen, and the second is the viral (V) or surface antigen. Complement-fixing antibodies against the S antigen are present within 2 to 3 days of onset, peak at about 10 days, and disappear in 8 to 9 months. Complement-fixing antibodies against the V antigen appear at about the 10th day of infection and persist for years. The transient nature of the S antibodies may sometimes assist in defining recent infection.

Special consideration should briefly be given to those persons commonly thought to be at increased risk from viral infections. The pregnant woman or her fetus is frequently considered at higher risk with regard to particular infections such as varicella and rubella. At this time, conflicting reports relating to mumps infection during pregnancy have been published. There is no confirmed evidence of the occurrence of congenital anomalies in the human fetus as a result of mumps infection sustained by the pregnant mother and transmitted to her fetus in utero. Although there have been documented cases of mumps in mothers during pregnancy and in the immediate partum period, neither clinical mumps nor excretion of virus has been documented in these infants.

Patients who have altered immunity, either humoral (e.g., the agammaglobulinemics) or cellular (the naturally occurring lymphopenic or the immunosuppressed patient) do not appear to be at increased risk with respect to mumps infection. Live virus vaccine is contraindicated for pregnant women and for the groups of persons with altered immunity, but to date no increased risk to these patients has been demonstrated.

The mumps intradermal skin test has been utilized in attempts to define an individual's previous experience with mumps antigen. There is no doubt that there can be a demonstrable delayed type of hypersensitivity response which is dependent on previous lymphocyte sensitization to the antigen, but this skin test is less reliable than antigens such as IPPD. There are a significant number of false-negative and false-positive tests, so in an individual case, the skin test will not be helpful.

Pathology

Examination of involved tissues is unusual because of the ordinarily benign nature of the illness. When salivary glands have been studied, there is no disruption of the general architecture of the gland after several days of illness. The involved salivary ducts demonstrate changes in the epithelial lining cells ranging from swelling to complete desquamation. The ducts are dilated and the lumen may be filled with cellular debris and polymorphonuclear cells. There is a moderate amount of periductal edema around the involved ducts, and the interstitial inflammatory cell is predominantly mononuclear.

The pathology of other involved tissues, e.g., the testes, is similar, with mononuclear infiltrates, edema, and no specific hallmarks allowing the diagnosis of mumps infection to be made solely on the basis of the observed pathology.

Epidemiology

Mumps virus infection is predominantly a disease of childhood, with the majority of clinically evident infections being seen between the ages of 5 to 10 years. It has been estimated that 90 percent of the population is immune by the time they reach 15 years of age. Although mumps virus infection is contagious, it is less communicable than measles and varicella. The degree of communicability is estimated most accurately by serologic surveys of exposed individuals, since as many as one-fourth of the infections with mumps virus occur without clinical symptoms.

Isolation of the patient within the hospital setting or in homes, when it is attempted, has not effectively curtailed spread of disease. This is usually attributed to the period of virus shedding which occurs prior to symptomatic onset of illness and thus precedes recognition of infection. As previously mentioned, one-fourth of patients have an asymptomatic infection, but they also excrete virus. Their infection is self-limited, and their immunity is comparable to those with symptomatic infection. To the best of our knowledge, there are no animal reservoirs or human "carriers" of mumps virus.

Diagnostic Approach

The work of Johnson and Goodpasture first established that mumps was caused by a filterable virus and demonstrated that rhesus monkeys could be experimentally infected. The description of the CF test and successful propagation of virus in chick embryos preceded the now generally employed standard tissue culture techniques. These methods employ monolayers of one of several cell types, including primary monkey kidney, human amnion, or human kidney and cell lines such as HeLa. With these techniques, virus has been isolated from such varied sources as blood, cerebrospinal fluid, urine, saliva, salivary gland tissue, and human milk.

In many academic and large hospital settings, viral diagnostic laboratories are available, and virus isolation can be attempted from clinical materials. The responsible laboratory will provide directions for submitting materials for culture. Saliva or urine can be collected at the time of clinical CNS symptoms and submitted for culture. Mumps isolation in tissue culture is usually not necessary for either diagnosis or patient care, but the techniques and facilities are available for defining the unusual or complicated situation.

For practical reasons, many diagnostic laboratories can offer more extensive serologic diagnosis than cultural facilities for virus isolation. They will evaluate sera for the presence of antibodies to mumps virus. The serum for evaluation should be obtained as early as possible in the illness, and a convalescent specimen should be obtained after an interval of 2 to 3 weeks. A pair of sera can determine whether a specific illness is mumps infection by demonstrating an increase in antibody titer. A single serum can determine whether a person has ever had mumps infection, but cannot define when it occurred. As indicated above, there are several types of antibody elicited by mumps infection. As with other virus infections, understanding the time sequence of mumps antibody formation and the specificity of a given type of antibody will allow determination of the significance of a laboratory-determined antibody titer. Such understanding depends on knowledge of the pathogenesis of the disease and the host response to mumps virus infection.

Treatment

There is no specific therapy available for mumps infection. Symptomatic management of patients includes adequate hydration and analgesic and antipyretic therapy, as well as local measures such as elevation and application of cold packs in orchitis.

Prophylaxis and Immunization

The problem repeatedly occurs of what to do after exposure to mumps infection. Usually the person concerned is an adult without previous symptomatic mumps infection. Hyperimmune globulin or pooled serum IgG has been administered after exposure without proven decrease in the number of patients acquiring illness or lessened severity of illness. However, there has been a controlled study purporting to demonstrate that the administration of hyperimmune globulin after the appearance of parotitis can decrease the incidence and severity of orchitis. For this reason, hyperimmune globulin may be administered to postpurbertal males who have parotitis.

Live attenuated mumps virus vaccine was licensed in January, 1968, and is available for prophylactic use. It is recommended for administration to children more than a year old and to young adults for induction of immunity parallel to that induced by natural infection. Vaccine should not be given to pregnant women because of the potential vulnerability of the fetus. Although no data exist which demonstrate transmission of attenuated virus to the fetus, placental infection has been documented after maternal immunization. There is only a single serologic strain of mumps virus; hence a single infection with either natural or attenu-

ated virus confers immunity. The vaccine is a live attenuated virus produced in tissue cultures of chick embryo fibroblasts and is administered parenterally. Virus is not shed by the vaccinee and immunization does not cause any side effects. The vaccine produces 95 to 100 percent serologic conversion from antibody negative to positive in vaccinated susceptibles. The antibody levels, although considerably lower, parallel those produced by natural infection and persist for the 9 to 10 years that vaccine has been available for study. Immunized children in contact with naturally occurring mumps in their families, institutional settings, or the community have been protected against clinical illness. The presence of detectable mumps antibody correlates with protection against clinical illness.

The vaccine will not offer protection against mumps if someone has already been exposed to natural infection and is in the incubation period of illness. On the other hand, no harmful effects have been noted after administration vaccine to an exposed susceptible.

FURTHER READING

Influenza Viruses

Barry DW et al: Comparative trial of influenza vaccine. I. Immunogenicity of whole virus and split product vaccine in man. Am J Epidemiol 104:34, 1976

Dowdle WR, Coleman MT, Gregg MB: Natural history of influenza type A in the United States, 1957–1972. Prog Med Virol 17:91, 1974

Galbraith et al: Therapeutic effect of 1-amantadine HC1 in naturally occurring influenza A$_2$ Hong

Gregg MB et al: Influenza related mortality. JAMA 239:115, 1978

Hall WJ, Douglas RG, et al: Pulmonary mechanics after uncomplicated influenza A infection. Am Rev Resp Dis 113:141, 1976

Influenza Vaccines. J Infect Dis Suppl, December 1977, pp S341–S742

Jackson GG, Muldoon RL: Viruses causing common respiratory infections in man. V. Influenza A (Asian). J Infect Dis 131:308, 1975

Kavet J: A perspective on the significance of pandemic influenza. Am J Public Health 67:1063, 1977

Kilbourne ED (ed): The Influenza Viruses and Influenza. New York, Academic 1975

Kilbourne ED: Influenza pandemics in perspective. JAMA 237:1225, 1977

Knight V (ed): Viral and Mycoplasmal Infections of the Respiratory Tract. Philadelphia, Lea & Febiger, 1973

Larson HE et al: Immunity to challenge in volunteers vaccinated with an inactivated current or earlier strain of influenza A (H3N2). J Hyg 80:243, 1978

Linnemann CC et al: Reye's syndrome: Epidemiologic and viral studies 1963–1974. Am J Epidemiol 101:517, 1975

Little JW, Hall WJ, Douglas RG, et al: Amantadine effect on peripheral airway abnormalities in influenza: study in 15 students on natural influenza A infection. Ann Intern Med 85:177, 1976

Louria DB et al: Studies on influenza in the pandemic of 1957–1958. II. Pulmonary complications of influenza. J Clin Invest 38:213, 1959

O'Donoghue JM, Ray CG, Terry DW, Beaty NH: Prevention of Nosocomial influenza infection with amantadine. Am J Epidemiol 97:276, 1973

Osborne D: Radiologic appearance of viral disease of the lower respiratory tract in infants and children. Am J Roentgenol 130:29, 1978

Perkins FT, Reganey RH: Influenza immunization. II. Developments in biological standardization, Vol 39. Basel, Karger, 1977

Petersdorff RG et al: Pulmonary infections complicating Asian influenza. Arch Int Med 103:106, 1959

Rogers DE, Louria DB, Kilbourne ED: The syndrome of fatal influenza virus pneumonia. Trans Assoc Am Physicians 71:260, 1959

Selby P (ed): Influenza: Virus, Vaccines and Strategy. New York, Academic, 1973

Stuart-Harris CH, Schild GC: Influenza: The Viruses and the Disease. Littleton, Mass, Publishing Sciences Group, 1976

Parainfluenza Viruses

Chanock RM, Parrot RH, Cook MK, et al: Newly recognized myxoviruses from children with respiratory disease. N Engl J Med 258:207, 1958

Gardner PS, McQuilin J, McDuckin R, Ditchburn RK: Observations on clinical and immunofluorescent diagnosis of parainfluenza virus infections. Br Med J 3:7, 1971

Glezen WP, Denny FW: Epidemiology of acute lower respiratory disease in children. N Engl J Med 288:498, 1973

Glezen WP, Fernald GW: Effective or passive antibody on parainfluenza virus type 3 pneumonia in hamsters. Infect Immunol 14:212, 1976

Gross PA, Green RH, Curnen MG: Persistent infections with parainfluenza type 3 virus in man. Am Rev Resp Dis 108:894, 1973

Gross PA, Green RH, Lerner E, Curnen MG: Immune response in persistent infection. Further studies on persistent respiratory infection in man with parainfluenza type 3 virus. Am Rev Resp Dis 110:676, 1974

Hall CB, Geiman JM, Breese BB, Douglas RG Jr: Parainfluenza viral infections in children: correlation of shedding with clinical manifestations. J Pediatr 91:194, 1977

Kim HW, Canchola JG, Vargosko AJ, et al: Immunogenicity of inactivated parainfluenza type 1, type 2, and type 3 vaccines in infants. JAMA 196:111, 1966

Klein JD, Collier AM: The pathogenesis of human parainfluenza type 3 virus infection in hamster tracheal organ culture. Infect Immunol 10:883, 1974

Musson MA, Mocega HE, Krause HE: Acquisition of parainfluenza 3 virus infection by hospitalized children. I. Frequencies, rates, and temporal data. J Infect Dis 128:141, 1973

Workshop on broncholitis, Pediatr Res 11:209, 1979

Mumps

Bang H, Bang J: Involvement of CNS in mumps. Acta Med Scand 113:487, 1943

Brunell PA, Brickman A, O'Hare D, Steinberg S: Ineffectiveness of isolation of patients as a method of preventing the spread of mumps. N Engl J Med 279:1357, 1968

Enders JF, Cohen S: Detection of antibody by complement fixation in sera of man and monkey convalescent from mumps. Proc Soc Exp Biol Med 50:180, 1942

Gellis SS, McGuiness AC, Peters M: A study of the prevention of mumps orchitis by gamma globulin. Am J Med Sci 210:661, 1945

Habel K: Cultivation of mumps virus in the developing chick embryo and its application to studies of immunity to mumps in man. Public Health Rep 60:201, 1945

Johnson CD, Goodpasture EW: An investigation of the etiology of mumps. J Exp Med 59:1, 1934

Johnson RT, Johnson KP, Edmonds CJ: Virus induced hydrocephalus: development of aqueductal stenosis in hamsters after mump infection. Science 157:1066, 1967

Levens JH, Enders JF: The hemagglutinative properties of amniotic fluid from embryonated eggs infected with mumps virus. Science 102:117, 1945

Modlin JF, Orenstein WA, Brandling-Bennett AD: Current status of mumps in the U.S. J Infect Dis 132:106, 1975

Prince GA, Jenson AB, Billups LC, Notkins AL: Infection of human pancreatic beta cell cultures with mumps virus. Nature 271:158, 1978

Weller TH, Craig JR: Isolation of mumps virus at autopsy. Am J Pathol 25:1105, 1949

Weibel RE, Buynak EB, McLean AA, Hilleman MR: Persistence of antibody after administration of monovalent and combined live attenuated measles, and rubella virus vaccines. Pediatrics 61:5, 1978

Witte JJ, Karchmer AW: Surveillance of mumps in U.S. as background for use of vaccine. Public Health Rep 83:95, 1968

Yamauchi T, Wilson C, St. Geme JW Jr: Transmission of live attenuated mumps virus to the human placenta. N Engl J Med 290:710, 1974

CHAPTER 18
Pseudomyxoviruses

Measles (rubeola)

Because of its distinctive clinical features, measles was recognized as a disease entity long before the demonstration of its viral etiology. Early medical writings by Hebrew and Arabic physicians include clear descriptions of the illness. In 1758, Home, a Scottish physician, first demonstrated the transmissibility of the disease by scarification of susceptible individuals with blood taken from infected patients.

Measles virus was first isolated in tissue culture in 1954 by Enders and his associates. With the demonstration that the agent replicated in renal cell cultures of human or simian origin, investigations of the virus and its properties were undertaken. The use of attenuated vaccines began in 1963 and has continued to extend throughout the world. A childhood disease which was accepted as inevitable may become a rarity. In the United States, mortality from measles in the twentieth century has resulted consistently in 1 death per 1000 cases. In nations with less well developed health services, the mortality rate has often exceeded 10 percent, especially among children suffering from nutritional disorders and other debilitating conditions.

Clinical Course

Measles has a regular incubation period varying from 10 to 14 days (Fig. 18-1). A prodromal stage is marked by catarrhal symptoms of cough, coryza, and conjunctivitis. Fever accompanies these symptoms and rises steadily each day, until the appearance of rash 2 to 4 days after the onset. Preceding the skin eruption, the pathognomonic Koplik spots may be found on the lateral buccal mucosa. These are pinpoint, grayish white spots surrounded by bright red inflammation. They are found over the lateral buccal mucosa, as well as the inner lips and may spread to involve the entire inner anterior mouth. They also may be detected in the palpebral conjunctiva. The exanthem begins on the head, behind the ears, on the forehead, and on the neck. Discrete macular and papular lesions progress downward to involve the trunk and upper extremities. Over a period of 3 days, the entire body becomes involved. When the lower extremities first show discrete lesions, those on the head and neck have begun to coalesce. With rash reaching the lower extremities, the high fever recedes dramatically. The bright red rash fades, to leave a brown discoloration which does not blanch with pressure and represents capillary leakage and hemorrhage into the skin.

The respiratory tract manifestations of measles vary in severity but include laryngitis, tracheobronchitis, bronchiolitis, and some degree of interstitial pneumonitis as manifestations of the primary virus infection that damages surface mucosal cells. With defervescence there is improvement, but secondary bacterial infections of the respiratory tract may complicate the recovery phase in 5 to 15 percent of patients. These include otitis media, sinusitis, mastoiditis, and pneumonia. The most dread complication of measles is an encephalomyelitis,

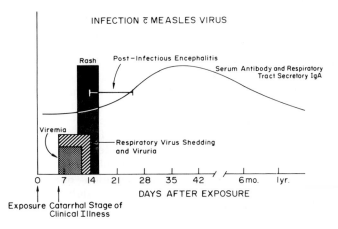

FIG. 18-1. The course of measles virus infection.

which occurs in approximately 0.1 percent of cases. Classically, this follows a period of 3 to 4 days' recovery from the acute illness and is marked by a sudden onset, with seizures, confusion, and coma. The mortality rate of central nervous system involvement approaches 25 percent. Additionally, nearly half of those who survive are left with some sequelae involving impaired intellectual, motor, or emotional development. More recently, the association of measles infection with a late central nervous system complication, subacute sclerosing panencephalitis (SSPE), has been established. This occurs in about 1 per 100,000 cases and is discussed in greater detail in Chapter 25.

Measles is acquired as an infection of the respiratory tract, and the principal damage is to the mucosal lining cells of respiratory tract surfaces. Secondary bacterial infections may be further augmented by a diminished function of alveolar macrophages. The virus spreads via the lymphatics and through the blood to give widespread involvement, particularly in lymphoid cells. The large, multinucleated giant cells which are found in these tissues are quite similar to those produced by measles virus when grown in vitro. Patients excrete large amounts of virus during the catarrhal phase prior to the appearance of the characteristic exanthem. Virus can be recovered from the blood, particularly from the white cell fraction, for several days before rash appears, but rarely thereafter. The urine may remain positive for virus up to 4 days after the onset of rash. The pathology of central nervous system involvement includes edema, congestion, and scattered petechial hemorrhages with perivascular cuffs of round cells. There may be some perivascular demyelination in the later stages.

The exact nature of the measles rash is uncertain, but involves a vasculitis. Viral microtubular aggregates have been observed by electron microscopy of nuclei and cytoplasm of skin biopsies. They are also found in the oral lesions (Koplik spots). Evidence of the interaction of measles infections and cellular immune mechanisms is the loss of delayed hypersensitivity to tuberculoprotein among tuberculin-positive patients with measles. This may persist for several weeks to months following the acute infection. Closely correlated is the repeated observation of the worsening or exacerbation of underlying tuberculosis in children or adults who acquire measles infection.

Epidemiology

Measles is one of the most highly communicable of all viral infections, so that nearly all susceptible children acquire the infection. In rural settings or among isolated communities, it has been possible for a population to reach adult life without experiencing the infection. Under such circumstances, the introduction of measles virus has produced devastating epidemics. Particularly noteworthy are those that have occurred among island populations. Of an epidemiologic class is Panum's description in 1846 of such an outbreak in the Faroe Islands. He showed the persistence of lifelong immunity among individuals who had acquired the infection six and seven decades previously.

In the temperate zones, measles has occurred in winter–spring epidemics at 2- or 3-year cycles, apparently related to the new groups of susceptible children born during the interval since the last outbreak. Maternal antibody is transplacentally acquired by the infant, so that infection under 6 months of age is rare. As shown in Figure 18-1, the catarrhal stage of the illness is marked by extensive respiratory virus excretion, so that infected droplet nuclei provide the usual mode of transmission within families, schoolrooms, or other crowded settings. A single attack confers lifelong immunity.

Although measles virus is closely related to the agents of canine distemper and rinderpest of cattle, there is no evidence of natural spread from one species to another. Their biologic similarities and serologic overlapping offer an interesting area for evolutionary speculation. Except for minor variations, only one distinct serotype of measles virus has been identified. Infection sustained in any part of the world confers uniform geographic protection.

Diagnosis

Diagnosis is clinical, based on the characteristic history and findings. Examination of the urinary sediment or of nasal smears will show characteristic inclusion-bearing syncytia, and there is peripheral leukopenia. Indirect immunofluorescent microscopy has been used to show measles antigen in nasopharyngeal cells. Although virus can be isolated from blood, respiratory tract secretions, conjunctival secretions, or urine, this is not ordinarily required.

In cell culture systems, the cytopathic effect is similar to that of the histology observed in vivo with measles infections.

A number of serologic tests are available, utilizing the antigens of the measles virion or its infectivity. These include virus neutralization, hemagglutination inhibition, complement fixation, and immunofluorescence. Antibodies appear very rapidly after the appearance of rash and rise to high titers in the next 30 to 60 days. A pair of sera obtained early in the course of illness and 7 to 14 days thereafter will show a marked rise in antibody titer by any of the methods described. Because of its ease and rapidity, the hemagglutination-inhibition test is most often utilized.

Treatment and Prevention

The primary viral disease is not amenable to any therapy. Supportive measures may be employed to reduce fever, ameliorate cough, and maintain hydration. Secondary bacterial complications are treated with antibiotics selected by culture of appropriate specimens. The use of antimicrobials prior to the appearance of secondary bacterial complications has not diminished their incidence, but has altered the flora, so that more resistant organisms have survived. The treatment of encephalomyelitis is also symptomatic, with careful attention to the maintenance of an airway, control of seizures, and provision of fluid, electrolyte, and caloric requirements. The administration of immune serum globulin early in the incubation period of measles may completely abort or modify the infection, depending on the amount employed. At a dose of 0.04 ml per kg body weight, gamma globulin will reliably modify measles so that a more benign course ensues, followed by lasting immunity. Using a dose of 0.2 ml per kg body weight, it is possible to abort the infection completely so that no clinical symptoms result, and the patient may remain susceptible to future infection after catabolism of the exogenous globulin.

The prevention of measles by proper use of the available attenuated active virus vaccines offers the most reliable and enduring protection against the infection. Vaccines are recommended for all healthy children shortly after 1 year of age. Several different vaccines have been available, but they all originate from the Edmonston strain of virus. The virus is propagated in chick embryo fibroblast cell cultures. It is given parenterally, with successful infection in at least 95 percent of susceptibles. The infection is usually occult but may cause fever in 15 percent of recipients. Rarely there is moderate, transient rash following the fever. This attenuated infection is noncommunicable and results in antibody responses somewhat lower than those which follow the natural infection. In patients studied to date, antibodies have persisted for periods up to 18 years after immunization. When exposed to natural infection, immunized children remain solidly protected. For several years an inactivated measles vaccine was available. It has been abandoned because chick embryo experiences that revealed a severe and unusual illness following the exposure of children to naturally occurring measles several years after the receipt of inactivated vaccine. They developed fever, pneumonitis, and an unusual rash with petechiae. It seemed to represent a hypersensitivity reaction in patients whose previous immunity had waned after inactivated vaccine. These inactivated vaccines have been withdrawn from the market.

In the initial 5 years after the onset of vaccine use in the United States, reported cases of measles were reduced by 90 percent. Figure 18-2 graphically demonstrates the striking decline in reported cases of measles in the United States since vaccine licensure in 1963 and widespread public health immunization programs beginning in 1966. In the next several years there was a small upswing of cases due to a failure to immunize the new crops of susceptible infants. In those instances in which outbreaks have occurred, epidemiologic studies revealed that the patients were susceptible to measles for one of the following reasons. The majority had never received measles vaccine and had not previously acquired natural immunity to the infection. Some had received inactivated vaccine in the past. Others had been given active attenuated vaccine prior to their first birthday, at a time when persistent maternally transmitted antibody aborted the vaccine infection. In some instances vaccine failures were traced to improper storage of the attenuated vaccine, so that heat and/or light had reduced the infectivity of the material injected. Continued attention to immunization of all children 1 year of age or older is needed in order to attain the goal of measles eradication.

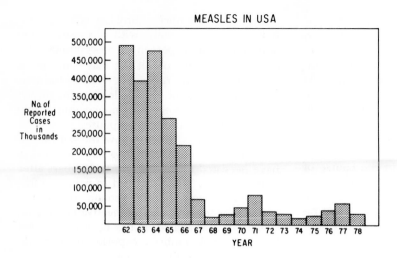

FIG. 18-2. Cases of measles reported in the United States.

Respiratory syncytial virus

Clinical Illness

Respiratory syncytial virus (RSV) was first described in 1956 and since that time has become recognized as the single most important respiratory tract pathogen of infants and young children. Studies of hospitalized young children with acute respiratory disease reveal 20 to 30 percent of their illnesses are attributable to RSV. The type and severity of illness which is produced by this agent varies markedly with the age of the person infected, although other determinants are also operative.

Infection had been experimentally produced by introduction of virus into the upper respiratory tract of volunteers and animals (chimpanzees). The communicability of disease among children is consistent with acquisition of virus by inhalation of contaminated secretions. Thus the virus comes in contact with the mucosal surfaces of the nasopharynx and virus replication occurs. Immunofluorescence of nasal secretions provides evidence that ciliated epithelial cells and macrophages contain RSV antigen and are therefore sites of viral replication. Viremia has not been documented with RSV infection. Infection appears to be limited to the respiratory tract, and virus has been isolated only from respiratory tissues and secretions.

The incubation period preceding clinical illness is ordinarily 1 to 4 days after exposure to virus. Virus excretion from the respiratory tract in infants may precede clinical symptoms for 1 to 3 days, and the mean duration of shedding is for 7 days after the onset of symptoms, with a range of 1 to 21 days (Fig. 18-3). The mean titer of virus detected in secretions is 5.0

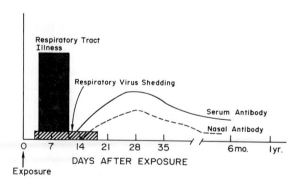

FIG. 18-3. The course of RSV infections.

$\log_{10}$ TCID$_{50}$. Children less than 2 years of age excrete virus longer than older children or adults.

Severe disease caused by RSV is most often manifest as bronchiolitis or pneumonitis and occurs predominantly in children less than 6 months of age. These very young children are severely ill, have fever, lethargy, and apnea, which is a prominent feature of their respiratory tract disease. Physical examination will reveal labored, rapid respirations, with radiologic changes ranging from hyperaeration to consolidation. Involvement of the lungs may be sufficiently extensive to necessitate respiratory assistance and, in a small percentage of children, is fatal.

Children between the ages of 1 and 3 years more often have symptomatic disease or illness classified as laryngotracheobronchitis and croup. Older children and adults may have "colds" with cough, nasal congestion, and fever or asymptomatic infection evidenced only by serologic or cultural evidence of infection. The symptomatic illness produces significant morbidity for periods up to 8 weeks. There is a measurable increase in respiratory tract resistance, apparently due to altered reactivity of small airways.

Reinfection with RSV may be either symptomatic or asymptomatic. Those patients less than 2 years of age are most frequently symptomatic, virus is excreted, and an antibody rise occurs. Antigenic variation of RSV does not seem to explain reinfection of humans, since the measured antibody response to different isolates of virus does not detect any antigenic difference. Antibody specificity has been used in animals to detect slight antigenic variation in separate isolates of RSV. For such practical purposes as immunization, only a single strain of RSV exists.

The severity and frequency of RSV infections in infants have led to rather extensive study of this virus in an attempt to guide the development of preventative procedures. Serum antibody against RSV is largely IgG and as such crosses the placenta to the fetus, gradually declining during the first months of life. Recent longitudinal studies of infants have shown that maternal antibody is protective and as it is catabolized infants become susceptible. Infants with absent or lowest titers develop disease earlier in life. This information has accrued through the meticulous prospective collection of cord bloods and subsequent serial evaluations of the same infants, including serum antibody measurement and virus isolations. Infants may have several RSV infections during the first years of life. They develop serum antibodies which appear in approximately 10 days to 2 weeks after the onset of infection, persist for long periods, and then show an anamnestic response with natural or artificial challenge with virus. These antibodies do not cross-react with any other known virus.

Antibodies to RSV of the IgA and IgG classes are found in respiratory secretions. Specific IgA antibody is usually detectable on respiratory epithelial cells from infants as soon as secretions are available (usually 1 to 2 days into hospitalization). Cell-free antibody is detectable by the sixth day of hospitalization. IgG cell-associated antibodies are less uniformly demonstrated and are detected after several days of illness. Cell-free specific IgG antibodies appear regularly during the illness.

Studies demonstrate a better correlation with protection against RSV infection by secretory antibodies than by serum IgG antibodies. It has been shown in virus challenge studies in adults that there is an inverse correlation of the titer of nasal wash antibody with the quantity of virus excreted and the antibody rise. Those persons with high levels of specific IgA antibody challenged with RSV excrete less virus and may show no antibody rise in secretions or serum. Those adults with low or absent levels of specific nasal IgA show excretion of increased amounts of virus and do develop rises in antibody. Clinical illnesses were observed only in the volunteers with low nasal levels of IgA who received the larger of two inoculation dosages of virus. The larger inoculum resulted in more rapid production of virus, as measured in the virus content of nasal secretions, but did not increase the total amount of virus excreted over one entire time period. Thus illness was associated with an altered pattern of virus excretion.

Pathology

Infants with RSV infection show severe changes in their lungs. Pulmonary tissues from children with pneumonia show marked inflammation, with mononuclear cell infiltrates in interstitial tissues, alveoli, small bronchioles, and alveolar ducts. There may be demonstrated syncytia formation and intracytoplasmic inclusions, which are consistent with, but not pathognomonic of, RSV infection. Infants with bronchiolitis show less extensive interstitial and alveolar involvement but instead have moderate to marked changes in the bronchioles. Epithelial necrosis and plugs consisting of cell debris and fibrin are seen, as well as peribronchiolar lymphocytic infiltrates. Virus has been isolated from tissues of children with both types of illness.

Epidemiology

RSV infection is worldwide in distribution and tends to cause yearly outbreaks of illness alternating from midwinter to late spring in occurrence. The increased frequency of severe disease in infants has been commented on, and during community studies, attack rates of 30 to 60 percent are estimated for exposed infants less than a year old. RSV infection seems to be introduced into a family by an older (2 years or more) sibling. The secondary attack rate is then greatest for infants less than 1 year of age, but appreciable attack rates occur at all ages, with approximately 50 percent of family members acquiring infection. Investigations conducted simultaneously on different populations in the same geographical setting pointed out that patients seen in private practice had RSV infection at an older age than either urban or rural clinical patients. These older infants with less severe disease were seldom hospitialized. We do not know whether the attack rates were the same in the groups studied, but features of the environment such as the number of siblings in the home and breast feeding may be associated with the altered epidemiologic pattern of disease in these children.

There is some epidemiologic evidence to suggest that breast feeding provides protection against RSV infection. Human colostrum has been shown to contain specific RSV IgA and IgG antibodies. Virus-neutralizing titer correlates with the quantity of specific IgA present. It is possible that ingestion of these antibodies provides some local protection.

A final epidemiologic consideration is that of nosocomial infections with RSV. The hospitalized infant excretes virus for days, and it is transmitted to ward personnel as well as to pa-

tients. The acquired illness also results in prolongation of hospitalization for infants and may produce fatal illness in those infants hospitalized with other underlying problems. Consideration should be given to patient isolation and careful handwashing in providing care for infected infants.

Diagnosis

Presumptive clinical diagnosis can be made in an infant with bronchiolitis or pneumonia with no demonstrable bacterial pathogens. A definite etiologic diagnosis of RSV infection can be made by detection of viral antigen or serologic tests. The virus can be isolated from nasal secretions utilizing tissue cultures of cells such as the Hep2 line. Although this information may assist in management of severely ill infants, all too frequently it is not known until the patient has completed the natural course of illness.

The virus is extremely labile, and respiratory secretions, throat swab material, lung biopsy, or postmortem material must be kept at 4C and immediately processed by a knowledgeable laboratory utilizing specific techniques in order to preserve cells for immunofluorescence or infectious virus for tissue culture inoculation. Several laboratories have demonstrated RSV antigens in cells present in nasal secretions by indirect fluorescent antibody techniques. The indirect immunofluorescent method employing bovine anti-RSV serum and fluorescein-conjugated anti-bovine globulin has been successful in the few laboratories familiar with the procedure. The technique affords a means of identification of RSV infection within hours but is not generally available. Later in disease, it can also identify viral antigen despite the presence of antibody, which may neutralize the infective virus in secretions.

Serologic techniques allow the recognition of RSV antibodies by complement-fixation, neutralization, or plaque-reduction techniques, with the latter being the most sensitive method. Experimentally, radioimmune and enzyme-linked immunosorbent assays are presently being developed. There is no hemagglutinating antigen of RSV recognized at present. Acute and convalescent sera are necessary to define the significance of measured antibody. As with other virus infections, a fourfold or greater antibody rise indicates recent infection, but a single value will indicate only previous experience with the virus at an undetermined time. Investigational use of a specific IgM radioimmunassay may prove helpful in defining acute illness.

Therapy

The therapy of RSV infection is entirely nonspecific and consists of support of the respiratory system, control of fever, adequate hydration and nutrition, and therapy of secondary bacterial infection.

Prevention

The prevalence of RSV infection has resulted in attempts to develop an effective vaccine. A formalin-inactivated, alum-precipitated virus vaccine was prepared and utilized in field trials. The vaccine produced a rise in serum antibodies in adults and children but no rise in secretory antibody. During naturally occurring outbreaks of RSV disease, the attack rate among infants immunized with inactivated vaccine was the same as that in the control (unimmunized) populations. This indicated that this immunization did not protect from subsequent infection. In addition, the clinical illness was more severe in vaccinees, and the age group suffering severe disease was extended to include older infants. Such unanticipated observations stopped the use of inactivated vaccine.

The pathogenesis of severe disease in the youngest infants and inactivated vaccine recipients remains to be determined. Present investigations include attempts to define specific RSV cell-mediated immunity as measured by in vitro lymphocyte-transformation assays. Natural infection and inactivated vaccine produce a measurable cell-mediated immune response. Further work is necessary to elucidate the role of cell-mediated immunity in protection against infection and/or pathogenesis of disease.

Attempts are being made to develop a live virus vaccine which will be protective. It is predicted that an effective vaccine will induce formation of respiratory tract IgA because of its recognized association with protection against disease. The first RSV vaccine administered intranasally was a cold-adapted strain which

was found to be attenuated in adults but produced mild disease and prolonged virus shedding in seronegative infants. Temperature-sensitive RSV mutants multiplying only at 33 to 35C have been isolated. It was hoped that such mutants would multiply only in the upper respiratory tract and thus reproduce the immunologic events of natural infection without causing illness in the infants. Administration of the temperature-sensitive mutant produced asymptomatic infection in seropositive children and infection with mild rhinitis in seronegative children. Using intranasal administration of 100 $TCID_{50}$, children with preexisting nasal antibody were not infected. Virus shedding was monitored, with recovery of a small population of genetically altered virus with partial loss of the temperature-sensitive property in a few instances. The vaccine virus was not transmissible to control children in intimate contact with vaccinees. Sequential studies have shown subsequent natural RSV infection occurred in vaccinees. Neither protection from infection nor potentiation of natural infection has been demonstrated to date. An effective immunizing material must be able to decrease the morbidity and mortality associated with natural infection without being detrimental to the infant, and so careful study of potential immunizing agents will continue.

FURTHER READING

Measles

Babbott FL Jr, Gordon JE: Modern measles. Am J Med Sci 228:334, 1954

Barkin RM: Measles mortality. Am J Dis Child 129:307, 1975

Enders JF, Katz SL, Milovanovic MJ, Holloway A: Studies on an attenuated measles virus vaccine. I. Development and preparation of the vaccine: techniques for assay of effects of vaccination. N Engl J Med 263:152, 1960

Fulton RE, Middleton PJ: Immunofluorescence in diagnosis of measles infections in children. J Pediatr 86:17, 1975

Krugman S, Ward R, Katz SL: Infectious Diseases of Children, 6th ed. St Louis, Mosby, 1977

Morgan EM, Rapp F: Measles virus and its associated diseases. Bacteriol Rev 41:636, 1977

Morley DC: Measles in the developing world. Proc R Soc Med 67:112, 1974

Nader PR, Horwitz MS, Rousseau J: Atypical exanthem following exposure to natural measles: 11

cases in children previously inoculated with killed vaccine. J Pediatr 72:22, 1968

Panum PL: Observations made during the epidemic of measles on the Faroe Islands in the year 1846. New York, American Public Health Association, 1940

Paule CL, Bean JA, Burmeister LF, Isacson P: Post-vaccine era measles epidemiology. JAMA 241:1474, 1979

Respiratory Syncytial Virus

Chanock RM, Kim HW, Vargasko AJ, et al: RSV. I. Virus recovery and other observations during 1960 outbreak of bronchiolitis, pneumonia, and minor respiratory diseases in children. JAMA 176:647, 1961

Chin J, Magoffin RL, Shearer LA, Schieble JB, Lennette EH: Field evaluation of a respiratory syncytial virus vaccine and a trivalent parainfluenza virus vaccine in a pediatric population. Am J Epidemiol 89:449, 1969

Fulginiti VA, Eller JJ, Sieber F, et al: Respiratory virus immunization. 1. A field trial of two inactivated respiratory virus vaccines; an aqueous trivalent parainfluenza virus vaccine and an alum precipitated respiratory syncytial virus vaccine. Am J Epidemiol 89:435, 1969

Gardner PS, McQuillin J: Application of immunofluorescent antibody technique in rapid diagnosis of RSV infection. Br Med J 3:340, 1968

Gharpure MA, Wright PF, Chanock RM: Temperature sensitive mutants of RSV. J Virol 3:414, 1969

Glezen WP, Denny FW: Epidemiology of acute lower respiratory disease in children. N Engl J Med 288:498, 1973

Glezen WP, Loda FA, Clyde WA Jr, et al: Epidemiologic patterns of acute lower respiratory disease of children in a pediatric group practice. J Pediatr 78:397, 1971

Hall CB, Douglas RG Jr, Geiman JM: Quantitative shedding patterns of RSV in infants. J Infect Dis 32:151, 1975

Hall CB, Douglas RG Jr, Geiman JM, Messner MK: Nosocomial RSV infections. N Engl J Med 293:1343, 1975

Hall CB, Geiman JM, Biggar R, et al: RSV infections within families. N Engl J Med 294:414, 1976

Hall WJ, Hall CB, Speers DM: Respiratory syncytial virus. Infection in adults. Clinical, virologic, and serial pulmonary function studies. Ann Intern Med 88:203, 1978

Kapikian AZ, Mitchell RH, Chanock RM, et al: An epidemiologic study of altered clinical reactivity to respiratory syncytial (RS) virus infection in children previously vaccinated with an inactivated RS Virus vaccine. Am J Epidemiol 89:405, 1969

Kim HW, Arrobio JO, Brandt CD, et al: Safety and antigenicity of temperature sensitivity (TS) mutant respiratory syncytial virus (RSV) in infants and children. Pediatrics 52:56, 1973

Kim HW, Arrobio JO, Pyles G, et al: Clinical and immunological response of infants and children to administration of low temperature adapted respiratory syncytial virus. Pediatrics 48:745, 1971

Kim HW, Canchola JG, Brandt CD, et al: Respiratory syncytial virus disease in infants despite prior administration of antigenic inactivated vaccine. Am J Epidemiol 89:422, 1969

Mills J, Van Kirk JE, Wright PF, Chanock RM: Experimental RSV infection of adults. Possible mechanism of resistance to infection and illness. J Immunol 107:123, 1971

Morris JA, Blount RE Jr, Savage RE: Recovery of cytopathogenic agent from chimpanzees with coryza. Proc Soc Exp Biol Med 92:544, 1956

Parrott RH, Vargosko AJ, Kim HW, et al: II. Serologic studies over a 34 month period of children with bronchiolitis, pneumonia and minor respiratory diseases. JAMA 176:653, 1961

Wright PF, Shinozaki T, Fleet W, et al: Evaluation of a live, attenuated respiratory syncytial virus in infants. J Pediatr 88:931, 1976

CHAPTER 19

Rubella (German Measles)

CLINICAL FEATURES
EPIDEMIOLOGY
DIAGNOSIS
TREATMENT AND PREVENTION

From the mid-nineteenth century until 1941, rubella was regarded as a benign childhood exanthem. When the Australian ophthalmologist, Sir Norman Gregg, reported the association of intrauterine rubella infection with congential cataracts, this attitude changed completely. Subsequently, deafness, congenital heart disease, and other malformations were found to result from maternal rubella during the first months of pregnancy. Despite heightened interest in all aspects of this infection, it was not until 1962 that Weller in Boston and Buescher in Washington were able simultaneously to report successful laboratory techniques for the isolation, propagation, and study of rubella virus. With the adaptation and modification of these techniques by other research groups, a number of investigators were prepared to study the effects of a pandemic of rubella in 1964 and 1965. From that single major outbreak, basic data about the epidemiologic, virologic, immunologic, and pathogenetic events were acquired. Stimulated by this new knowledge, several laboratories mounted major efforts to derive immunizing antigens which would effectively prevent rubella. By 1969 attenuated active rubella virus preparations were licensed in the United States and Western Europe. They offered the first real promise of controlling this major cause of congenital malformations.

Earlier investigators had demonstrated the viral etiology of rubella by transmission studies in monkeys and later in children, using filtered respiratory tract secretions from infected patients. The absence of any serologic tests made reliable clinical and epidemiologic studies very difficult. With newer techniques after 1962, it became apparent that many diseases that had previously been diagnosed clinically as rubella were not. Studies of the enteroviruses and adenoviruses revealed that many cases of mild rash diseases that were clinically indistinguishable from rubella were actually caused by members of these large virus groups. Since these agents are not known to be teratogenetic, the importance of laboratory differentiation became crucial. Rubella virus is classified as a togavirus.

Clinical Features

Rubella is a mild rash disease which occurs principally in children but is seen at all ages. As shown in Figure 19-1, the incubation period is approximately 16 to 18 days, with minimal prodromal signs or symptoms. Most often the first awareness of illness is mild fever and respiratory signs immediately preceding the onset of rash. The exanthem is pink in color with macules and papules. They appear at first on the face and then spread to the neck, trunk, and extremities, where they remain discrete and rarely coalesce. The rash is shorter lived than that of measles and has ordinarily disappeared by the third day. Preceding and accompanying the rash there is lymphadenopathy, which may involve the postauricular, suboccipital, and cervical nodes. Rash is observed commonly among children, but infection may be occult or present only as febrile pharyngitis in as many as one-third of adult patients. Although major complications are rare (thrombocytopenic purpura and encephalitis), the incidence of arthralgia and arthritis is much greater than generally

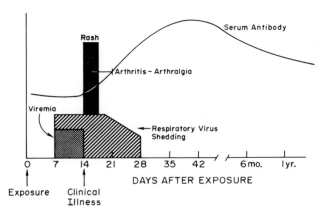

FIG. 19-1.   The course of rubella virus infection.

appreciated. The frequency of joint involvement is directly correlated with increasing age and appears also to be more common among women.

The route of infection is via the respiratory tract, with spread to lymphatic tissues and then to the blood. Both viremia and respiratory tract shedding of virus may precede the rash by 1 week, and the latter may follow it for another several weeks. Because the major excretion of virus occurs prior to the recognition of illness, secondary infection of intimate contacts has usually transpired before the primary patient has been diagnosed. Little is known of the actual pathology of the postnatal disease because it is not a fatal one. However, the pathogenesis of congenital infection has been well studied during and since the 1964 outbreak. Maternal viremia is followed by infection of the placenta with the characteristic microscopic lesions described. Placental infection may then lead to virus invasion of the fetus. Multiple tissues and organs support the replication of virus, which continues to multiply throughout the remainder of pregnancy and in the postnatal period. A large percentage of maternal infections that occur in the first 3 months of pregnancy result in fetal illness. There is a diminishing number in the fourth month, and it is uncertain whether any fetal infections have resulted from maternal rubella in later pregnancy. Although the exact mechanism of the damage to fetal organs is not clear, rubella infection of human embryonic cells in vitro is associated with both chromosomal breakage and inhibition of normal mitosis. Infants with congenital rubella have a subnormal number of cells in some infected organs, which suggests that these same features may also occur in vivo.

Congenital rubella infection may result in a large variety of abnormalities, including deafness, congenital heart disease, eye defects (cataracts, glaucoma, retinitis, microphthalmia), growth retardation, thrombocytopenic purpura, osteitis, hepatitis, pneumonitis, encephalitis, and cerebral damage with mental retardation. In contrast to the postnatal infection, intrauterine disease is marked by a continued replication and excretion of virus, which may persist throughout the first year of life and has been demonstrated in selected tissues such as the lens of the eye as long as 3 or 4 years postnatally. In addition to malformations compatible with life, intrauterine rubella infection may also produce fetal death, abortion, and neonatal death. Although many of the acute neonatal manifestations resolve over the first months of life, long-term sequelae result in multiple developmental handicaps. Recent studies also indicate a significant increase in diabetes mellitus, chronic pneumonitis, thyroiditis, and degenerative encephalitis among long-term survivors of intrauterine infection.

The immunity which follows naturally acquired postnatal rubella is similar to that observed with other common virus infections such as measles and mumps. Only one serologic type of virus has been identified, and a single attack apparently confers lifelong immunity. The immunologic events that accompany intrauterine infection differ strikingly. The antibody response in utero is one of IgM rather than IgG. Although IgG specific for rubella is found in fetal circulation, it is mainly of maternal origin and transplacentally acquired. Despite the presence of specific IgM and IgG in the fetal circulation, chronic infection of cells continues, as already mentioned above. Postnatally, the infant infected in utero synthesizes rubella-specific IgG. Virus replication in infected cells diminishes gradually over ensuing months. A small number of congenitally infected infants have developed unusual forms of hypogammaglobulinemia in the first year of life. Another small group have lost all their detectable rubella antibody postnatally but have proven resistant to challenge with attenuated rubella viruses. The explanations for a number of these apparent immunologic paradoxes are not yet known.

Epidemiology

In most urban communities, rubella infection was acquired during early childhood, principally in the school years. Because it is not as highly communicable as measles or varicella, as many as 15 to 20 percent of women reach childbearing age without having acquired natural immunity. In the United States, large epidemics have occurred at 6- to 8-year intervals, with smaller numbers of cases in the intervening years. This cycle was interrupted in 1969 by the onset of rubella vaccine programs, and no large epidemic has developed in the United States in the 15 years since the 1964–65 pandemic. Usual transmission is by the respiratory route, but the prolonged viruria of the congenitally infected infant may be of great importance in spread to close contacts. In situations where adolescents and young adults have been placed in crowded living conditions, rubella outbreaks have recurrently been observed. Examples are regularly seen among Armed Forces recruits, preparatory school and college groups, and summer camps. Deliberate efforts to expose susceptibles in the hopes of their acquiring natural immunity have not been uniformly successful. The reasons for these failures include the lack of clinical reliability of the diagnosis of rubella, so that many of the alleged outbreaks were due to other agents and, second, the apparent necessity for fairly intimate and prolonged contact to ensure transmission. With widespread use of rubella vaccines, childhood outbreaks have disappeared, and sporadic clusters of cases are reported now on college campuses and among other teen-age groups who have escaped immunization and natural infection.

Diagnosis

Until 1962, the diagnosis was entirely a clinical one and, therefore, sometimes unreliable. Current virus isolation techniques are reliable but not readily available. They involve the use of cell culture systems susceptible to the virus, with direct observation for cytopathic effect, or the use of an interfering agent as an indicator of viral replication. Because these are time-consuming and somewhat fastidious, serologic tests are more commonly used. The antigens of rubella virus are responsible for induction of a number of antibody types that can be assayed in serum. These include complement-fixing (CF), virus-neutralizing, hemagglutination-inhibiting (HI), immunodiffusion, and immunofluorescent antibodies. The rapidity, reliability, reproducibility, and low cost of the HI test have made it the most commonly employed. Paired serum specimens obtained early in the course of illness and 14 to 21 days thereafter will demonstrate a rise in the particular antibody tested. The presence in many sera of non-specific inhibitors of hemagglutination makes the HI test susceptible to some error unless care is taken first to treat the sera with kaolin, heparin, or manganese chloride to remove inhibitors. The HI test is also valuable for screening populations to determine serosusceptibility or immunity and in the consideration of the use and efficacy of vaccines. The diagnosis of intrauterine rubella is more complicated but may be accomplished on a single serum specimen if neonatal blood is assayed for IgM antibodies specific to rubella virus. If such a technique is not available, it may be necessary to assay paired specimens obtained over a period of several months in order to determine whether there is active postnatal antibody synthesis by the infant or merely a decline of transplacentally acquired antibody.

Treatment and Prevention

There is no specific therapy for rubella virus infection. In the case of documented maternal infection during the first trimester of pregnancy, therapeutic abortion is commonly employed. Even though rubella is confirmed virologically and/or serologically in the mother, it is not possible to be certain that fetal infection has occurred. It is estimated that maternal rubella in the first month of gestation carries a 30 to 50 percent incidence of fetal infection, in the second month a gestation of 25 percent risk, and in the third month a 10 percent incidence. After the third month there are insufficient numbers to quantitate the minimal risk of fetal involvement. Amniocentesis has been employed in a few cases to examine cells and fluid for evidence of rubella virus infection. However, there are insufficient data at present to evaluate the reliability of this method.

The use of immune serum globulin to prevent fetal rubella in an exposed pregnant woman has been the subject of long, continued controversy. With the availability of virologic and serologic testing, the lack of efficacy of ordinary gamma globulin has been repeatedly demonstrated. Studies utilizing a high-titered convalescent rubella gamma globulin preparation have led to encouraging results, but this material remains investigational and not available. In most situations, it has been difficult to ascertain for how long a mother has been exposed to the virus. As depicted in Figure 19-1, it is conceivable that she may have had a week's exposure to respiratory shedding from a family member by the time he or she develops rash disease. With this in mind, the expectation of globulin prevention of secondary infection is not very high.

Since 1969 several attenuated active rubella virus vaccines have become available. Studies to date suggest that these vaccines confer lasting effective immunity. They have been prepared in cell cultures of duck embryo, canine kidney, rabbit kidney, and a human diploid line (WI-38). The use of all these vaccines is followed by some respiratory shedding of the vaccine virus. In contrast to the shedding which follows natural infection, this is not transmissible. A serologically detectable HI antibody response is produced in more than 95 percent of susceptible vaccine recipients within 4 to 6 weeks of immunization. As with acquired rubella, vaccination may be followed by joint complaints but in a much smaller percentage of recipients than is noted after natural disease. Whether the attenuated strains of virus might also be teratogenic for the fetus or embryo is uncertain. Vaccine virus can cross the placenta to reach the products of conception. The risk of any teratogenesis is estimated at far less than 5 percent. Of 66 infants born to seronegative mothers who inadvertently received rubella vaccine in early pregnancy, none had a detectable congenital malformation. The use of vaccine in pregnant women, is however, strictly contraindicated. Before administering attenuated rubella vaccine to a susceptible woman of childbearing age, it is essential to be certain that she is not pregnant and that she will follow an acceptable method of pregnancy prevention for 2 to 3 months thereafter. Reinfection with "wild" rubella virus has been demonstrated to occur following either naturally acquired or vaccine-induced immunity. This usually happens in individuals whose antibody titers have fallen to low levels. It is more likely to occur in vaccinated individuals because they may lack detectable rubella-specific respiratory tract secretory immunoglobulins (IgA). The virologic events of such reinfections are markedly abbreviated, and it has not been possible to demonstrate viremia, but there is brief respiratory shedding of virus and a rapid secondary type antibody response with no overt illness. There is no evidence to suggest that such a reinfection phenomenon threatens the fetus of a woman in early pregnancy. Continued study will be required to answer fully all the questions raised by the vaccines. In 10 years of utilization in the United States with administration of more than 83 million doses, they have produced a striking reduction in reported cases of intrauterine and postnatal rubella.

FURTHER READING

Blattner RJ, Williamson AP, Heyes FM: Role of viruses in the etiology of congenital malformations, Prog Med Virol 51:1, 1973

Cooper LA: Congenital rubella in the United States. Infections of the fetus and the newborn infant. In Krugman S, Gershon AA (eds): Progress in Clinical and Biological Research, vol 3. New York, Alan R Liss, 1975

Fuccillo DA, Sever JL: Viral teratology. Bacterial Rev 37:19, 1973

Gregg NM: Congenital cataract following German measles in the mother. Trans Ophthalmol Soc Aust 3:35, 1942

Johnson RT: Progressive rubella encephalitis. N Engl J Med 292:1023, 1975

Krugman S, Ward R, Katz SL: Infectious Diseases of Children, 7th ed. St. Louis, Mosby, 1977

Krugman S, Rubella immunization: progress, problems and potential solutions. Am J Public Health 69:217, 1979

Modlin JF, Herrmann K, Brandling-Bennett AD, Eddins DL, Hayden GF. Risk of congenital abnormality after inadvertent rubella vaccination of pregnant women. N Engl J Med 294:972, 1976

Peckham CS, Martin JAM, Marshall, WC, Dudgeon JA: Congenital rubella deafness: a preventable disease. Lancet 1:258, 1979

Ziring PR, Florman AL, Cooper LA: The diagnosis of rubella. Pediatr Clin North Am 18:87, 1971

CHAPTER 20

Arboviruses

Arboviruses (*ar*thropod*bo*rne) are maintained in nature usually by cyclical transmission between susceptible vertebrates and blood-sucking arthropods, with the virus multiplying in both. In vertebrates, arboviruses produce, after an incubation period, a viremia of sufficient titer and duration to permit infection of arthropods. In arthropods, virus multiplication results in the capacity to transmit virus by bite to new vertebrates after a period of time known as the "extrinsic incubation period." This biologic transmission, which depends on the multiplication of virus in the arthropod, is in contrast to mechanical transmission, which occurs as the result of rapid carriage of virus by an arthropod from one host to another. The consequences of infection of the vertebrate range from total absence of disease to major illness and death; with rare exceptions, no effect on the arthropod has been detected.

Most arboviruses are members of the families Togaviridae and Bunyaviridae; some are Reoviridae and Rhabdoviridae. A small number, generally of no significance in human disease, may be found in other families. Proof of biologic transmission by arthropods has not been obtained for many arboviruses, but is inferred from epidemiologic and taxonomic characteristics. The discipline of arbovirology often deals, in addition, with some viruses such as arenaviruses that probably are not arthropod-borne and for which other modes of transmission have been demonstrated. This is because of an historical, initial presumption of arthropod transmission based on clinical and epidemiologic similarities to known arbovirus disease.

The year 1900 marked the beginning of arbovirology, when Walter Reed demonstrated the biologic transmission of yellow fever by the mosquito *Aedes aegypti*. The facts, so convincingly elucidated by the commission of which Reed was chairman, proved the remarkable perception of Dr. Carlos Finlay and the essentials of the basic definition of arboviruses.

During ensuing decades, other viruses, many of which are human pathogens, were recognized to possess life cycles that involve arthropod–vertebrate transmission. By the end of the 1930s there were approximately 30 arboviruses. Since World War II hundreds more have been discovered, so that now there are in excess of 200 known arboviruses. Some 90 may infect man, and most produce disease.

Beginning in the 1920s with the yellow fever program of its International Health Division, the Rockefeller Foundation has provided the major physical and intellectual base of arbovirology. In 1964 the Rockefeller Foundation Virus Program became the Yale Arbovirus Research Unit. Besides developing the conceptual base, much of the basic technology, and most of the facts of arbovirology, the Foundation's conquest of yellow fever by the development of the first tissue culture attenuated, live virus vaccine stands as a major achievement.

CLASSIFICATION AND NOMENCLATURE

Initially, arboviruses were named for the disease that they caused, for example, dengue and yellow fever. Later, combined names were devised that comprised both geography and disease such as St. Louis encephalitis. Names currently derive from the place of collection of the specimen from which the first isolation was made, especially for those agents for which no disease is recognized. Most arboviruses fall into this group.

Arobviruses are classified according to morphologic and serologic relationships based on tests employing complement fixation, hemagglutination inhibition, and infectivity neutralization. At first, serogroups were designated A, B, and C. More recently, groups are given the name of the first virus isolated. More than 40 serogroups have been identified, and more than 60 viruses are as yet ungrouped.

There are several other ways of grouping arboviruses that may be useful in certain circumstances. These include classification according to vector, geographic range, type of disease, or combinations of these such as "tickborne hemorrhagic fevers." Because this text is intended primarily for medical students, it seems appropriate to use a clinically oriented classification (Tables 20-1 to 20-3).

TECHNOLOGY

The epidemiologic and clinical study of arboviruses involves two basic technologies, virus isolation and measurement of antibodies. The major culture medium for isolation is the infant mouse, in which nearly all arboviruses are pathogenic when inoculated intracerebrally.

TABLE 20-1   SUMMARY OF SELECTED HUMAN ARBOVIRUS DISEASES CHARACTERIZED BY FEVER WITH OR WITHOUT RASH

Virus			Maintenance Hosts		Clinical (Other than Fever) and Epidemiologic Features
Name*	Antiserogroup	Geography	Vertebrate	Arthropod	
Dengue	B	Tropics and sub-tropics through-out world	Man	Mosquito	Headache, myalgia, rash, arthralgia; endemic and epidemic
Colorado tick	Ungrouped	Western North America	Small mammals	Tick	Headache, myalgia, rash, biphasic rash; spring and summer focal endemnicity
West Nile	B	Mediterranean, Middle East Russia, and Asia	Birds	Mosquito, ticks	Headache, lymphadenopathy, rash; endemic and epidemic
Sandfly	Phlebotomus fever group	Mediterranean, Asia, tropical America	Man, monkeys, small mammals	Phlebotomus flies	Myalgia, rash, conjunctivitis; seasonally epidemic and sporadically endemic
Rift Valley	Ungrouped	Africa	Large mammals	Mosquito	Headache, myalgia, photophobia, retinal damage may occur; sporadic from mosquito or by contact with tissue from infected animals
O'nyong-nyong	A	Africa	Man	Mosquito	Like dengue; epidemic
Chikungunya	A	Africa	Man	Mosquito	Like dengue; epidemic
Mayaro	A	South and Central America	Monkeys and marsupials	Mosquito	Like dengue; forest-associated endemic
Ross River	A	Australia	Mouse	Mosquito	Like dengue, arthritis; endemic and seasonal epidemic

*Disease name is generally the viral name plus fever, e.g., Colorado tick fever

TABLE 20-2 SUMMARY OF SELECTED HUMAN ARBOVIRAL DISEASES CHARACTERIZED BY ENCEPHALITIS

Name*	Antiserogroup	Geography	Maintenance Hosts — Vertebrate	Maintenance Hosts — Arthropod	Clinical (Other than Encephalitis) and Epidemiologic Features — Case Fatality (F) and Sequelae (S)
Western equine	A	North and South America	Birds	Mosquito	Summer epidemics North America; equine epizootics; F = high, S = severe
Eastern equine	A	Eastern and North America and Caribbean	Birds	Mosquito	Summer epidemics and equine and bird epizootics; F = high, S = severe
Venezuelan	A	South and Central America, U.S.A., Caribbean	Rodents, birds, equines	Mosquito	Influenzalike, CNS involvement far less common. Encephalitis epizootics in equines; F = low, S = moderate
St. Louis	B	Western hemisphere, especially U.S.A.	Birds	Mosquito	Summer epidemics in U.S.A.; F = high, S = severe
California (LaCrosse)	California	U.S.A.	Small mammals	Mosquito	Summer outbreaks and sporadic cases; F = low, S = ?
Japanese B	B	Asia, Japan, Pacific Islands	Birds, pigs	Mosquito	Summer and fall epidemics in temperate areas; F = high, S = severe
Murray Valley	B	Australia and New Guinea	Bird	Mosquito	Epidemic; F = high
Powassan	B	U.S.A. and Canada	Small mammals	Tick	Two cases of human disease recognized, one fatal
Russian spring-summer	B	Russia	Birds, mammals	Tick†	Diphasic pattern, bulbospinal paralysis; F = high, S = severe
Central European	B	Europe	Birds, mammals	Tick†	Diphasic pattern
Kyasanur Forest disease	B	India	Birds, mammals	Tick†	Hemorrhagic manifestations more common; F = low, S = 0
Louping ill	B	British Isles	Birds, mammals	Tick†	Diphasic pattern; F = 0, S = 0

*Disease and virus name is name listed plus, usually, encephalitis, e.g., St. Louis encephalitis
†Transovarially transmitted

TABLE 20-3   HEMORRHAGIC FEVERS DUE TO ARBOVIRUSES, ARENAVIRUSES, AND RHABDOVIRUSES

Fever	Causative Agents	Vector(s)	Vertebrate Host(s)	Geographical Distribution	Epidemiologic Features of Involvement of Man	Control	Remarks
Yellow fever (urban)	YF virus, a group B arbovirus	*Aedes aegypti* in cities	Man	Human populations (usually urban) in tropics of South and Central America	Person-to-person passage by *Aedes*	*Aedes aegypti* control; vaccination	Sylvan YF can spread to cities
Yellow fever (sylvan)	YF virus, a group B arbovirus	*Haemagogus* mosquitoes in New World; *Aedes* species in Africa	Monkeys of several genera and species	Forests and jungles of South and Central America and West, Central and East Africa	Man infected by exposure in jungles (i.e., woodcutters, hunters, etc.)	Vaccination	Human cases sporadic and unpredictable; disease often a "silent" endemic in forests
Dengue hemorrhagic fever	Dengue viruses of four types; group B arboviruses	*Aedes aegypti*	Man (involvement of other primates has been postulated)	Tropical and subtropical cities of Southeast Asia and Philippines	Small children usually involved in cities where *Aedes aegypti* densities are high	*Aedes aegypti* control: mosquito repellent, screens, etc.	Disease may represent an immunologic over-response to a sequential infection with a different dengue strain
Omsk hemorrhagic fever	Two distinguishable subtypes arboviruses	Ticks of genus *Dermacentor*	Small rodents and muskrats	Omsk region of USSR; northern Romania	People exposed in fields and wooded lands	Tick repellents and protective clothing	
Kyasanur forest disease	KFD virus, a group B arbovirus	Ticks of several species in genus *Haemaphysalis*	Monkeys (rhesus and langur) and small rodents and birds	Mysore State, India	People exposed in fields and wooded lands	Tick control: tick repellents and protective clothing	Monkey mortality signals epidemic

Disease	Etiology	Vector	Reservoir	Geographic distribution	Persons at risk	Control	Comments
Argentinian hemorrhagic fever	Junin virus, an arenavirus	None proved; mites suspected	Small rodents: *Akodon*, *Calomys*	Argentina: NW of Buenos Aires extending west to Province of Cordoba	Field workers at harvest time are particularly at risk	Rodent control in fields	Infected rodents contaminate environment with urine
Bolivian hemorrhagic fever	Machupo virus, an arenavirus	None recognized	Small rodent, *Calomys callosus*	Beni Province of Bolivia	Residents of small rodent-infested villages and homes; 1971 nosocomial outbreak in Cochabamba, Bolivia	Rodent control in fields	High mortality in man
Lassa fever	Lassa virus, an arenavirus	None required	Small rodent; *Mastomys natalensis*	West Africa; Nigeria, Liberia, Sierra Leone	Residents of small rodent-infested villages; dramatic nosocomial outbreaks	None known; possibly rodent control	High mortality in man
Crimean hemorrhagic fever	CHF Congo virus, an ungrouped arbovirus	Ticks of several genera	Larger domestic animals implicated; also African hedgehog	Southern USSR, Bulgaria, East and West Africa	Cowhands and field workers in USSR; nosocomial outbreaks reported	Tick control relating to livestock; full isolation in patient care	Human disease important in USSR; importance to man in Africa not known
Korean hemorrhagic fever (hemorrhagic fever with renal syndrome)	Not known; virus suspected	Not known	Possibly small mammals	Korea; northern Eurasia to and including Scandinavia	Rural or sylvan exposure (military, forest occupations, farmers)	None	A baffling epidemiologic and clinical entity
African hemorrhagic fever (Marburg disease)	Marburg and Ebola viruses, rhabdoviruses	None required	Unknown	Africa	Person- or monkey-to-person	Clinical isolation techniques	Highly fatal

(Adapted from Johnson: In Beeson and McDermott (eds): Textbook of Medicine, 14th ed., 1975, p. 239. Courtesy of W. B. Saunders Co.)

311

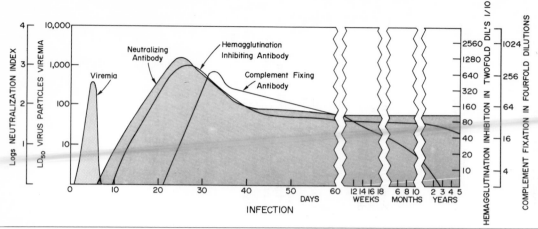

FIG. 20-1. Hypothetical diagnostic virologic and serologic features of arbovirus infections of man. (From Work: Hunter, Swartzwelder JC, Clyde DF (eds): Tropical Medicine, 5th ed., 1976. Courtesy of W. B. Saunders Co.)

Cell cultures are the next most useful medium, especially for the study of many established isolates. In the usual search for arboviruses, serum or tissue specimens from man or other vertebrates, as well as aqueous suspensions of ground-up arthropods, are routinely inoculated intracerebrally into each infant mouse in a litter. The mice are observed for signs of illness over the next month. Brain tissue from mice that become sick during this time is used for further passage. Serial passages are made until mouse pathogenicity is constant, at which time a large pool of infected brain tissue is stored frozen for use in biologic, physical, chemical, and serologic characterization of the agent.

Serum antibody measurements are usually essential in the diagnosis of individual patients (Fig. 20-1). Antibodies also provide a record of past infection that permits detection of the involvement of specific populations of man and other vertebrates in the transmission cycle. These seroepidemiologic investigations are of great value to the understanding of the geographic distribution of arboviruses. Geographic distribution is also delineated through the use of sentinel animals such as caged mice or chickens placed in the forest; presence of virus in these animals is revealed by disease or by a rise in serum antibodies.

The techniques of virus isolation and antibody measurement may be applied to captured or colonized vertebrates and arthropods to de-termine their potential as natural hosts for a given virus. Experts in field and laboratory zoology and entomology are therefore essential. As a result, the arbovirus laboratory is staffed by a broad-based medical zoology group.

ECOLOGY, EPIDEMIOLOGY, AND THE PUBLIC HEALTH

Arboviruses are distributed throughout the world. While most exist in the tropics and subtropics, some have also been isolated from arthropods and vertebrates in climatically less hospitable areas such as Finland, Canada, Siberia, and Alaska. Their epidemic potential continues to surprise us. Outbreaks of major and often new clinical disease in the savannahs, forests, and swamps of the tropical world provide episodes of romance and high adventure as well as tragedy and misery.

The conceptual and technical base of arbovirus ecology grew from the work in the late 1920s and 1930s on yellow fever. Yellow fever initially was thought to be entirely a problem of man-to-man transmission by the relatively domestic mosquito *A. aegypti.* When the entirely nonhuman jungle cycle was discovered and the study broadened zoologically, several arthropod-transmitted viruses were discovered that were not yellow fever. After World War II, an extraordinary global effort was made to study

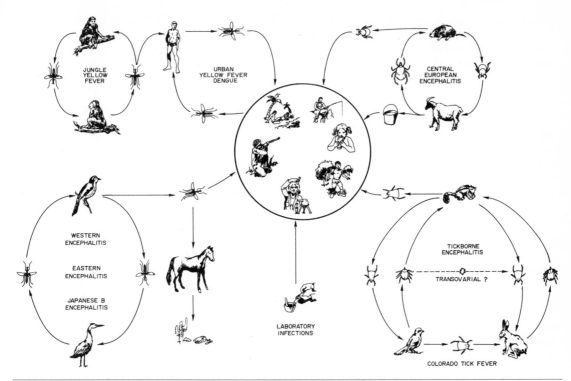

FIG. 20-2. Some basic arbovirus transmission cycles showing their relationship to man. (From Work: In Hunter, Swartzwelder JC, Clyde DF (eds): Tropical Medicine, 5th ed., 1976. Courtesy of W. B. Saunders Co.)

these agents. Arbovirus laboratories were established around the tropical world in such places as Jamaica, Trinidad, Egypt, India, Malaya, Taiwan, Brazil, Panama, Nigeria, and Uganda. The hundreds of arboviruses that we know today were discovered in this effort.

Arboviruses generally maintain themselves in nature by constant and fairly frequent transmission between an arthropod and a vertebrate host (Fig. 20-2). Many variables influence the likelihood and frequency of successful transmission. For example, the frequency of feeding of an arthropod on a vertebrate host depends on numerous innate and behavioral factors of each animal that may be environmentally determined. These factors may differ from time to time and from place to place. Arthropod longevity, a major determinant, depends in turn on climatic factors such as humidity and temperature, availability of food, and the prevalence of predators, as well as on noxious physical, chemical, and biologic agents. Many other complex factors similarly influence transmis-

sion. Clearly, a systems analysis approach is needed to assess the effect of any change in any variable. Such an approach in malariology has been nicely expressed both in words and mathematically by the late Professor George Macdonald. Because of the large number of agents, the numerous nonhuman hosts, and the great gaps in our knowledge, this approach has not yet come into its own.

Although viremia is usually of sufficient intensity for transmission only for a matter of days, there is evidence that long-lasting infections in some vertebrates, such as turtles infected with Eastern equine encephalitis virus, may permit a virus to remain in an area during a period of unsuitable climate, such as the New England winter, and to be transmitted again when arthropod feeding resumes. Transovarial transmission through generations of arthropods occurs with some viruses and provides the prospect of indefinite survival of virus without cyclical transmission. The important subject of how arboviruses survive periods when cyclical transmission cannot occur has been reviewed recently by Reeves. Arthropod-to-arthropod transmission probably does not occur other than transovarially. However, direct transmission may occur between vertebrate hosts, as in the case of Eastern equine encephalitis virus, which is passed from pheasant to pheasant under crowded conditions of commercial pheasant farms. The importance in nature of these alternate maintenance cycles is an area of active, current research.

Man is generally an incidental host and a dead end for virus transmission. A notable exception is the man–mosquito cycle of dengue and urban yellow fevers. As a rule, arboviruses are clinically silent as they cycle in nature. Disease occurs in unnatural hosts such as man and equines in Eastern equine encephalitis virus, and these hosts are of no consequence to viral transmission.

There are two basic ecologic mechanisms that ensnare man and other animals: either man enters the geographic area of the transmission cycle, or the cycle changes and moves closer to man. Yellow fever ecology provides examples of both. The former occurs when a person enters a forest where yellow fever virus is cycling between monkeys and the tree-top mosquitoes. Without the intrusion, these mosquitoes might never feed on man. When the infected person returns to town, he may introduce yellow fever virus into a cycle involving man and the peridomestic mosquito *A. aegypti*, and an urban epidemic may result. Some viruses may come to man by another mechanism, the amplification cycle. These viruses have a basic cycle away from man that involves arthropods that feed almost exclusively on animals. With seasonal buildup of populations of different arthropods and, let us say, birds, another cycle is started that carries the virus closer to man and arthropods that are more likely to bite man.

From time to time an arbovirus disease appears where it has been previously unknown. Hemorrhagic fever of southeast Asia, Kyasanur Forest disease in India, Chikungunya and O'nyong-nyong fevers in East Africa, and Rift Valley Fever in Egypt are all examples of new disease situations that have been recognized within the past two decades. Did O'nyong-nyong fever virus become more pathogenic for man as a result of passage in anopheline mosquitoes? This is a possibility, since all other

group A arboviruses (Alphaviruses) are transmitted by culicenes. Was pickup and transmission by a certain genus of tick responsible for human pathogenicity of Kyasanur Forest disease virus? It is difficult to prove that pathogenicity for man may be enhanced by a change in the transmission cycle, but the question is of great importance, since we persist in meddling with our environment. The possibility that the clinical consequences of infection may be affected by sensitization from prior infection by a related virus has been raised to explain the hemorrhagic fever syndromes in the Philippines and southeast Asia caused by dengue viruses. Before the 1950s, dengue virus was known only in connection with classical dengue, a relatively benign disease characterized by fever and nonhemorrhagic rash. Thus one disease may change as a result of a change in the ecology of another infection. Whatever the reasons prove to be, we are clearly in an unstable situation, with some surprises in the future.

Arboviruses cause disease and death in thousands of persons each year. From year to year and decade to decade, the epidemic potential results in wide swings in incidence in a given geographic area. In the United States, the number of known and reported cases of arboviral encephalitis varies from tens to thousands each year. In the 1960s, there was an epidemic of yellow fever in remote areas of Ethiopia that is believed to have killed 10,000 persons. Dengue hemorrhagic fever in the Philippines and southeast Asia affects thousands each year, with substantial mortality. Laboratories must not be overlooked in a discussion of epidemiology. Man's curiosity about arboviruses has resulted in more than 400 reported laboratory infections, of which 18 caused the deaths of investigators.

HUMAN DISEASE

Arboviruses cause acute febrile illnesses that range from minimal severity to highly fatal diseases with devastating sequelae for the survivors. There are three general types of disease. One type is characterized by undifferentiated fever that may or may not be accompanied by rash and arthralgia (Table 20-1). In another type, there is viral invasion of the central nervous system productive of an often severely damaging encephalitis (Table 20-2).

In the third category are those arboviruses that cause hemorrhagic fever, as in yellow fever, hemorrhagic dengue fever, and a number of tickborne hemorrhagic fevers. Hemorrhagic fevers (Table 20-3) are also caused by a number of arena and rhabdoviruses; thus additional material on this category of diseases is to be found in the appropriate chapters.

There follow brief descriptions of respresentatives of each of these categories, somewhat arbitrarily selected because they occur within the United States or because of their significance to the public health elsewhere.

Arboviruses that Cause Fevers/Rash/Arthralgia

Undifferentiated arbovirus fevers with or without rash and arthralgia are generally of low mortality, and recovery is complete with rare exceptions. Classical dengue fever is the prototype, and many of the other members of the group often are described as denguelike. Members of the group that are transmitted in the United States include Colorado tick fever and vesicular stomatitis viruses.

Dengue fever is a disease of rapid onset, with the temperature rising suddenly to high levels and persisting for several days. Accompanying the fever are headache and aching in the muscles and joints. There is an associated macular rash that blanches with pressure; later in the disease it may become papular and morbilliform, sparing only the palms and soles. After 5 to 10 days, the acute symptoms disappear, but generalized weakness and lassitude may delay complete recovery for weeks. In the midst of the acute disease there may be one or several days when manifestations of illness disappear. This provides the biphasic course that characterizes many arbovirus diseases. Leukopenia is the most striking and consistent laboratory abnormality. Etiologic diagnosis is confirmed by virus isolation from the blood or by the demonstration of a rise in serum antibody titer. Death is rare, and most patients recover completely. The disease occurs throughout the tropical world. Epidemics may be widespread, affecting a very high proportion of the population. Typically, the virus cycles between man and *Aedes* mosquitoes, especially *A. aegypti*. In some areas there may be a jungle cycle involving primates.

O'nyong-nyong, Chickungunya, Mayaro, and

Ross River are four related, mosquitoborne viruses that cause denguelike disease in Africa, southern Asia, South and Central America, and Australia.

Colorado tick fever is characterized by fever, headache, myalgia, and lethargy. Rash and central nervous system signs and symptoms occur less commonly. The rash is maculopapular, sometimes petechial, and does not have any particular distribution. A biphasic fever pattern occurs in about half of the cases and does not occur in any other infectious disease that is endemic in the United States. Disease begins approximately five days from tick bite and lasts for about a week, although malaise may continue for several more weeks, especially in the older patient. Approximately 20 percent of patients require hospitalization; neither death nor permanent clinical sequelae occur. Diagnosis is confirmed by laboratory methods that detect antibodies in serum, infectious virus in blood, or viral antigen in erythrocytes. Curiously, infectious virus and/or viral antigens may be detected in the blood for up to several weeks after onset of disease.

The disease is transmitted throughout the Rocky Mountain states from March to October, with peak incidence in May and June. During 1973 and 1974, active search revealed 228 cases in Colorado alone. The arthropod host is *Dermacentor andersoni*, in which transovarial passage occurs.

Arboviruses that Cause Encephalitis

The arthropodborne viral encephalitides are a group of serious illnesses that occur sporadically or in epidemic form throughout the tropical and subtropical world. Infected persons are at varying risk of death or severe central nervous system damage depending on their age and on the type of virus. In the United States since 1960, there have been an average of 300 arbovirus encephalitis cases reported to the Center for Disease Control annually, ranging from a low of 45 in 1960 to a high of 2113 cases in 1975. In recent years, California and St. Louis encephalitis viruses account for most of these cases, followed by Western, Eastern, and Venezuelan equine encephalitis viruses, with Powassan virus making its appearance in human disease for the first time in single cases in 1971 and 1972.

In Britain, Louping ill is a tick-transmitted encephalitis of sheep and man. Central European and Russian spring–summer encephalitides are found in Europe and Russia, while Japanese B encephalitis occurs in Japan, Asia, and the Pacific. Kyasanur Forest disease in India entails signs of central nervous system involvement, as well as abnormal bleeding, and is thus classified both as an encephalitis and a hemorrhagic fever. Murray Valley encephalitis is found in Australia.

Western equine encephalitis is an acute generalized illness characterized by fever, meningeal signs, somnolence, coma, convulsions, and paralysis. Fatality is high, and permanent brain damage is common, especially in children. The virus cycles in nature between a variety of small birds and the mosquito *Culex tarsalis.* Man and equines are dead-end hosts because viremia adequate for mosquito infection does not occur. Overwintering may occur as a result of chronic viremia in snakes, but annual return of the virus to temperate areas is probably more often by way of migratory birds. Diagnosis depends on demonstration of a rise in antibodies between acute and convalescent serum specimens. Virus isolation attempts from human tissue are rarely successful. The disease is confined to the western hemisphere and in North America occurs in epidemic form in spring and late summer.

Eastern equine encephalitis is an acute generalized illness characterized by fever and central nervous system signs and symptoms, including those of meningeal irritation, lethargy, coma, pareses, and convulsions. Death is frequent, and in survivors residual central nervous system damage is common and severe. A striking polymorphonuclear leukocytosis occurs in blood and in cerebrospinal fluid. Etiologic diagnosis depends on virus isolation from brain tissue or the demonstration of an increase in serum antibody. Mosquitoborne epidemics and equine epizootics occur in late summer along the eastern strip of North America from Florida to Canada and the Caribbean.

California encephalitis is an acute febrile illness of 7 to 10 days duration, characterized by headache, vomiting, lethargy, disorientation, seizures, and focal neurologic signs. Although convalescence may be prolonged, fatality and morbidity are low. The cerebrospinal fluid contains abnormally high protein and increased numbers of white blood cells during acute phase. Etiologic diagnosis depends on either virus isolation or demonstration of a rise of specific antibodies. It is a disease of children and occurs during summer months in wide areas of midwestern and southern United States. Mosquitoes of the genus *Aedes* transmit the virus to man. Vertebrate hosts include small animals.

St. Louis encephalitis is an acute febrile illness that occurs in a small percentage of those infected. It is usually characterized by fever and headache, and abnormalities of coordination and motor cranial nerve function may be common, especially in children. Signs of urinary tract inflammation may occur. Fatality may reach 25 percent, and mental and emotional sequelae are common. Cerebrospinal fluid changes are consistent with virus infections. Diagnosis depends on demonstration of serum antibody rise, and virus isolation is rarely achieved from human specimens. The virus cycles between birds and *Culex* mosquitoes. Large urban and suburban outbreaks occur in the United States.

Powassan virus encephalitis has been identified as the cause of encephalitis in an occasional patient for the past 20 years in northern United States and Canada. While human infection does not appear to be very common, clinical consequences can be severe, with fever, headache, seizures, coma, and paralysis. Permanent neurologic damage may follow. Powassan virus is transmitted by ticks, although history of tick bite is frequently lacking. In one report of three cases, there was evidence for concurrent enterovirus infection. This raises the fascinating possibility that here may be found the first example of human disease enhancement as a result of concurrent infection with more than one virus.

Venezuelan equine encephalitis is characterized by fever, severe headaches, chills, and gastrointestinal symptoms. In a few children, serious encephalitis occurs, with convulsions, coma, paralysis, and abnormal reflexes. Myalgia, conjunctivitis, and sore throat commonly appear as the disease progresses. Leukopenia is typical. Diagnosis is suspected in patients who have been in appropriate geographic areas during epidemics. Demonstration of a rise in antibodies in serum samples obtained in the acute and convalescent stages of disease is necessary.

The disease occurs as equine epizootics in South and Central America, the Caribbean, Mexico, and southern United States. Strains of the virus that are responsible for endemic cases cycle in mosquitoes, small rodents, and birds. Epidemic strains cycle in equines and a variety of mosquitoes; it is this poorly understood amplification cycle that results in humans becoming infected. A live attenuated vaccine (strain TC-83) has been shown capable of immunizing man and horses. While it may be too dangerous for widespread use in people, this vaccine can be applied on a mass basis to horses with prompt cessation of both equine and human cases. Insecticides can be used concurrently.

Japanese B encephalitis is an acute illness that occurs in a small percentage of those infected with the virus. The usual consequence of infection is either inapparent or mild, undifferentiated disease. The encephalitis disease is similar to St. Louis encephalitis, with more severe neurologic involvement. Fatality is high, especially in older persons. Diagnosis is confirmed serologically, and virus isolation from human specimens is unusual. The virus cycles annually between *Culex* mosquitoes and birds or pigs in temperate areas including Japan, Siberia, Korea, Taiwan, Pacific islands, and in tropical areas of southeast Asia.

The tickborne group B viruses comprise a group of related agents that are found, each in a circumscribed domain, in North America, Britain, Europe, Russia, and Asia. The most common ones are Powassan, Russian spring–summer, central European, and Louping ill encephalitides, and Omsk hemorrhagic fever and Kyasanur Forest disease. The clinical spectrum includes hemorrhagic fever or encephalitis or a combination of the two. They are listed in both Tables 20-2 and 20-3. The diseases vary greatly in fatality and in the severity of permanent neurologic damage. Diagnosis is by virus isolation from blood or brain tissue and by serologic tests.

Arboviruses that Cause Hemorrhagic Fevers

The viral hemorrhagic fevers comprise 10 distinct diseases with widely differing epidemiologies and geographic loci. They are caused by viruses representing several taxonomic groups (Table 20-3). Some are mosquitoborne, others are tickborne, and the arthropod vectors of a third group are unknown or nonexistent. One

disease is only presumed to have a viral etiology, and its mode of transmission is unknown. In addition to man himself, vertebrate hosts generally include primates or small mammals or rodents. The diseases are severe, with high fatality, but are self-limited, and survivors usually recover completely. None of these diseases is transmitted within the United States.

Because of their importance, two, yellow fever and hemorrhagic dengue, are described here. See also Arenaviruses (Chap. 22).

Yellow Fever In spite of the presence of a safe, effective, attenuated vaccine that has been available since the 1930s, yellow fever remains today a real or potential public health problem for many peoples of the tropical world. Clinically, the disease is marked by the sudden onset of fever associated with severe headache, myalgia, back pain, conjunctivitis, and photophobia. As the disease progresses, prostration increases, and there are signs of involvement of the liver, kidneys, and heart with hepatomegaly, albuminuria, and a striking slowing of the heart rate. A hypovolemic shock phase associated with bleeding gums, hematemesis, oliguria, and jaundice follows. Case fatality is high; recovery is complete. Diagnosis is confirmed by the isolation of the virus from the blood or from the liver, or by demonstration of a specific serologic response.

The characteristic histopathology of midzonal necrosis in the liver lobule has given rise in some areas to an epidemiologic method wherein mandatory postmortem corings of the liver are obtained from all deaths and sent in fixative to a central examining point, thus permitting an assessment of the proportion of deaths that are due to yellow fever.

Yellow fever has two distinct cycles in nature. One is an *A. aegypti*–man cycle that is operative in the large urban epidemics that are now, fortunately, history. The virus also cycles in the forest, involving monkeys and different genera of mosquitoes. The virus is carried from the forest cycle to the urban cycle by infected people. Transmission can be effectively interrupted, in the case of the urban cycle, by *A. aegypti* control measures. Such measures, used prior to the recognition of the sylvan cycle, rid the urban centers of the Western hemisphere of dreadful epidemics that affected cities as far north as Philadelphia into the first decade of the present century. These same measures permitted completion of the Panama Canal. They are, however, relatively useless when one is dealing with zoophilic forest mosquitoes. As if in defiance of man's best effort, yellow fever epidemics occur today in areas where lack of basic logistical underpinnings of health services prevent vector control or immunization programs.

Hemorrhagic Dengue This acute febrile illness begins abruptly like classical dengue with fever, nausea, vomiting, and cough. However, after a day or two, hemorrhagic manifestations appear, first as petechiae followed by purpura and signs of gastrointestinal hemorrhage. The clinical picture of shock ensues. Hypoproteinemia, rising hematocrit, and hemostatic abnormalities, including thrombocytopenia, are present. During the next days the patient either dies or begins a complete recovery. Fatality can be as high as 50 percent. Diagnosis is confirmed by demonstration of rise in antibodies or by isolation of virus from specimens of blood.

This hemorrhagic form of dengue was first seen in the 1950s in the Philippines and Southeast Asia and subsequently in India, Indonesia, and Oceania. Children comprise the vast majority of those affected. Although a hypersensitivity state resulting from prior infection with a related virus has been postulated, the answer to the question of why dengue virus causes this serious disease in certain areas and age groups, and not in others, remains a mystery. The vector is *A. aegypti*, as it is for classical dengue. Accordingly, the frequency of transmission is governed by climate and numerous domestic factors (such as those relating to collections of standing water) that affect breeding and feeding of this mosquito. Control measures entail all those measures that reduce the frequency of the vector mosquito feeding on potential patients. No vaccine is available.

TREATMENT, CONTROL, AND PREVENTION

There are no therapeutic agents that are specific for arboviruses. The management of patients with disease involves measures designed to restore and maintain nutrition and reasonably normal physiology. The latter might include, for example, restoration of intravascular volume during the hypovolemic shock phase of hemorrhagic fever or measures to reduce destructive degrees of cerebral edema in patients with encephalitis.

With the exception of the highly effective, safe, attenuated, tissue-culture vaccine that has been in use since the 1930s for yellow fever, there are no arbovirus vaccines in general use.

Venezeulan equine virus vaccine has a restricted use for laboratory workers and can protect man indirectly by preventing the development of a large reservoir of infected horses. Many experimental vaccines are under trial or laboratory development.

For arboviruses with limited host range and an accessible vector, there have been instances of effective prevention based on vector control. Urban yellow fever was eradicated in the early decades of this century by measures against *A. aegypti*. The task becomes infinitely more complex when one considers arboviruses that cycle silently in a variety of natural settings in birds, mammals, and arthropods. In these situations, where no ecologically acceptable vector control is available, the possibilities may be restricted to lessening man's exposure to the ecosystem in question. Such a measure may fail because of anticipated and unacceptable consequences as, for example, might beset a woodcutter denied access to the forest.

FURTHER READING

Berge TO: International Catalogue of Arboviruses, 2nd ed. Dept. of Health, Education, and Welfare, Washington DC, US Gov Printing Office, 1975. See also Karabatsos N (ed): Supplement to International Catalogue of Araboviruses Including Certain Other Viruses of Vertebrates Am J Trop Med Hyg (Suppl): 27 372, 1978

Blaskovic D, Nosek J: The ecological approach to the study of tick-borne encephalitis. Prog Med Virol 14:275, 1972

Casals J: Arboviruses, arenaviruses and hepatitis. In Hellman A, Oxman MN, Pollack R (eds): Biohazards in Biological Research. Cold Spring Habor Laboratory, New York 1973, p 223

Dietz WH Jr, Alvarez O Jr, Martin DH, Walton TE, Ackerman LJ, Johnson KM: Enzootic and epizootic Venezuelan equine encephalomyelitis virus in horses infected by peripheral and intrathecal routes. J Infect Dis 137:227, 1978

Downs WG: Arboviruses. In Evans AS (ed): Viral Infections of Human—Epidemiology and Control. New York, Plenum, 1976, p 71

Goodpasture HC, Poland JD, Francy DB, Bowen GS, Horne KA: Colorado tick fever: Clinical epidemiologic, and laboratory aspects of 228 cases in Colorado in 1973–1974. Ann Intern Med 88:303, 1978

Hammon W McD, Suther GE: Arboviruses. In Lennette EH, Schmidt NJ (eds): Diagnostic Procedures for Viral and Richettsial Infection. American Public Health Association, 1969, p 227

Henderson PE, Coleman PH: The growing importance of California arboviruses in the etiology of human disease. Prog Med Virol 13:404, 1971

Johnson KM: Arthropod-borne viral fevers, In Beeson and McDermott (eds): Textbook of Medicine, 15th ed. Saunders, Philadelphia, 1979, p 276

Lee HW, Lee PW, Johnson KM: Isolation of the etiologic agent of Korean hemorrhagic fever. J Infect Dis 137:298, 1978

McDonald G: The Epidemiology and Control of Malaria. London, Oxford Univ Press, 1957

Reeves WC: Overwintering of arboviruses. Prog Med Virol 17:193, 1974

Smith R, Woodall JP, Whitney E, et al: Powassan virus infection, a report of three human cases of encephalitis, Am J Dis Child 127:691, 1974

Strode GK: Yellow Fever. New York, McGraw-Hill, 1951

Theiler M, Downs WG: The Arthropod-Borne Viruses of Vertebrates. New Haven, Yale Univ Press, 1973 (see also review of this book, Work TH: Science 182:273, 1973)

Traub R, Wisseman CL JR: Korean hemorrhagic fever. J Infect Dis 138:267, 1978

US Public Health Service: Morbidity and Mortality Weekly Report. Atlanta Ga, Center for Disease Control

US Public Health Service: Neurotropic Viral Diseases Surveillance, Annual Summary. Atlanta Ga, Center for Disease Control

Woodruff AW, Bowen ETW, Platt GS: Viral infections in travellers from tropical Africa, Br Med J 1:956, 1978

Work TH: The expanding role of arthropod-borne viruses in tropical medicine. In: Industry and Tropical Health, Vol 4. Boston, Harvard School of Public Health, 1961, p 225

Work TH: Exotic virus diseases. In Hunter GW, Swartzwelder JC, Clyde DF (eds). Tropical Medicine, 5th ed. Philadelphia, Saunders, 1976, p 1

CHAPTER 21
Rhabdoviruses

But Africa has had a nasty habit recently of turning up new and dangerous virus diseases such as Lassa, Marburg, and Ebola. . . .
British Medical Journal, 6112:529, 1978, 1.

Rabies

There are more than 60 rhabdoviruses. Of these, 8 are known to cause disease in man (Tables 21-1 and 20-3). Some of these, such as rabies, Marburg, and Ebola, cause serious illnesses, often with very high fatality rates. Rabies is nearly 100 percent fatal, while for Marburg and Ebola, fatality rates ranging from 20 to 90 percent have been observed. Other members of the group such as Isfahan virus, isolated in Iran in 1975, probably infect humans, as evidenced by a high prevalence of antibodies; no disease has been observed, however. Rabies was the first rhabdovirus isolated, in the early years of this century, while the newest member of the group was isolated in the 1970s; clearly there are more to come.

Transmission from man to man or animal to animal is clearly by arthropod in some instances with certain viruses, while others are transmitted mainly through animal-to-animal contact. For Marburg and Ebola, in addition to person-to-person transmission, man probably acquires infection as a result of environmental contamination by chronically infected, viruric rodents.

Serologic relationships and morphology define the subgroups of the Rhabdoviridae and individual members of subgroups. Arbovirologists often have been drawn into the study of rhabdoviruses either because of definite arthropod transmission or epidemiologic features that were initially highly suggestive of arthropod-borne transmission. In addition, rhabdoviruses possess complex life cycles and transmission patterns, and these aspects in themselves attract the attention of medical scientists and clinicians whose interests are broadly based zoologically. Other than support of physiologic processes in the sick individual, the only protection that man and animals have from rhabdoviruses are natural or induced immunity, and environmental manipulations based on an understanding of the zoonotic details of the life cycles.

In this chapter, attention will be paid only to those agents known to cause human disease, i.e., rabies, Marburg, Ebola, and certain members of the vesicular stomatitis group.

Rabies has been recognized for centuries. In the late 1800s Pasteur sorted out the confusion concerning the etiologic agent and pathogenesis of the disease. He then crowned these extraordinary achievements with the development of a method of active immunization, which, although never tested in a controlled trial in man, is generally believed to have been an effective vaccine. Today, rabies continues to frighten man, and it poses a distinct and highly fatal hazard to him as well as his domestic animals. In the 1960s the physical and chemical properties, as well as the morphology of the rabies virus, were established.

Epidemiology

Rabies virus persists in nature by passage from animal to animal, usually during direct contact. Adequate contact generally requires that virus-containing saliva be inoculated through broken skin by bite or scratch into a susceptible animal. The infected animal's illness alters behavior in favor of transmission by making the animal deranged and aggressive and thus more likely to attack or bite more frequently and to less purpose than normally.

In some situations, rabies virus may be transmitted by aerosol. Such situations occur in caves that contain large populations of insectivorous bats. At certain times of the year, when bats are being born and are developing and when their mothers are lactating, sufficient quantities of rabies virus may be present in the air of the cave to infect susceptible animals placed in such a fashion as to be exposed only to cave air. Beyond such observations as these, however, the ecology of bat-cave rabies is poorly understood.

Some biologic determinants of transmission include the following. The saliva must contain an adequate titer of virus. In this connection, infected skunks have been found to develop the highest titers. The bitten animal must be of a susceptible species. In the laboratory, the route of innoculation is important, with intracerebral being among the most sensitive

TABLE 21-1   RHABDOVIRUSES THAT CAUSE ILLNESS IN MAN

Group	Serotype	Year of Original Isolation	Natural Geographic Distribution	Disease in Man
Rabies	Rabies	1903	Worldwide	Meningo-encephalitis
Marburg	Marburg	1967	Africa	Hemorrhagic fever
	Ebola	1976	Africa	Hemorrhagic fever
Vesicular stomatitis	Indiana	1925	Western hemisphere	Fever and vesicles
	New Jersey	1952	Western hemisphere	Fever and vesicles
	Alagoas	1964	Western hemisphere	Fever
	Piry	1960	Western hemisphere	Fever
	Chandipura	1965	Asia and Africa	Fever

routes, whereas intraperitoneal injection is less likely to transmit infection. Transmission transplacentally or via the gastrointestinal tract by drinking milk of infected animals or by arthropod vector does not seem to occur to any significant degree in nature. The age of the receiving animal can be important; generally, the younger the animal, the more susceptible.

The site of the inoculation is clearly important. In man, bites about the head and neck are more likely to result in disease sooner than are bites at sites more peripheral to the central nervous system. This is thought to be because there is a greater density of nerve endings about the head and face, providing more neurologic pathways for the virus to migrate to the central nervous system. Also, the distance to the brain is shorter from the head and neck area.

Sources of infection for man and his economically important domestic animals vary to some extent, geographically. For example, in North America, foxes, skunks, raccoons, and bats are the usual reservoir and vector hosts. In Latin America, bats of various species seem to be the most important, while in the Mediterranean area the wolf, in India the jackal, and in certain Caribbean areas, such as Cuba, Puerto Rico, and Granada, the mongoose is the most significant reservoir/vector animal. In the United States, due to various protective measures, chiefly immunization, rabies in dogs, cats, and

farm animals such as cattle has fallen to a low level in recent years. For poorly understood reasons, however, rabies in skunks and bats seems to be on the increase. In the United States and in Europe nearly all reported rabies occurs in wild animals, with skunks and bats most common in this country, while foxes are now the most likely animal in Europe.

Bat rabies may be transmitted in two ways. In South America, vampire bats transmit rabies to domestic cattle while taking their blood meal. This results in enormous economic and nutritional losses to man in these areas. As mentioned above, aerosol transmission can occur in some insectivorous cave-dwelling bats of North America. Presumably this mechanism, as well as transmission by bite, can be operative in vampire populations.

The epi- and enzootic behavior of rabies in wild animals may display swings of incidence in one or another species and from one place to another that are often poorly understood. Spread of animal rabies into a geographic area typically involves but one species. Evidence that strains of wild rabies may differ in their pathogenicity for one species as compared to others may explain such epizootic behavior.

Density of susceptible and infected animal populations, the biology of the virus in the infected host, the degree to which rabies can become latent, and factors that influence aggressive behavior, other than the rabies infection

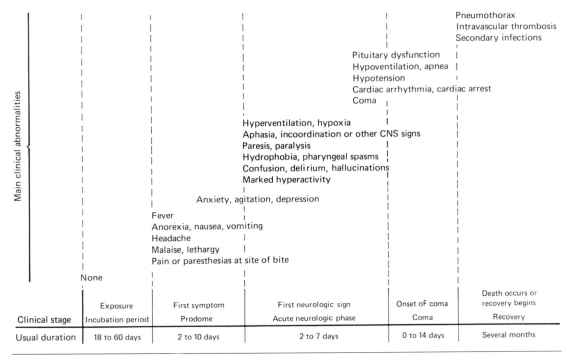

Main clinical abnormalities

Pneumothorax
Intravascular thrombosis
Secondary infections

Pituitary dysfunction
Hypoventilation, apnea
Hypotension
Cardiac arrhythmia, cardiac arrest
Coma

Hyperventilation, hypoxia
Aphasia, incoordination or other CNS signs
Paresis, paralysis
Hydrophobia, pharyngeal spasms
Confusion, delirium, hallucinations
Marked hyperactivity

Anxiety, agitation, depression

Fever
Anorexia, nausea, vomiting
Headache
Malaise, lethargy
Pain or paresthesias at site of bite

None

Clinical stage	Exposure Incubation period	First symptom Prodome	First neurologic sign Acute neurologic phase	Onset of coma Coma	Death occurs or recovery begins Recovery
Usual duration	18 to 60 days	2 to 10 days	2 to 7 days	0 to 14 days	Several months

FIG. 21-1. The natural history of clinical rabies in man; hypothetical composite case. All clinical abnormalities need not be present in each case. (From Hatwick MAW, Gregg MB: In Baer GM (ed): The Natural History of Rabies, Vol 2. Academic, New York, 1975, p 289.

itself, all require further understanding before the determinants of rabies transmission are completely elucidated.

Serologic surveys for the purpose of detecting rabies antibodies indicative of nonfatal, past infection has not been helpful in rabies because no adequately sensitive, specific, and inexpensive test is available. Accordingly, the question of whether latent or clinically silent infections are epidemiologically significant is not yet answered.

There is seasonal variation in rabies in animals as well as man. In animals the greatest number of cases occur in the spring and early summer during the breeding season. In man, most cases occur in the summer, possibly as a result of more outdoor contact with rabid animals.

Disease in Man

Generally, disease begins within 2 to 8 weeks of exposure, with extremes of the incubation period ranging from 9 days to more than a year, and even many years in very rare instances

(Fig. 21-1). Early symptoms include loss of appetite, nausea, vomiting, and fever associated with feeling poorly, pain, and paresthesias at the site of the bite. Anxiety, depression, and agitation ensue. As the disease progresses, spasms begin to occur. When they involve the muscles of deglutition, the inability to drink liquids or to swallow one's own saliva properly gives rise to the picture of hydrophobia. Soon, a wide variety of central nervous system signs appear, including hyperactivity, confusion, delirium, hallucinations, disorders of coordination, and paralyses. Finally, coma develops, and by this time secondary infection, nutritional deficiencies, and respiratory problems bring on death.

Recovery from rabies is considered to be extremely rare. There is one definite, well-documented case of recovery reported in the literature.

Laboratory findings include an elevated white cell count of 20 to 30 thousand cells per cu mm associated with increased percentage of polymorphonuclear leukocytes. Cerebral spinal fluid is usually normal, but a mild pleocytosis may be observed consisting of less than 100 mononuclear cells per cu mm. Protein, sugar, and hyaline casts may be seen in the urinary sediment.

The clinical features described above following a bite by a rabid animal is the major criterion for diagnosis. Virus isolation attempts from saliva, cerebrospinal fluid, tears, urine, and nasopharyngeal secretions may be successful, but they are meaningless if negative. A rise in serum antibodies is generally not detectable by the time the patient dies. Following death, virus can be demonstrated in the central nervous system by one or more of the three methods used to detect rabies virus in animal brain specimens. One method, the most rapid, involves the detection of intracellular rabies virus antigen by the use of fluorescent antibody. A second method is the identification of typical intracytoplasmic, eosinophilic inclusions (Negri bodies, see color plate I) in nerve cells in certain areas of the central nervous system. Third, intracerebral inoculation of mice with suspensions of brain tissue is followed by identification of rabies virus in brain tissue after a suitable period of time for virus multiplication to occur.

The pathologic findings in rabies are principally inflammatory necrosis that chiefly involves the brain. Grossly, the brain is swollen and congested. Microscopically nerve cells are seen to be destroyed and perivascular, mononuclear cuffing is present in addition to meningeal mononuclear cell infiltrates. Negri bodies are found. The spinal cord may show similar changes.

Prevention and Control

Other than avoidance of rabid animals and their bite, prevention of human disease is based on immunization. Pre-exposure immunization is recommended for certain people at high risk of exposure because of their animal-handling occupation. Because of the long incubation period, active immunization to rabies *after* exposure (inoculation) is considered to be effective for preventing disease and has been the basic measure in the management of human rabies since it was first performed by Pasteur.

The decision to immunize depends on evidence that the biting animal is rabid. Such evidence results from examination of the animal's brain for rabies virus by one or more of the methods mentioned above. Presently in the United States, animals such as skunk, fox, coyote, raccoon, or bat that carry out unprovoked attacks on man may be presumed rabid.

Immunizing agents include killed virus vaccine prepared from virus grown in duck embryos or human diploid cell cultures. Antiserum derived from hyperimmunized horses or people is used concurrently with vaccine. Details of active and passive rabies immunization procedures for animals and man are constantly changing as products exhibiting fewer undesirable side effects and better antigenicity are developed. Groups such as the Advisory Committee on Immunization Practices of the U. S. Public Health Service publish their recommendations periodically in the Morbidity and Mortality Weekly Report of the Center for Disease Control.

The control of rabies in wild animals is considerably more difficult than for domestic animals and man. Human rabies in the United States has decreased to less than three cases per year for the past 19 years, while cases of wild animal rabies continue to be recognized by the thousands. Measures for control of wild animal rabies include quarantine or slaughter. More recently some innovative work employing baited food with live virus vaccine designed to attract specific species such as foxes is being tried in certain parts of the world.

Marburg and Ebola disease

In 1967 in Marburg, Germany, and in Yugoslavia, 25 persons became ill (See Table 20-3). There were 6 secondary cases among persons in direct contact with the primary 25 cases. Seven of these 31 patients died. All of the primary cases were laboratory workers who had been in contact with tissue or body fluid of green monkeys *(Cerpithicus aethiops)*. The monkeys were from two shipments from Uganda. On route there was an overnight stay at Heathrow Airport in London. A new virus was isolated from the patients and subsequently named Marburg virus. On the basis of its physical and chemical properties, Marburg

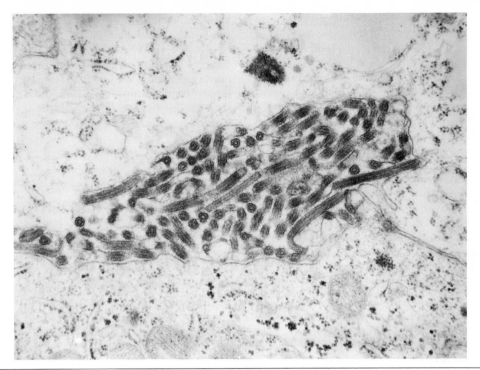

FIG. 21-2. The Marburg agent (1975 isolate from South Africa). ×75,000. (Courtesy of Dr. Erskine Palmer, Center for Disease Control, Atlanta, Georgia.)

virus is now classified as a rhabdovirus (Fig. 21-2).

During early February 1975, a young Australian, accompanied by his girlfriend, travelled extensively in Rhodesia. He then became ill and died in Johannesburg. Marburg virus was isolated from blood and throat specimens. His girlfriend and a nurse who assisted her in caring for him during his fatal illness suffered a similar illness but recovered. Intensive epidemiologic investigations have failed to reveal the source of their infections.

Between August and November 1976, in two towns 600 miles apart, one in southern Sudan and the other in northeastern Zaire, hundreds of persons became ill, and most died. From these patients a virus was isolated which, when compared with Marburg virus, was found to be antigenically distinct but morphologically identical. This agent was named Ebola, for the river that runs near the location of the outbreak in Zaire. In Zaire, except for the index case, all 235 cases were linked by intimate contact or by syringes and needles used for medicinal purposes. Of these cases, 201 died, including all of those infected by syringe and needle transfer.

Thus Marburg virus has struck at man on three occasions in the past 12 years with devastating effect. While the method of spread within a given outbreak has become evident quickly, the natural life cycle and the source of the infection for the index cases remain a mystery except in the original outbreak involving green monkeys from Uganda. But even in that instance, the epidemiologic determinants of the monkey infection are not at all clear. Investigations in Uganda revealed some serologic evidence for unusually high infection rates in the monkeys at the time of the Marburg outbreak. But these rates have not persisted and the thousands of green monkeys shipped from Uganda during the years before the Marburg incident were apparently not infected. As one prominent worker in this field has stated, "the green monkey is a red herring."

Exhaustive studies of the 1975 Rhodesian cases provide evidence suggestive of arthropod transmission of infection to the index case, but no proof could be obtained. Similarly, the sources of infection in the index cases of the two simultaneous 1976 outbreaks are unknown. Serologic surveys show that between 1 and 20 percent of persons, depending on the geographic area surveyed, may show antibodies, and it is thus likely that considerable human transmission occurs in clinical silence. Beyond that, however, routes of transmission, reservoir hosts, and all of the other essentials that permit rational defense are totally lacking.

Transmission routes within outbreaks, however, are much better understood. Close personal contact during acute illness clearly results in transmission. Handling of secretions, tissues, and body fluids from acutely ill patients is another source. In one instance, venereal transmission occurred through infected semen long after the acute disease. In Zaire, improper use of syringes and needles spread the infection nosocomially in a uniformly fatal manner among hundreds of persons.

Clinically, Marburg disease begins with fever, headache, nausea, vomiting, myalgia, and diarrhea. After a few days, sore throat, conjunctivitis, abdominal pain, and rash appear. The rash is fine, maculopapular, and may appear initially in relation to hair follicles. It begins either on the trunk or limbs, spreading thereafter to involve both. After five to seven days, severe bleeding begins involving chiefly the gastrointestinal tract. Shock and death often follow soon thereafter.

Laboratory findings include leukopenia, thrombocytopenia, and often the findings of disseminated intravascular coagulation. proteinuria and markedly elevated serum amylase and transaminases are also found.

The major pathologic finding is extensive hepatocellular necrosis. Eosinophilic cytoplasmic inclusions are seen, and on electron microscopy numerous virus particles are seen in the liver.

Diagnosis rests on isolation of virus in Vero cell cultures or guinea pigs from blood or oropharyngeal secretions from acute-phase specimens. After approximately two weeks, serum antibodies may be best detected by fluorescent antibody techniques.

Case fatality approaches 100 percent under some circumstances, but this has been held to 20 percent when first-class intensive care facilities are available. The roles of heparin to combat intravascular coagulation, specific antibody-containing plasma, and interferon in the treatment of the acute disease are not fully defined, but all have been used.

Index cases cannot be prevented, for lack of knowledge. Secondary cases resulting from person-to-person contact or the handling of patient's secretions or tissues can be controlled by rigorous isolation and protective practices. Laboratory work on Marburg viruses must be confined to maximum containment laboratories such as the facility at the Center for Disease Control at Atlanta.

Vesicular stomatitis virus

The vesicular stomatitis virus group is generally considered to comprise seven serologically related but distinct viruses; they are the Indiana, New Jersey, Cocal, Alagoas, Piry, Chandipura, and Isfahan strains of VSV. The first of these was isolated in 1925 and the latest in 1975. All but two have been found solely in the western hemisphere, while the remainder have been found in Africa, Asia, and Iran.

As a group, the vesicular stomatitis viruses have been found in nature in a variety of animals ranging from insects to man. Evidence for arthropodborne transmission includes the fact that VSV has often been isolated from naturally infected insects, including sand flies, mites, black flies, and mosquitoes. With some serotypes, insect transmission has been achieved experimentally. For other serotypes, the above information is lacking. In addition, the seasonal pattern of transmission of some of these agents strongly suggests arthropod vector.

The disease vesicular stomatitis is most frequently seen in cattle, pigs, and horses, where it occurs in outbreaks of illness associated with vesicular lesions in the mouth and elsewhere on the body. Lesions may also be seen on the feet, thus mimicking foot and mouth disease, a much more severe condition. Although three-quarters of a herd may become ill, fatality is less than 5 percent. Transmission among domestic animals seems to require contact with very sharply circumscribed pasture areas, epidemiologic conditions that are difficult to explain on the basis of arthropod transmission.

The disease in man is an acute, self-limited, benign, febrile illness associated with general

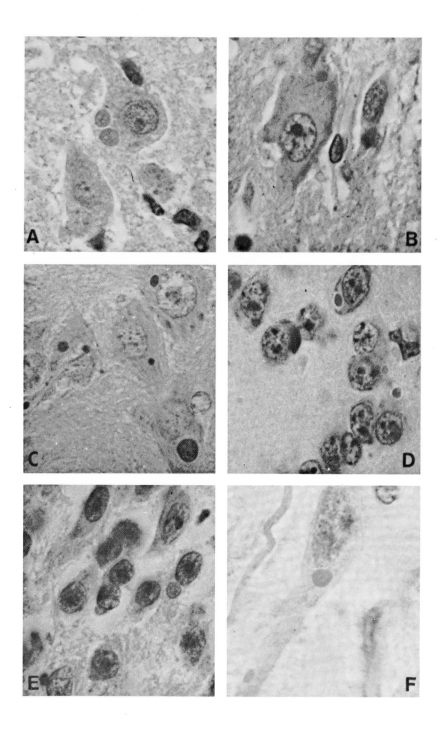

PLATE I. Negri bodies are shown here as the round eosinophilic cytoplasmic inclusions in brain cells from a variety of animals. A, Human; B, Cat; C, Dog; D, Mouse; E, Bovine; F, Fox. All sections are magnified approximately 1000–1400 ×. (From Atanasiu P, Sisman J: In Baer GM (ed): The Natural History of Rabies, Vol 1. Academic, New York, 1975, p 374.)

malaise, myalgia, nausea and vomiting, and headache. No deaths have been reported. In some instances vesicles may be seen on the gums and buccal and pharyngeal mucosa; herpeslike lesions on the lips may also be seen. Illness is generally of less than a week's duration. Many laboratory infections have occurred, and the illnesses resulting from these infections seem to be identical to those occurring as a result of natural transmission.

In the United States, small outbreaks have been detected wherein direct transmission from animal to man is the most probable route of infection. Several serologic surveys in a variety of populations throughout the world suggest that infections with these viruses may be quite widespread. No serious public health problem, however, in the form of human disease at least, has seemed to occur as a result of these widespread infections. Because of the nonspecific nature of the illness, it is possible that many cases of human disease go undetected.

No characteristic clinical laboratory or histopathologic features have been recognized. Diagnosis depends on the clinical picture and the knowledge of close contact with diseased animals. Specific, complement-fixing, and neutralizing antibodies are detectable approximately two weeks after onset of illness.

Except for avoidance of contact with diseased domestic animals, no preventive measures are known. No specific treatment of the sick individual is available.

FURTHER READING

General

Knudson DL: Rhabdoviruses. J Gen Virol (Supplt) 21: 105, 1973

RABIES
Baer GM (ed): The Natural History of Rabies, (Vols 1 and 2), Academic, New York, 1975
Center for Disease Control: Recommendations of the public health service advisory committee on immunization practices, rabies. In Morbidity and Mortality Weekly Report 25:403, 1976
Center for Disease Control: Rabies surveillance report, annual summary for 1977. U.S. Dept. HEW Public Health Service, Sept. 1978
Hattwick MAW et al: Recovery from rabies: a case report. Ann Intern Med 76:931, 1972
Shope RE: Rabies. In Evans AS (ed): Viral Infections of Humans, Epidemiology and Control. Plenum, 1976, p 351
Sikes RK: Rabies. In Hubbert WT et al (eds): Diseases Transmitted from Animals to Man, 6th ed. Thomas, Springfield, Ill, 1975, p 871
Steele JH: The epidemiology and control of rabies. Scand J Infect Dis 5:299, 1973

MARBURG AND EBOLA FEVER
Bowen ETW et al: Viral hemorrhagic fever in southern Sudan and northern Zaire. Lancet x:581, 1977
Conrad LJ et al: Epidemiologic investigation of Marburg virus disease, Southern Africa, 1975. Am J Trop Med Hyg 27:1210, 1978
Editorial: After Marburg, Ebola . . . . Lancet 1:581, 1977
Editorial: Ebola virus infection. Br Med J, 1:539, 1977
Esmond RTD et al: A case of Ebola virus infection. Br Med J 1:541, 1977
Gear JSS et al: Outbreak of Marburg virus disease in Johannesburg. Br Med J 4:489, 1975
Johnson KM et al: Isolation and partial characterization of a new virus causing acute hemorrhagic fever in Zaire. Lancet 1:569, 1977
Kissling RE: Marburg virus. In Hubbert WT et al (eds): Diseases Transmitted from Animals to Man, 6th ed. Thomas, Springfield, Ill, 1975, Chap 71
Pattyn SR (ed): Ebola Virus Hemorrhagic Fever. Elsevier, Amsterdam, 1978

VESICULAR STOMATITIS
Bhatt PN, Rodrigues FM: Chandipura: A new arbovirus isolated in India from patients with febrile illness. Ind J Med Res 55:1295, 1967
Fields BN, Hawkins K: Human infection with the virus of vesicular stomatitis during an epizootic. N Engl J Med 277:989, 1967
Tesh RB, Johnson KM: Vesicular stomatitis In: Hubbert WT et al (eds): Diseases Transmitted from Animals to Man, 6th ed. Thomas, Springfield, Ill, 1975, Chap 73
Tesh RB, Saidi S, Javadian E, Loh P, Nadim A: Isfahan virus, a new vesiculovirus infecting humans, gerbils and sand flies in Iran. Am J Trop Med Hyg 26:299, 1977

CHAPTER 22
Arenaviruses

Thomas Robson, 65, an engineer who died in a hospital after returning from Nigeria, was ordered cremated in secret with his family banned from the ceremony. Robson died of Lassa fever, a disease so dangerous that authorities refused to order an autopsy.

Southall, England
(Associated Press, November 2, 1978)

Arenaviruses have in common a similar morphology, related antigens, and the capacity to cause chronic infections in a variety of rodents (Fig. 22-1). Three of them—Junin, Machupo, and Lassa—can cause serious, highly fatal, hemorrhagic fever in man; one, lymphocytic choriomeningitis (LCM), causes a sporadic, relatively benign meningoencephalitis (Tables 22-1 and 20-3).

Perhaps the most impressive feature of arenaviruses is their capacity to surprise us. Recent Lassa fever outbreaks warn us in the 1970s, as Machupo and Junin did in the 1960s, of the dire consequences of our ignorance and of our constant meddling with the balance of nature.

Arenaviruses establish chronic infections in a variety of laboratory and wild rodent species. When the young animal is inoculated perinatally, a chronic, lifetime infection can result with continuous viremia, as well as high titer viruria. There may be a total lack of immunologic response or disease. It is this sort of rodent infection that is of great importance in the perpetuation of the virus in nature, as well as the main source of human infection. Inoculation of adult rodents may cause severe acute, often fatal disease and a sharp, conventional immunologic response. Isolation techniques are designed to take advantage of this latter circumstance. By varying such factors as dose, route of inoculation, strain, and age of animal, interme-

FIG. 22-1. Approximate distribution of isolations of arenaviruses in the New World (LCM excluded). (From Arata AA, Gratz NG: The structure of rodent faunas associated with arenaviral infections. Bull WHO 52:621, 1975.)

TABLE 22-1   ARENAVIRUSES

Virus	Date and source of specimen yielding initial isolate			Disease in humans
Lymphocytic choriomeningitis	1934	United States	Human (male)	Meningoencephalitis
Tacaribe	1956	Trinidad	Bats	None
Junin	1958	Argentina	Man	Hemorrhagic fever
Machupo	1963	Bolivia	Man	Hemorrhagic fever
Amapari	1964	Brazil	Rodent	None
Latino	1965	Bolivia	Rodent	None
Parana	1965	Paraguay	Rodent	None
Pichinde	1965	Columbia	Rodent	None
Tamiami	1965	USA	Rodent	None
Lassa	1969	Nigeria	Human (female)	Hemorrhagic fever

diate results may be produced with, for example, chronic immune complex disease being the result of the infection. Transmission by arthropods has been demonstrated experimentally for some of these agents, but such transmission does not seem to be of importance in nature.

The human diseases

When a person is infected, one of a spectrum of consequences ensues. On the one hand, no clinical disease appears, with infection detectable only by a rise in titer of serum antibody; on the other hand, virulent disease with an even chance of death may result. In many instances during the acute stages of infection, virus may be isolated readily from a variety of body fluids or tissues by the use of laboratory rodents or cell cultures. The methodology of arbovirology has served well to elucidate the enzootic features of arenaviruses during the past two decades.

LYMPHOCYTIC CHORIOMENINGITIS

LCM virus was isolated initially in the 1930s from patients with acute aseptic meningitis. From the very beginning it has been recognized that LCM virus causes but a fraction of a percent of the overall number of these cases.

In nature the virus is found in house mice, where, as a result of natural perinatal inocula-

tion, chronic infection with no clinical manifestations is established in a high proportion of animals. Contact with mice or their excreta brings the virus to man, who, from the virus's point of view, is a dead end, as person-to-person or person-to-animal spread does not seem to occur. More recently, captive Syrian hamsters have been found to be a source of human infection. There have been sizable outbreaks in laboratory personnel who have been in contact with infected hamsters that were being used in cancer research. In one instance, the infected hamsters were widely dispersed to many cancer laboratories from a single provider's colony.

Disease begins with fever, headache, myalgia, and malaise. Conjunctival injection and cough may be present. Within a few days the disease may subside, with rapid return of well being. In some, however, it may worsen, with the addition of confusion and somnulence. Stiff neck, vomiting, and marked increase in the severity of the headache may become prominent. Gross disorientation may be a feature. Physical examination reveals fever, meningeal, and encephalitic signs. Children tend to have less severe disease than adults.

The total and differential white blood counts are often normal, as is the urine. Cerebrospinal fluid, especially from those persons showing a meningeal or encephalitic clinical picture, is generally under abnormally high pressure, with an elevated number of lymphocytes and increased protein concentration. Low spinal fluid sugar, hypoglycorrhachia, is often present.

The disease is rarely, if ever, fatal, and it generally subsides after a few days so that the acute febrile stage of the illness is usually over within 7 to 14 days. In older and more severely affected patients, convalescence may be prolonged.

Clinically, there is often little to suggest LCM viral etiology as being any more probable than many of the far more usual causes of viral meningitis such as entero- or myxoviruses. Diagnosis is made by isolation of virus and/or demonstration of high titer, complement-fixing antibodies in serum obtained during the acute disease, or by a rise in titer between samples obtained early in the acute stages and again in convalescence. Adult white mice from colonies known to be free of LCM virus are the medium of choice for isolation. Following intracerebral inoculation, the mice become ill at about 10 days, at which time antigen can be detected in brain tissue by immunofluorescent techniques.

Human deaths are very rare. Consequently, the pathology of LCM infection must be inferred from that observed in experimental animals such as mice and monkeys, where lymphocytic infiltrates of the meninges and choroid plexus are prominent features and are the features from which the name of the virus is derived.

Treatment is supportive; there are no specific measures available.

HEMORRHAGIC FEVERS

Argentinian and Bolivian hemorrhagic fever and Lassa are similar diseases.

Junin Virus in Argentina

Argentinian hemorrhagic fever occurs in certain agricultural areas of northern Argentina. The disease was recognized in the 1940s, and in the 1950s definitive epidemiologic and clinical studies culminated in the identification of Junin virus as the causative agent. The disease occurs predominantly in male field workers, who in the course of their work contact the virus which contaminates the environment from the excreta of chronically infected wild rodents of the genus *Calomys*. During the period from February to May when agriculture, especially of maize, is intense, thousands of cases a year may occur, with a 10 to 20 percent case fatality. Concurrent transmission of LCM and arboviruses may confound the clinical and epidemiologic features of the annual outbreaks.

Machupo Virus in Bolivia

Bolivian hemorrhagic fever was recognized in 1959 as a disease of agricultural workers in specific areas of Bolivia. In 1962 and 1963 the disease moved into the town of San Joaquin. Intensive investigation of this outbreak revealed Machupo virus to be the etiologic agent and the source of man's infection to be the contamination of the household environment with the excreta of chronically infected rodents of the species *Calomys*. These mice, normally found in the fields, invaded the houses of the town and thereby brought the virus to family groups in the general population. The reasons for this ecologic shift are not well understood, but serve to remind us of the complexity of the balances of nature. The efficacy of domestic rodent control was established in the San Joaquin outbreak. In 1971 a small, highly fatal outbreak in an area where there are no *Calomys* was due to person-to-person spread of the infection.

Lassa Fever in West Africa

Lassa fever was first encountered in 1969 in a small outbreak of serious disease among the hospital personnel from missions in the towns of Lassa and Jos in northeastern and central Nigeria (Figures 22-2, 22-3, 22-4). Subsequent to the startling events surrounding this initial outbreak, a more balanced picture of this zoonosis has emerged. Additional nosocomial outbreaks occurred in the subsequent years, and considerable concern spread to Europe and America as the result of the importation of seven cases from West Africa. Fortunately, very little spread, if any, resulted from these importations, and it can be definitively stated now that careful isolation procedures are adequate to protect hospital staff and visitors. Handling of specimens and tissues in the laboratory, however, must be done with extreme care, as fatal laboratory infections have occurred. Presently, maximum security laboratory facilities such as the one available at the Center for Disease Control in Atlanta are the only laboratories where work with this virus should be carried on.

Studies in West Africa, guided by knowledge gained from studies of LCM, of Junin and Machupo viruses resulted in the recognition that the rodent *Mastomys natalensis* carries the agent in nature with the lifelong infection that

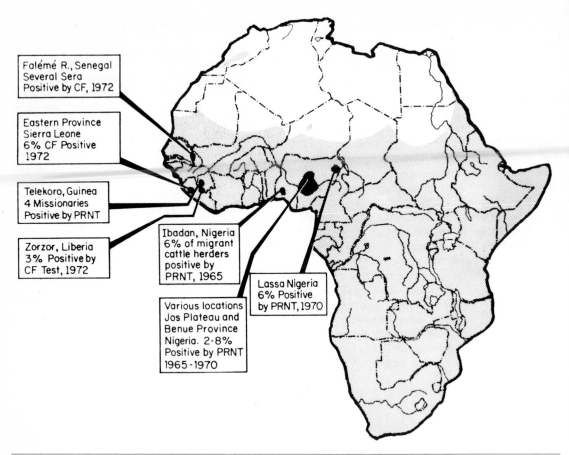

Faléné R., Senegal
Several Sera
Positive by CF, 1972

Eastern Province
Sierra Leone
6% CF Positive
1972

Telekoro, Guinea
4 Missionaries
Positive by PRNT

Zorzor, Liberia
3% Positive by
CF Test, 1972

Ibadan, Nigeria
6% of migrant
cattle herders
positive by
PRNT, 1965

Lassa Nigeria
6% Positive
by PRNT, 1970

Various locations
Jos Plateau and
Benue Province
Nigeria. 2-8%
Positive by PRNT
1965-1970

FIG. 22-2. Summary of serologic surveys for Lassa fever antibodies. Shaded area represents the distribution of the rodent reservoir-host *Mastomys natalensis.* PRNT, plaque reduction neutralization test. (From Monath TP: Lassa fever: Review of epidemiology and epizootiology. Bull WHO 52:587, 1975)

characterizes arenaviruses. Man probably acquires infection from contamination by these rodents of the environment with infectious urine and excreta. Human practices that entail field burning and capture of rodents fleeing therefrom contribute to the intimate man–rodent contact necessary for transmission.

Man provides the only overt indication that virus is present (indicator host) and is of no significance in the perpetuation of this agent in nature. For all three hemorrhagic fever arenaviruses, man-to-man transmission has been documented but is relatively unusual.

The virus, as is so far known, seems to be present in several countries of West Africa, including Liberia, Nigeria, and Sierra Leone. There is a recent isolation of a Lassa-like virus from Mozambique on the other side of the continent.

The clinical consequences of infection with the three viruses range from no disease and simple seroconversion to a fulminant and highly fatal condition. Following an incubation period of one to two weeks, disease begins in a nonspecific manner with fever, headache, sore throat, and myalgia. Toward the end of the first week, flushing, edema, and petechial rash of the face and neck appear in association with adenopathy and worsening of the above symptoms. Mouth lesions, exudative pharyngitis, and pulmonary symptoms may predominate in Lassa fever. During the second week, serious complications relating to loss of intravascular mass through generalized capillary leakage cause a shock syndrome characterized by peripheral vasoconstriction, generalized edema, renal failure, and metabolic acidosis. Myocardial involvement may be evident. There is

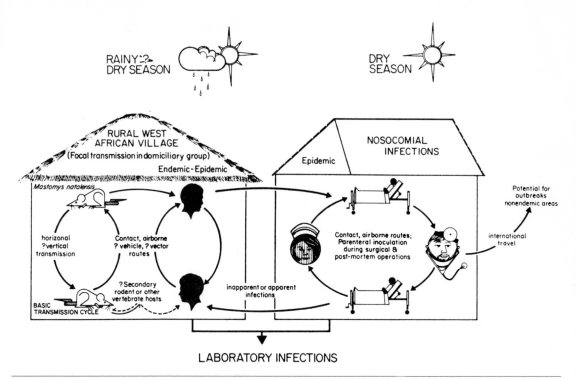

FIG. 22-3. Diagramatic presentation of the epidemiologic cycles of Lassa virus transmission. (From Monath TP: Lassa fever: Review of epidemiology and epizootiology. Bull WHO 52:579, 1975)

moderate hemorrhage in the form of petechiae and bleeding from the gastrointestinal and genitourinary tracts. The bleeding is not sufficient to produce the shock syndrome. After a few days of the shock syndrome, the patient dies or recovery begins. During convalescence there may be loss of hair and neurosequelae, including nystagmus, tremulousness, and episodic dizziness. In spite of considerable damage to the liver seen histopathologically, no jaundice or other signs of liver failure occur. The spleen does not enlarge.

Laboratory findings include leukopenia, decreased platelets, increased prothrombin time, and evidence of disseminated intravascular coagulation. Protein, cells, and casts are found in the urine. The cerebrospinal fluid is normal. A chest x-ray may reveal a patchy pneumonia, and there may be electrocardiographic changes consistent with diffuse myocardial damage. During the shock syndrome, hemoconcentration may be evident.

Pathologic findings include little inflammatory response other than modest monocytic in-

filtrates. There is widespread edema indicative of the capillary damage that permits leakage of plasma from the intravascular space. Cellular damage may be found in many organs, including heart, muscles, lungs, liver, adrenals, kidneys, spleen, lymph nodes, and brain.

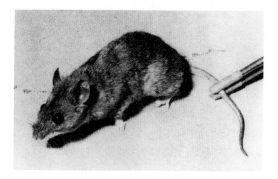

FIG. 22-4. Praomys (Mastomys) natalensis: chronically infected reservoir host Lassa fever virus. [From Isaacson M: The ecology of Praomys (Mastomys) natalensis in Southern Africa. Bull WHO 52:631, 1975]

Diagnosis depends on isolation of virus from either body fluids or from tissue specimens obtained after death. Mice, hamsters, or cell cultures are inoculated, depending on the virus suspected. The presence of virus in the laboratory system is detectable by a variety of means, including disease, cytopathology, and immunofluorescent techniques. Antibody response in man occurs within a few weeks of onset and is generally detected by testing for a rising titer of group-specific complement-fixing antibodies or type-specific neutralizing antibodies. Immunofluorescent techniques are also available for detection of antibody and in some situations may provide the earliest detection of antibody response.

The general theme of clinical management of arenavirus disease calls for efforts to support basic physiologic processes long enough for the body to cure itself. There are no specific therapeutic agents except humoral antibodies derived from persons who have survived natural infection. Such immunotherapy has seemed to improve dramatically a few patients who appeared to be in imminent danger of dying. No vaccine is currently available, although experimental vaccines have been prepared and tested.

Prevention of person-to-person transmission can be achieved in any hospital by standard strict isolation techniques. Outbreaks can be controlled with measures aimed at breaking the intimate contact between the excreta of certain rodents and man. A protocol for the management and containment of imported cases of Lassa fever has been used and is available through the Center for Disease Control of the U. S. Public Health Service.

FURTHER READING

Arenaviruses in perspective. Br Med J 1:529, 1978

Berge TO: International Catalogue of Arboviruses, 2nd ed., Dept. of Health, Education and Welfare, Washington, DC, US Government Printing Office, 1975. See also Karabatsos N (ed): Supplement to International Catalogue of Arboviruses Including Certain Other Viruses of Vertebrates. Am J Trop Med Hyg 27:(Suppl): 372, 1978

Buckley SM, Casals J: Pathobiology of Lassa fever. Intl Rev Exp Pathol 18:97, 1978

Cassals J: Arenaviruses. In Evans AS (ed): Viral Infections of Humans — Epidemiology and Control. New York, Plenum, 1976, p 103

deBracco MME, Rinoldi MT, Cossio PM, et al: Argentine hemorrhagic fever — Alterations of the complement system and anti-Junin-virus humoral response. N Engl J Med 299:216, 1978

Fuller JG: Fever! The hunt for a new killer virus. New York, Readers Digest Press, 1974

Hirsch MS, Moellering RC Jr, Pope HG, et al: Lymphocytic chorimeningitis virus infection traced to a pet hamster. N Engl J Med 291:610, 1974

International Symposium on Arenaviral Infections of Public Health Importance 14–16, July 1975, Atlanta, Georgia. Bull WHO 52:381, 1975

Johnson KM: Hemorrhagic diseases caused by arenaviruses. In Beeson, McDermott (eds): Textbook of Medicine, 15th ed, Philadelphia, Saunders, 1979, p 276

Rowe WP, Murphy FA, Bergold GH, et al: Arenoviruses: Proposed name for a newly defined virus group. J Virol 5:651, 1970

Weissenbacker MC, Grela ME, Sabattani MS, et al: Inapparent infection with Junin virus among laboratory workers. J. Infect Dis 137:309, 1978

Zweightaft RM, Fraser DW, Hattwick MAW, et al: Lassa fever: Response to an imported case. N Engl J Med 297:803, 1977

Hepatitis

Classically two forms of hepatitis have been distinguished: one which may be orally or parenterally transmitted and is characterized by short incubation, while the other appears after a much longer incubation period and was thought in the past to require parenteral transmission of virus. Short-incubation hepatitis has also been termed "epidemic jaundice," "catarrhal jaundice," "infectious hepatitis," "virus A," or "MS-1 hepatitis." The long-incubation disease has at various times been called "homologous serum jaundice," "serum hepatitis," "virus B," or "MS-2 hepatitis." Although the present chapter deals exclusively with these classic forms of hepatitis, it is important to remember that the clinical manifestations of liver inflammation comprise a syndrome and that nearly identical symptoms and signs may result from infections caused by *Leptospira*, treponemes, protozoa, bacteria, and a variety of viruses.

The most evident manifestation of hepatic dysfunction in hepatitis is the yellow discoloration of skin and sclerae, which reflects the presence of elevated levels of circulating bilirubin and is termed "jaundice" (from the French *jaunice*) or icterus (Greek derivation). For thousands of years, observers noted that occasional clusters of jaundice were seen, suggesting the presence of a common factor—dietary, atmospheric, or a transmissible agent. During military campaigns, these epidemics of jaundice were described in association with the crowding and poor sanitation prevalent in wartime. Although the more severe illnesses may have reflected the presence of yellow fever, leptospirosis, or malaria, it is certain that some of these cases were manifestations of virus hepatitis.

Sporadic cases of hepatitis were recognized early in the twentieth century, and a viral etiology had been postulated. However, attempts to transmit recognizable infection to a variety of animals were unsuccessful. Even after it became apparent that hepatitis was a transmissible disease, the existence of two epidemiologically distinct entities was not initially appreciated.

As early as 1885 an observant German doctor reported an outbreak of what was apparently long-incubation hepatitis among several hundred factory workers in Bremen. Although this physician remained uncertain about the etiology of the disease, he traced it epidemiologically to the administration two to six months previously of smallpox vaccine, which at that time was prepared with human lymph and serum. In the 1930s, yellow fever vaccine virus, grown in eggs, was stabilized with pooled human serum. Subsequently, clusters of hepatitis were recognized in vaccinees, although the problem was not adequately recognized until later, when mass yellow fever immunization of military personnel was followed by thousands of cases of hepatitis.

The importance of this debilitating, transmissible disease to the military led to a major investigative effort in the United States during World War II. Hepatitis was transmitted experimentally by filtrates to man. Much of our information concerning these diseases was accumulated during volunteer studies performed on military personnel and prisoners. These studies suggested that two distinct infectious agents existed: one (virus A) responsible for a short-incubation illness, and another (virus B) responsible for a disease with a long incubation period. These early investigations defined the stability and resistance of the viruses to chemicals and temperature and, very significantly, demonstrated that many inoculated individuals exhibited chemical evidence of hepatic dysfunction although they remained in apparent good health and did not become jaundiced (anicteric hepatitis).

CHARACTERIZATION OF THE AGENTS OF VIRUS A AND B HEPATITIS

In 1965 Blumberg and associates were seeking to define different allotypes of beta-lipoproteins in connection with genetic studies. They employed an Ouchterlony agar gel double-diffusion technique and used the serum of a multiply transfused individual (presumably carrying antibodies to a variety of circulating antigens) to screen various test serums. An antigen was recognized in the serum of an Australian Aborigine (called Australia antigen or Au) which appeared rarely in serum from North Americans (0.1 percent), though more commonly in residents of the tropics and frequently in association with cases of hepatitis. Although the antigen was most consistently related to hepatitis, it was also found frequently in association with Down's syndrome, lepromatous leprosy, and certain forms of leukemia and chronic renal disease. It was uncertain whether this association related to the increased possibilities for transmission of hepatitis to these patients, to their genetic susceptibility to the infection, or to a combination of these and other factors.

Electron-microscopic studies performed on the antigen which was aggregated by antibody revealed many spherical and filamentous particles with a mean diameter of 20 nm and some large spherical double-shelled particles, described first by Dane, with a mean diameter of 42 nm. On the basis of human inoculation studies, it was then definitely ascertained that the antigen was associated exclusively with the long-incubation hepatitis (B). The 20-nm particles were identified as noninfectious surface components of the 42-nm Dane particles, which probably represent the complete virions of virus B hepatitis (HBV). Cores of the virus, which are present in infected liver, contain double-stranded DNA, a DNA polymerase, and core antigens termed HB_c; the surface antigens (responsible for the original immunodiffusion reactions) are designated HB_sAg. The presence of HB_sAg in clinical specimens coincides reliably with the presence of infectious HBV.

The surface antigens of hepatitis B have been further characterized and subtypes have been identified. All hepatitis B surface antigens manifest a group-specific determinant, a, as well as specific subtypes related to mutually exclusive alleles (for example d or y; w or r). The prominence and distribution of these subtypes have been found to vary in different countries and social groups. The clinical significance, if any, of these variants is as yet uncertain. The e antigen described by Magnius and Espmark in 1972 is probably closely associated with the intact HB virion and, using new radioimmunoassays, can be detected in most cases during acute hepatitis B illness and, in some cases, in carriers. Beyond the acute period, the presence of e antigen coexisting with HB_sAg seems to correlate directly with the morbidity of the illness, as well as with the infectivity. The absence of the e antigen (and the presence of anti-e antibody) is associated with reduced (but not necessarily absent) infectivity. A core antigen (HB_cAg) has also been identified, which, as the name implies, is also closely associated with the intact hepatitis B virion.

Anti-HB_s appears and persists for days or even weeks in most cases as the clinical symptoms resolve. Antibody to core antigen (anti-HB_c) is detectable just before the onset of the icteric phase of the disease, subsides gradually after the disease resolves, and may persist at a low level for years or indefinitely. In carriers, anti-HB_c usually persists at elevated levels, while anti-HB_s is not detectable. Current hepatitis B nomenclature is summarized in Table 23-1.

Various techniques have been developed to test for HB_s. These include complement fixation, indirect hemagglutination, high-voltage crossover immunoelectrophoresis (CIE), and radioimmunoassay (RIA). HB_s RIA, which requires 24 hours, is more sensitive than the more rapid (2 to 4 hours) CIE technique, but its use is limited to those institutions that possess the more elaborate and expensive radiation-detection equipment. An enzyme-linked immunosorbent assay (ELISA) comparable in its sensitivity to the RIA for HB_sAg has been developed. This test obviates the need for isotopes and a radioactive counter but requires about an equal number of manipulations and time. Almost the same degree of sensitivity reached with the RIA test may be achieved with the more rapid and much simpler reversed passive hemagglutination test, which employs a suspension of turkey erythrocytes coated with purified antibody to HB_sAg. Table 23-2 compares most currently available methods for the detection of HB_sAg. Modifications of these immunologic procedures (Table 23-3) have been developed for the assessment of anti-HB_s. Sensitive detection procedures (including RIA) have been developed for HBV core and e antigen–antibody systems, though at this writing these procedures are not routinely available.

Long-incubation hepatitis was thought for some time to be transmissible only by the parenteral route. Using the HB marker, it was demonstrated that the virus could be transmitted by the oral route and that it probably exists in infectious form in feces as well as in blood. Antigen also occasionally appears in saliva, and sexual transmission has been implied by some epidemiologic observations. The dosage required for induction of infection following ingestion of virus may be 50 times as high as that necessary for parenteral transmission.

Certain individuals become permanent carriers following infection with long-incubation hepatitis, and recognition of these persons has

TABLE 23-1 CURRENT HEPATITIS NOMENCLATURE

Hepatitis B (long-incubation hepatitis, serum hepatitis).

HB_sAg	Hepatitis B surface antigen (pleomorphic spheres and filaments 16–27-nm diameter)
HB_cAg	Hepatitis B core antigen
HBV	Hepatitis B virus
Dane particle	Large (42-nm) spherical particles with 27-nm core, probably the intact virion
$HB_sAg/a \ \dfrac{d}{y} \ \dfrac{w}{r}$	Hepatitis B surface antigen subtypes ($\dfrac{d}{y}$ and $\dfrac{w}{r}$ are mutually exclusive determinants)
anti-HB_s	Antibody to hepatitis B surface antigen
anti-HB_c	Antibody to hepatitis B core antigen
e antigen	
anti-e	
HBAg-associated DNA polymerase	

been sought to avoid their use as blood donors. The presence of the HB antigen provides a convenient marker. The antigen is usually detected two weeks to two months prior to the onset of jaundice and persists for a mean duration of one to two months. If it remains demonstrable four months after the onset of acute disease, it is likely to persist indefinitely. In rare cases, HB_sAg disappears after an even more prolonged persistence (1 to 6 years). Apparently a chronic carrier state is induced more frequently following mild cases of virus B hepatitis. Thus the persistent carrier is unlikely to be aware of his own previous illness. Mosley estimates that 5 to 10 percent of infected persons become carriers.

With the development of sensitive detection methods for hepatitis B, persistent antigenemia was found associated with several conditions, including some cases of chronic hepatitis and some of glomerulonephritis. Persistent antigenemia occurs in a variety of individuals with depressed or altered immunologic reactivity, including some patients with chronic renal disease and others with certain malignancies. The early recognition by Blumberg of antigenemia in a proportion of individuals with Down's syndrome, certain forms of leukemia, and patients with lepromatous leprosy probably reflects the varying degrees of immunologic compromise associated with these conditions. The HBV carrier state is found commonly among immunosuppressed transplant recipients, though HBV infection and the carrier state does not interfere with graft survival. Interestingly, it has been shown that the response of renal allograft recipients to HBV infection pretransplant and the sex of the kidney donor

TABLE 23-2 METHODS OF DETECTION OF HB_SAg

Relative Sensitivity	Methods	Ease of Performance	Time Required for Completion (hours)
1	Immunodiffusion (ID, AGD)	Simple	24–72
2–10	Counterimmunoelectrophoresis (CIE)	Simple	2
	Complement fixation (CF)	Moderate	18–24
	Reversed passive latex agglutination	Simple	0.1–0.2
	Passive hemagglutination inhibition (HAI)	Moderate	2
	Immune adherence hemagglutination	Moderate	2
100	Radioimmunoassay (RIA)		
	solid phase	Moderate	4–24
	double antibody (DA-RIA, RIP)	Complex	24–72
	Reversed passive hemagglutination (RPHA)	Simple	1–4

(Adapted from Howard CR, Burrell CJ: Structure and nature of hepatitis B antigen. Prog Med Virol 22:36, 1976)

TABLE 23-3   METHODS FOR DETECTING ANTI-HB$_S$

Technique	Relative Sensitivity
Immunodiffusion	1
Counterimmunoelectrophoresis	1–4
Complement fixation	2–10
Passive hemagglutination	1000–10,000
Solid-phase radioimmunoassay	1000–10,000
Radioimmunoprecipitation	10,000–100,000

(Adapted from Howard CR, Burrell CJ: Structure and nature of hepatitis B antigen. Prog Med Virol 22:36, 1976)

affect graft survival. Grafts from male donors lasted significantly longer both in uninfected recipients and in those who were chronic carriers of AB$_s$Ag. Kidneys from male donors transplanted into HBV-immune recipients had short survival. It has also been reported that the response of parents to hepatitis B is related to the sex ratio of offspring, and a cross-reactivity between HB$_s$Ag and a male-associated antigen has been suggested.

Hepatitis B has proven to be particularly troublesome in renal dialysis units. Dialysis patients are at high risk of acquisition of hepatitis B infection (frequently subclinical), and a persistent carrier state is frequently established. Staff workers are thus heavily exposed and are also at high risk.

Hepatitis B can be transmitted congenitally or during the perinatal period. If acute hepatitis B occurs during gestation, the virus can be transmitted transplacentally. The risk of this transmission increases as gestation progresses. In contrast, maternal HBV carriers rarely transmit the virus prenatally, although a significant proportion of the babies of long-term carriers acquire hepatitis B infection during the first year of life. Many babies who acquire HBV infection in the perinatal period become chronic carriers, and it is presently presumed that perinatal infection with HBV is responsible for the existence of many (possibly most) asymptomatic carriers. A variable proportion of these infants manifest some degree of hepatic dysfunction. The high prevalence of HBV carriers in some parts of the world may be perpetuated by these circumstances, There is, in addition, some evidence that genetic factors are relevant to the establishment of prolonged or permanent HBV carriage.

The mortality of long-incubation hepatitis varies from 0 to 20 percent and increases with age. Approximately 6 to 8 per 1000 transfused develop this illness. Plasma, serum, fibrinogen, and thrombin may transmit the infection. Washed packed red blood cells (and frozen glycerinized blood) seem less likely to be contaminated with the virus.

Long-incubation hepatitis appears to be a sporadic disease, reflecting poor transmissibility under ordinary circumstances. This form of hepatitis is probably more frequent among adults. In contrast, short-incubation hepatitis is more often a disease of children. An attack of either disease apparently conveys permanent immunity. Since hepatitis A and B viruses are quite unrelated, no cross-immunity is conferred.

In 1973 Feinstone et al visualized 27-nm particles by immune electron microscopy in the stools of patients with acute virus A hepatitis. Using similar methods they were able to detect specific antibody to these particles, indicating that they were very likely the etiologic agents of virus A hepatitis. Morphologically they resemble picornaviruses or parvoviruses, and they probably contain RNA. Hepatitis A virus can be transmitted to marmosets, and virus-specific complement-fixation and immune-adherence serologic tests have been developed. The virus of hepatitis A has recently been propagated in vitro in primary explant cultures of marmoset liver, as well as in fetal rhesus kidney cells. This development, if confirmed, will greatly assist the further characterization of the virus of hepatitis A.

There remains a significant (perhaps greater than 50 percent) cluster of hepatitis cases that are for the moment best termed "non A-non B." It remains to be determined whether these cases will eventually prove to be related etiologically to additional viruses. At present, some non A-non B associated antigens are being evaluated, which may lead to the designation and study of additional specific etiologic agents of viral hepatitis.

CLINICAL COURSE

Table 23-4 summarizes many distinguishing features (clinical, virologic, and epidemiologic) of virus A and virus B hepatitis. Patients with short-incubation hepatitis remain asymptomatic after exposure for 15 to 50 days (Fig. 23-1). The onset of symptoms is abrupt, ushered in

TABLE 23-4   LONG-INCUBATION AND SHORT-INCUBATION HEPATITIS

Property	Short Incubation	Long Incubation
Etiologic agent	Virus A	Virus B
Size of virus	Approximately 27 nm	Dane particle, approximately 42 nm; HB_s, 20-nm spheres and filaments
Resistance to heat	Survives 56 C for 30 minutes	Survives 60 C for 4 hours
Route of infection	Oral and parenteral	Parenteral; oral and sexual transmission probably also occur
Virus in blood	Late incubation and early acute illness	Incubation period and acute phase (may persist)
Virus in stools	Incubation period and acute phase	Probably present
Virus in urine	Acute phase	Uncertain, may be present
Duration of carrier state:		
blood	Not beyond acute illness	Protracted (indefinite) carrier state may occur
stools	Several weeks or months	Present, duration uncertain
Incubation period	15–50 days	45–160 days
Type of onset	Sudden	Insidious
Seasonal incidence	Autumn and winter	All year
Fever	Common during pre-jaundice prodome	Less common
Age group	Children and young adults	All ages
Jaundice	More common in adults	More common in adults
SGOT elevation	Transient	More prolonged
Thymol turbidity	Usually elevated	Usually normal
IGM	Usually elevated	Usually normal
HB_sAg	Not present	Present during incubation and acute phase, may persist
ISG prophylaxis	Excellent	Fair to poor
High-titered globulin prophylaxis (HBIG)	Not applicable	May protect when given promptly

with anorexia, fever, nausea, vomiting, lassitude, and occasionally right upper quadrant abdominal pain. Smokers often lose their taste for cigarettes. Symptoms persist for several days to a week, after which jaundice, if any, becomes manifest. During the preicteric phase, rashes may occur transiently. With the appearance of jaundice, the fever and associated symptoms usually subside. Icterus is accompanied by the appearance of dark urine and light-colored or white stools. During the acute phase of hepatitis, functional intrahepatic obstruction

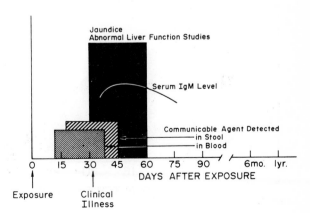

FIG. 23-1. The course of hepatitis A virus infection.

of biliary flow may be as complete as with mechanical blockage of the common bile duct. With the appearance of jaundice most patients begin to recover. The feces and blood of these patients are highly infectious during late incubation and throughout the period of acute illness. During convalescence, which may be quite prolonged, patients often remain weak, anorexic, easily fatigued, and depressed (posthepatitis asthenia).

The usual course of hepatitis is benign, and recovery is generally complete. However, in some cases the disease is fulminant and fatal. The illness is liable to be more severe during pregnancy and in postmenopausal females. Persistence of symptoms and relapses over a two-month period are usually designated subacute and, at an indefinite period thereafter, chronic hepatitis.

Virus B hepatitis, previously called "serum hepatitis" when it was thought to be transmitted only with blood products, is characterized by a long incubation period (45 to 160 days). Figure 23-2 summarizes the relationship of the exposure to HBV, the incubation period, appearance and persistence of various associated antigens and antibodies and clinical disease. It should be noted that tests for HB_sAg and anti-HB_s may both be negative late in the course of the illness. The appearance of HB_sAg and the occurrrence of an immune response to the surface antigen is recognized in about 70 percent of cases. In 25 percent the presence of HBV infection may be recognized only by an anti-HB_s response. This may reflect the sensitivity or the timing of testing or both. The remaining 5 percent of HBV infections result in the estab-

lishment of a carrier state. A secondary immune response can be detected following exposure to HBV of an antibody-positive individual. The onset of clinical symptoms in virus B hepatitis is usually insidious, and patients are often afebrile. Frequently, the initial evidence of disease is the appearance of jaundice. Although the onset is subtle, the overall course of this form of hepatitis is usually more severe than that of short-incubation disease. A significant mortality rate accompanies this infection, occasionally rising to as high as 10 to 20 percent.

The long incubation and the late appearance of symptoms associated with hepatitis B infections, the occurrence of articular and cutaneous prodromal manifestations, and the association of this infection with the presence of immune complexes, vasculitis, cryoglobulinemia, glomerulonephritis, and polyarteritis suggests that some or all of the manifestations of hepatitis reflect the operation of immunopathologic mechanisms. A large body of evidence suggests that hepatitis B infection plays a direct or indirect role in the genesis of hepatocellular carcinoma. In all of these conditions and illnesses, genetic factors are probably critical cofactors determining and interacting with immunopathogenic mechanisms.

The pathologic findings on gross and routine microscopic examination in virus A and virus B hepatitis are indistinguishable. Both illnesses are characterized by the development of inflammatory infiltrates in the liver parenchyma, variable degrees of distorted architecture, and bile stasis. Fatal fulminant hepatitis is accompanied by severe and widespread necrosis of pa-

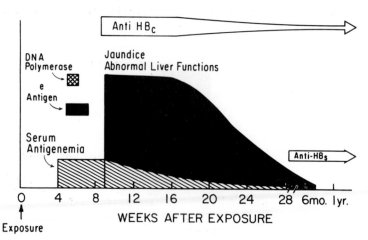

FIG. 23-2. The course of a case of hepatitis B infection.

renchymal cells. The extrahepatic pathology is nonspecific, being manifest primarily as a mild lymphoid hyperplasia, accompanied, in some cases, by a moderate nephritis.

During the incubation period and prodromal phase, patients may exhibit a leukopenia with the associated presence of some atypical lymphocytes. Liver function tests (particularly serum transaminase) become abnormal during late incubation. These chemical changes may persist for a prolonged period following long-incubation hepatitis.

EPIDEMIOLOGY

Short-incubation hepatitis is primarily a disease of children, with maximum incidence between the ages of 5 and 14. Infection is widespread, transmitted primarily by the fecal–oral route, and apparently conveys lasting protection to the host against reinfection by the same agent. The effectiveness of pooled gamma globulin in ameliorating or preventing this infection implies that most adults have had prior infection with this virus.

It is evident that since few people are aware of having had hepatitis, many infections must be anicteric and unrecognized. This supposition has been borne out by volunteer and careful epidemiologic studies in which the ratio of anicteric to icteric cases has been estimated to be as high as 10 to 1 or greater. As with many other childhood diseases, morbidity increases with the occurrrence of disease in older individuals. Army studies have indicated that morbidity associated with hepatitis may be twice as high at age 40 as it is in individuals of 20. It is paradoxical that improved sanitation and the resultant postponement of contact with hepatitis may be associated with the appearance of increased overall morbidity.

Short-incubation hepatitis spreads rapidly from person to person. Thus there is a high risk of infection among secondary contacts in households and within closed communities. Epidemics are frequent in homes for the retarded, army units, and childrens' homes. Common-source explosive outbreaks are usually traced to contaminated water supplies, milk, or food. Shellfish taken from sewage-contaminated estuaries may carry and transmit infection. The incidence of short-incubation hepatitis varies inversely with population density, perhaps reflecting the existence of more crude sanitary facilities in rural communities. In general, there is a seasonal pattern to the occurrence of this disease, with a rise demonstrable in late summer, autumn, and early winter. Occasionally huge outbreaks occur. Thirty thousand cases of hepatitis were recognized in New Delhi during a six-week period in 1955 following floods which disrupted water supplies.

DIAGNOSIS

Table 23-4 lists some distinguishing features of long-incubation and short-incubation hepatitis. Clinical differentiation of these conditions may be difficult.

In both illnesses, liver function may become abnormal during the preicteric phase. The most sensitive measures of hepatic dysfunction are provided by the levels of serum glutamic-oxaloacetic and pyruvic transaminase (SGOT and SGPT). If jaundice develops, bile appears in the urine, and the stools become acholic. There may be a leukopenia, and a few atypical lymphocytes usually appear among the peripheral circulating white blood cells.

The diagnosis of hepatitis is supported by epidemiologic data. However, it is well to remember that aside from the specific tests described for hepatitis A and B, no single feature of these diseases absolutely distinguishes them from carcinoma (primary or metastatic) affecting the liver, other infectious entities, or even obstructive jaundice due to a calculus or tumor.

TREATMENT

The administration of high-titered leukocyte interferon has been shown to eliminate the HBV carrier state in some patients with chronic hepatitis. There is some preliminary evidence that a combination of interferon and adenine arabinoside may be effective in the therapy of some patients who are resistant to interferon alone. Other than in these experimental studies, the treatment of uncomplicated hepatitis is nonspecific. In the past much attention has been paid to the control of activity and diet. It now appears quite certain that the activity of patients need not be restricted. The persistence of residual hepatic dysfunction and the development of chronic hepatitis are apparently unaffected by early ambulation. Activity should be regulated simply by the patient's state of well-being. If nausea and anorexia are prominent, intravenous fluids are indicated. Subsequently, diet may be managed without specific restrictions. However, if symptoms suggestive of hepatic coma supervene, control of diet becomes much more important. In this instance the level of blood ammonia must be controlled, and thus protein intake should be restricted and intestinal antibiosis instituted to reduce the bacterial breakdown of nitrogenous products. If the prothrombin time is prolonged, vitamin K may be administered. Exchange transfusion or hemodialysis may be employed in order to reduce the blood ammonia and other toxic products. Although of symptomatic importance, these measures probably will not alter the final outcome of the specific infection.

Steroids have been used in the management of hepatitis, but although the use of these agents may be associated with a more rapid fall in temperature, as well as of serum transaminase and bilirubin values, the ultimate time of recovery and the risk of development of chronic hepatitis are not significantly influenced. Since a variety of steroid-associated complications may occur, there does not appear to be a place for this therapy in acute hepatitis.

PREVENTION

It was anticipated that with the characterization of the hepatitis viruses, a major effort would be made to develop a vaccine. It has been shown that inoculation of volunteers with heat-killed short-incubation hepatitis virus is followed by the development of a protective immune response.

Pooled gamma globulin protects against short-incubation hepatitis effectively when administered prior to or shortly after exposure in a dosage of 0.02 to 0.04 ml per kg of body weight. In the face of intensive and chronic exposure, 0.06 ml per kg may be given. If exposure will be chronic (in an endemic area), this dose may be repeated after six months. Thereafter the development of passive-active (permanent) immunity is assumed. It is noteworthy that gamma globulin inoculation may reduce the clinical manifestations of short-incubation hepatitis without actually preventing infection.

Unaware of the presence of an active process, patients who have received gamma globulin may still be infectious and are capable of spreading the virus.

The prevention of serum hepatitis is more complex. All blood to be used in clinical medicine should be screened for the presence of HB_s. Donors with a past history of hepatitis should be excluded. Blood and pooled plasma should be used sparingly. There is virtually never an indication to administer a single-unit transfusion. Whenever possible, washed packed red blood cells should be employed. Needles, dental equipment, and syringes should be carefully sterilized or disposable.

Large doses of pooled gamma globulin given before or simultaneously with blood transfusions probably reduce the incidence of virus B hepatitis. There is evidence that hyperimmune anti-HB_s globulin (HBIG) administered to exposed individuals reduces the severity or even the likelihood of HBV infection without increasing the risk of the establishment of the carrier state. Since the effectiveness of HBIG (limited in supply and expensive) is still being evaluated, it is essential to restrict its use to cases in which definite parenteral or mucous membrane exposure of a susceptible individual to HBV has occurred. This type of exposure to HB_sAg-positive material must therefore be followed immediately by tests of the exposed individual to determine his HB_sAg and anti-HB_s status rapidly enough to permit administration of HBIG as soon as possible. In some cases the appearance of hepatitis B has only been postponed (as long as six months) by the administration of HBIG. Whether the clinical presentation has been ameliorated in these delayed cases is not known. In case of what is judged to be a very significant exposure, some individuals recommend the readministration of HBIG one month later. Recent studies by Krugman and colleagues have shown that by using sensitive techniques HB_sAg may be detectable in as short a time as six days after infection. With this background Krugman suggests that HBIG be administered within 48 hours of the exposure. The use of hyperimmune globulin in neonates born to HBV carrier mothers is presently recommended to reduce the likelihood of early infection with the associated high risk of the establishment of the carrier state.

The management of HBV carriers presents very significant medical and moral dilemmas. It has been suggested by some that medical personnel and food handlers (for example) be tested and, if HB$_s$ positive, change their employment or take other drastic measures to reduce the likelihood of HBV transmission. However, transmission from health worker HBV carriers has only rarely been reported, and there is no necessity to change employment of patient care personnel who are carriers. Simple measures of personal hygiene (equally applicable to all personnel) are advised. In the United States at present, there are no specific recommendations for routine testing other than of blood donors, workers, and patients in dialysis units, other patients considered at high risk, or those under diagnostic study.

Krugman and coworkers inoculated volunteers with heat-inactivated infectious serum, which they found induced an immune response to HBV and afforded protection to recipients subsequently challenged with infective material. Although HBV has not been replicated in vitro, large amounts of virus and viral antigens have been prepared from carrier plasma. Purified DNA-free subunit preparations of HB$_s$Ag derived from this material have been shown to be immunogenic and protective in chimpanzees and safe when inoculated into volunteers. Trials of this unique vaccine are currently underway in several high-risk groups (homosexual males and dialysis personnel).

FURTHER READING

Books And Reviews

Blumberg BS: Australia antigen and the biology of hepatitis B. Science 197:17, 1977

MacCallum FO (ed): Viral hepatitis (a collection of reviews). Br Med Bull 28:103, 1972

Symposium on Viral Hepatitis, Washington, D.C., National Academy of Sciences, March 17-19, 1975. Am J Med Sci 270:2, 1975

Selected Papers

Alter HJ, Chalmers TC, Freeman BM, et al: Health-care workers positive for hepatitis B surface antigen: Are their contacts at risk? N Engl J Med 292:454, 1975

Bunyak EB, Roehm RR, Tytell AA, et al: Development and chimpanzee testing of a vaccine against human hepatitis. Proc Soc Exp Biol Med 151:694, 1976

Diagnostic methods in viral hepatitis—Proceedings of a symposium. J Med Virol 3:1, 1978

Drew JS, London WT, Lustbader ED, et al: Hepatitis B virus and sex ratio of offspring. Science 201:687, 1978

Feinstone SM, Kapikian AZ, Purcell RH: Hepatitis A: Detection by immune electron microscopy of a viruslike antigen associated with acute illness. Science 182:1026, 1973

Feinstone SM, Purcell RH: Non-A, non-B hepatitis. Ann Rev Med 9:359, 1978

Greenberg HB, Pollard RB, Lutwick LI, et al: Effect of human leukocyte interferon on hepatitis B virus infection in patients with chronic active hepatitis. N Engl J Med 295:517, 1976

Isenberg JN: The infant and hepatitis B virus infection. Adv Pediatr 24:455, 1977

Krugman S, Overby LR, Mushahwar IK, et al: Natural history and prevention of hepatitis B reexamined. N Engl Med 300:101, 1979

Mosley JW: HBV carrier—A new kind of leper? N Engl J Med 292:477, 1975

Prince AM: Use of hepatitis B immune globulin: Reassessment needed. N Engl J Med 299:198, 1978

Provost PJ, Hilleman MR: Propagation of human hepatitis A virus in cell culture in vitro. Proc Soc Exp Biol Med 160:213, 1979

Szmuness W: Hepatocellular carcinoma and the hepatitis B virus: Evidence for a causal association. Prog Med Virol 24:40, 1978

CHAPTER 24

Miscellaneous Virus Infections

Rhinoviruses

Rhinoviruses (nose viruses) are important in causing acute respiratory infections with predominant involvement of the upper airway passages. Over 100 different serotypes have been recognized, the first one being recovered in 1956 and initially classified as ECHO virus type 28.

Clinical Illness

The usual symptoms of a rhinovirus common cold are nasal obstruction and discharge, sneezing, scratchy throat, mild cough, and malaise. Severe tracheobronchitis and even atypical pneumonia may occur occasionally in adults, but fever and significant lower respiratory tract involvement are more likely with rhinovirus infection in children. Rhinovirus infection has been associated in a few studies with acute exacerbations of chronic obstructive pulmonary disease in adults.

Epidemiology

Rhinovirus infections have been documented in all populations studied. They occur throughout the year with a tendency toward increased incidence in the fall. Occasional epidemics with a single serotype have been described, but more often multiple serotypes appear to be circulating at the same time. The viruses spread among close associates by uncertain means. Aerosol spread has been shown to be possible. Children frequently serve to introduce rhinoviruses into the family unit, with subsequent illnesses occurring within two or three days to a week.

Diagnosis

Acute upper respiratory infections can be caused and/or simulated by multiple viruses, bacteria, allergies, and so on. To distinguish rhinovirus infection from other etiologic factors requires virus isolation and demonstration of a rise in antibody titer between acute and convalescent sera. Assays for antibody must be done with neutralization tests against the specific rhinovirus causing the infection. Specific serologic identification requires multiple cross-neutralization tests and is impractical for general use.

Rhinoviruses are generally recovered by inoculating specimens from the nose and/or throat into human embryonic fibroblast or human embryonic kidney tissue cell cultures. They are almost never recovered from stool or rectal swabs. An isolate can be identified as a rhinovirus by its physicochemical properties, especially acid lability, which serve to distinguish rhinoviruses from enteroviruses. Rhinoviruses can be further subdivided into H strains that grow well only in tissues of human origin, and M strains that will also grow in monkey kidney tissue.

Treatment

Treatment of rhinovirus infection is aimed at prevention of secondary bacterial infections and relief of symptoms. Hydration and preventing obstruction of airways, paranasal sinuses, and eustachian tubes are the mainstays of therapy.

Prevention

Resistance to reinfection with the same rhinovirus serotype can be shown following an initial infection. An important element of this resistance appears to be secretory antibody, which is not induced well by parenteral killed-virus vaccines. For this reason, as well as because of the existence of a multiplicity of rhinovirus serotypes, vaccination appears unlikely to be a successful method of prophylaxis against the rhinovirus common cold.

Nonspecific resistance to reinfection with virus lasting one to two months appears to follow acute respiratory virus infection. The mechanism responsible for this resistance is uncertain, but attempts to simulate it using interferon inducers or other chemotherapeutic agents are underway. High doses of vitamin C have been advocated as a prophylactic measure, but the efficacy is unproven.

Coronaviruses

Coronaviruses are RNA-containing viruses with widely spaced 20-nm club-shaped surface projections. Tyrrell and Bynoe first described the isolation of one of these viruses in 1965 (B814) from a patient with a cold. This agent grew only in human tracheal organ culture and was later characterized by electron microscopy as being morphologically similar to avian infectious bronchitis (AIB) virus. The prototype strain of coronavirus (229E) was first isolated in 1962 and reported in 1966 by Hamre. This initial isolate was obtained in human kidney cell culture. Vacuolization of the cytoplasm of infected cells was noted in stained preparations after a second blind passage of the inoculated cultures. Subsequent isolations of these viruses were made in human embryonic tracheal organ cultures. The isolations may require multiple passages of the agent before the definitive observation, immobilization of cilia, is made.

Coronaviruses have been etiologically associated with upper respiratory tract disease in adults and lower respiratory tract disease in hospitalized children. Five to 10 percent of such illness appears to have coronavirus associated with it. Diagnosis can be made by demonstrating a rise in antibody titer (complement fixation, neutralization). This information has accrued as a result of serologic surveys, attempted virus isolations, and challenge studies in volunteers. These viruses will produce colds in adult volunteers. In addition, when very careful studies are done with hospitalized asthmatic children, it is found that coronaviruses may be responsible for an exacerbation of the respiratory symptomatology.

The epidemiology of these infections is incompletely characterized, but it is known that coronavirus infections occur in small outbreaks which tend to take place during late winter and early spring. Sporadic outbreaks at other times of the year can also occur. Periods of high incidence with one specific serotype have been reported to occur in cycles of two to three years. Reinfection with coronaviruses is a frequent event as shown by the presence of infection in persons having pre-existing neutralizing antibody.

In recent years particles with the morphol-ogy of the coronaviruses have been visualized by electron microscopy in the stool specimens of patients with symptomatic gastroenteritis. One of these coronaviruses has been propagated in primary embryonic kidney monolayers and human embryo intestinal organ cultures. Additional studies are necessary to determine their etiologic association with the observed gastroenteritis. It is presently accepted that coronaviruses cause gastroenteritis in pigs and cows.

Work with coronaviruses isolated from man is difficult because they will generally not grow in cultured cell strains or lines, but only in organ cultures; thus recovery of coronavirus from respiratory secretions requires the use of ciliated epithelium organ cultures. Significant advances concerning the coronaviruses that are human pathogens will have to await the development of convenient cell culture systems in which these agents will grow efficiently. It should be noted however that variants of human coronaviruses are known that can multiply in such systems, and significant advances in our understanding of the structure and mode of multiplication of coronaviruses are being made (see Chap. 2).

Reoviruses

Reoviruses are ubiquitous viruses that have been recovered worldwide from animals as well as from man. Three serotypes are recognized that share a common complement-fixing antigen. Some strains were initially classified as ECHO virus type 10. However, as the multiple dissimilarities from picornaviruses were recognized, establishment of a new virus group was suggested in 1959, with the first two letters of the name, reoviruses, serving to emphasize an association with the respiratory and enteric tracts.

Human infection with reoviruses is common. Reovirus has been recovered from healthy persons, as well as from persons with nonspecific fever, encephalitis, upper respiratory illness, pneumonia, diarrhea, steatorrhea, hepatitis, meningoencephalitis, and Burkitt lymphoma. However, little evidence exists to prove that reoviruses are the etiologic agent of these associated conditions. Reovirus infection of infants may be associated with a mild febrile respiratory illness, diarrhea, and on occasion, an exanthem. Studies of the viruses in human volunteers have documented that infection, as judged by recovery of virus, involves both the gastrointestinal and respiratory tracts. The most characteristic pattern of adult infection is that of an afebrile coryzal illness occurring in the winter. It is clear that asymptomatic infections, as documented by virus isolation, are common. The presence of neutralizing antibody in the serum seems to be relatively protective against infection with homologous serotype. The reoviruses have not been of major epidemiologic importance.

Reoviruses produce a spectrum of experimental illness in mice. Disease may be acute and disseminated, or focal and chronic. The age of the animal, serotype of virus, and route of inoculation contribute to the manifestations of the infection. For example, weanling mice are infected without detectable illness, whereas newborn mice sustain a viremic infection with high mortality. Reovirus type 3 inoculated intracerebrally into newborn mice results in a fatal encephalitis with neuronal destruction but sparing of ependymal cells. In contrast, reovirus type 1 produces a nonfatal infection and virus replicates in ependymal cells causing the subsequent development of hydrocephalus. The tropism for ependymal cells or neurons seems to be regulated by a single gene (the S_1 genome segment), which codes for the sigma 1 outer capsid protein. It has therefore been proposed that the S_1 gene of reovirus type 1 is responsible for the virus interaction with ependymal cells, and that the S_1 gene of reovirus type 3 is responsible for the virus interaction with neurons. This single protein specificity for the "receptors" of target cells provides a fascinating insight into the pathogenesis of these infections.

Diagnosis

Mammalian reoviruses replicate and produce visible cytopathic effects in a wide variety of cell cultures. Primary rhesus kidney cell cultures are satisfactory for routine isolation, but primary human kidney cell cultures have been used when it is essential to exclude the possibility of endogenous infection of monkey kidneys. The cytopathic effect consists of a granular change in the cytoplasm, and the cells do not slough off the glass quickly; often they remain attached to the surface by a single cytoplasmic process and may appear to be undergoing nonspecific degeneration. Stained preparations will show intracytoplasmic inclusions. Isolates are obtained primarily from the respiratory tract and fecal specimens, and occasionally from other sources such as urine, cerebrospinal fluid, and various tissues obtained at autopsy.

Reoviruses are usually identified as to serotype by the hemagglutination-inhibition technique. Human infection is associated with a demonstrable antibody rise to the homotypic serotype. The prevalence of reovirus antibodies makes it mandatory that an antibody titer rise be demonstrated in order to document the presence of acute infection.

Rotaviruses

Rotaviruses constitute another genus within the family Reoviridae. In 1943 an epidemic diarrhea of the newborn was described by Light and Hodes. They isolated a filtrable agent which caused diarrhea in calves, but no tissue culture verification of a viral agent could be obtained (Nebraska calf diarrhea virus). Naturally occurring diarrhea in other animal species has also been attributed to rotavirus agents; for example, epidemic diarrhea of infant mice (EDIM) was described in 1947 and the etiologic agent demonstrated in 1963. At the present time, rotavirus particles have been detected in association with diarrhea in foals, lambs, piglets, rabbits, deer, monkeys, and other species. Several animal models are now being utilized to study the illness and host response to these agents.

Clinical Illness

Pathologic studies suggest that the virus initially invades the epithelium of the duodenum and upper small intestine. In a few fatal cases, infection has extended the entire length of the small bowel. Villi are transiently denuded and shortened. Epithelial cells are rapidly regenerated, which accounts for the appearance of virus and virus-infected cells in the feces within a few days after the onset of symptoms. There is no proof of infection beyond the gastrointestinal tract, although some patients have had associated respiratory symptoms; but virus has not been demonstrated in respiratory tissues or gastric contents.

The incubation period of rotavirus-associated gastroenteritis in man is 48 to 72 hours. Characteristically, the onset of the disease is heralded by one or two episodes of vomiting with subsequent appearance of diarrhea. The stools are usually without blood or mucus, and there is a variable degree of fever associated with the illness. Uncomplicated illness lasts approximately one week.

Studies of household contacts suggest that a high rate of infection occurs, although symptomatic illness is commonest in young children. Rotavirus infection is communicable on infant wards and has been responsible for nosocomial outbreaks of diarrhea. Adults have been described to have diarrheal illness, with virus detected in their stools; but this occurs significantly less frequently than illness in infants. Clinical evidence suggests that reinfection does occur.

Serologic data show that rotavirus antibody can be detected in 90 percent of the sample population after two years of age. There are two serotypes of human rotavirus, the relative proportions of which tend to fluctuate in the seasonal outbreaks of rotavirus infection that occur every winter.

Acute gastroenteritis caused by these viruses is a major health problem and is certainly responsible for the majority of hospitalized cases of acute diarrhea among infants less than three years of age on all continents and among all races. Rotavirus-caused acute gastroenteritis is a major cause of death in infants in many developing countries, where its effects are compounded by malnutrition and inadequate medical care. In particular, weaning from breast milk is associated with the onset of the diarrhea–malnutrition cycle which contributes greatly to the huge infant mortality rates in such parts of the world. Unfortunately, despite the fact that breast milk has been shown to have preventive activity against undifferenti-ated infantile diarrhea, breast-feeding is declining in popularity in parts of the world with the greatest incidence of diarrheal disease.

Diagnosis

Human rotaviruses isolated from patients will not grow in any cultured cell. The presence of rotaviruses can be demonstrated by electron microscopic examination of fecal extracts obtained during the acute stage of gastroenteritis, particularly when specific antiserum is used to agglutinate virus particles and to surround them with a layer of antibody molecules, which renders them easily recognizable (immune electron microscopy). Enzyme-linked immunosorbent assay (ELISA) and radioimmunoassay (RIA) techniques are now becoming available for the detection of rotavirus infection. Serologic techniques employing complement fixation and immunofluorescence are being used to define the epidemiology of rotavirus infection. Further significant advances concerning human rotaviruses will have to await the development of cell culture systems capable of supporting their multiplication. Recent evidence suggests that sophisticated adaptation procedures may lead to the isolation of rotavirus variants capable of growing in cultured cells.

FURTHER READING

Rhinoviruses

Cate TR: Rhinoviruses. In Knight V (ed): Viral and Mycoplasmal Infections of the Respiratory Tract. Philadephia, Lea & Febiger, 1973, p 141

Fridy WW Jr, Ingram RH, Hierholzer JC, Coleman MT: Airway function during mild viral respiratory illness. The effect of rhinovirus infection in cigarette smokers. Ann Intern Med 80:150, 1974

Gwaltney JM Jr: Rhinoviruses. Yale J Biol Med 48:17, 1975

Gwaltney JM Jr, Moskalski PB, Hendley JO: Hand-to-hand transmission of rhinovirus colds. Ann Intern Med 88:463, 1978

Hendley JO, Wenzel RP, Gwaltney JM Jr: Transmission of rhinovirus colds by self-inoculation. N Engl J Med 288:1361, 1973

Jackson GG, Muldoon RL: Viruses causing common respiratory infections in man. J Infect Dis 127:328, 1973

Stanley ED, Jackson GG, Panusarn C, Rubenis M, Dirda V: Increased virus shedding with aspirin treatment of rhinovirus infection.

Coronaviruses

Bradburne AF, Bynoe ML, Tyrrell DA: Effects of a "new" human respiratory virus in volunteers. Br Med J 3:767, 1967

Caul EO, Clarke SKR: Coronavirus propagated from patients with nonbacterial gastroenteritis. Lancet X:953

Hamre D, Beem M: Virologic studies of acute respiratory disease in young adults. V. Coronavirus 229E infections during six years of surveillance. Am J Epidemiol 96:94, 1972

Hamre D, Procknow JJ: A new virus isolated from the human respiratory tract. Proc Soc Exp Biol Med 121:190, 1966

Jackson GG, Muldoon RL: Viruses causing common respiratory infection in man. III. Respiratory syncytial viruses and coronaviruses. J Infect Dis 128:674, 1973

McIntosh K, Chao RK, Kluse HE, et al: Coronavirus infection in acute lower respiratory tract disease in infants. J Infect Dis 130:502, 1974

McIntosh K, Dees JH, Becker WB, Kapikian AZ, Chanock RM: Recovery in tracheal organ cultures of novel viruses from patients with respiratory disease. Proc Natl Acad Sci USA 57:933, 1967

McIntosh K, Kapikian AZ, Turner HC, et al: Seroepidemiologic studies of coronavirus infection in adults and children. Am J Epidemiol 91:585, 1970

Mathan M, Swaminathan SP, Mathan VR, Yesudoss S, Baker SJ: Pleomorphic viruslike particles in human feces. Lancet X:1068

Tyrrell DAJ, Bynoe ML: Br Med J, 1:1467, 1965

Wenzel RP, Hendley JO, Davies JA, Gwaltney JM Jr: Coronavirus infections in military recruits. Three-year study with coronavirus strains 0C43 and 229E. Am Rev Respir Dis 109:621, 1962

Reoviruses

Fields BN: Genetic manipulation of reovirus—A model for modification of disease? N Engl J Med 287:1026, 1972

Jackson GG, Muldoon RL: Viruses causing common respiratory infection in man. IV. Reoviruses and adenoviruses. J Infect Dis 128:811, 1973

Lerner AM, Cherry JD, Klein JO, Finland M: Infections with reoviruses. N Engl J Med 267:947, 1962

Stanley NF: The reovirus murine models. Prog Med Virol 18:257, 1974

Weiner HL, Drayna D, Averill DR Jr, Fields BN: Molecular basis of reovirus virulence: Role of the S1 gene. Proc Natl Acad Sci USA 74:5744, 1977

Rotaviruses

Bishop RF, Davidson GP, Holmes IH, Ruck BJ: Virus particles in epithelial cells of duodenal mucosa from children with acute non-bacterial gastroenteritis. Lancet X:1281

Bishop RF, Davidson GP, Holmes IH, Ruck BJ: Detection of a new virus by electron microscopy of fecal extracts from children with acute gastroenteritis. Lancet 1:149, 1974

Davidson GP, Townley RRW, Bishop RF, Holmes IH: Importance of a new virus in acute sporadic enteritis in children. Lancet X:242

Flewett TH, Bryden AS, Davies H, et al: Relation between viruses from acute gastroenteritis of children and newborn calves. Lancet 2:61, 1974

Light JS, Hodes HL: Studies on epidemic diarrhea of the newborn: Isolation of a filtrable agent causing diarrhea in calves. Am J Public Health 33:1451, 1943

Light JS, Hodes HL: Isolation from cases of infantile diarrhea of a filtrable agent causing diarrhea in calves. J Exp Med 90:113, 1949

Middleton PJ, Holdaway MD, Petric M, Szymanski MT, Tam JS: Solid-phase radioimmunoassay for the detection of rotavirus. Infect Immun 16:439, 1977

Tallett S, MacKenzie C, Middleton P, Kerzner B, Hamilton R: Clinical, laboratory, and epidemiologic features of a viral gastroenteritis in infants and children. Pediatrics 60:217, 1977

Wyatt RG, Kapikian AZ, Thornhill TS, et al: In vitro cultivation in human fetal intestinal organ culture of a reovirus-like agent associated with nonbacterial gastroenteritis in infants and children. J Infect Dis 130:523, 1974

Yolken RH, Kim HW, Clem T, et al: Enzyme-linked immunosorbent assay (ELISA) for detection of human reoviruslike agent of infantile gastroenteritis. Lancet X:263

CHAPTER 25

Slow Virus Infections

Unconventional viruses: subacute spongiform (virus) encephalopathies

Definition

In simplest terms, slow virus infections refer to transmissible conditions associated with a very slow evolution of clinical symptomatology. Defined in this manner, consideration of slow virus infections must include a diverse assortment of conventional infectious agents, clinical conditions, and pathogenetic mechanisms. The slow progress of the effects of a virus infection may reflect properties inherent in the virus and the degree of genetic adaptation of host and parasite as reflected, for example, in the effects of latency and periodic reactivation of herpesviruses. Alternatively, the protracted course of a virus disease may be determined largely or even solely by host factors. In any case the progress of disease associated with known viruses is best taken up in the chapters devoted to consideration of the relevant taxonomic groups. Here we have chosen to limit the definition and consideration of slow virus infections to a group of progressive degenerative conditions of the central nervous system (CNS) with a similar histopathology, associated with unconventional transmissible agents, and termed the subacute spongiform virus encephalopathies. When, in the future, the nature of the unconventional transmissible agents has been defined, it may become apparent that similar infections are responsible for certain chronic diseases of other organ systems.

Examples of slowly progressive virus infections of the CNS caused by conventional viruses are listed in Table 25-1. Some of these conditions, as well as the responsible viruses, are discussed in other chapters in this text.

The four known naturally occurring diseases of the CNS caused by unconventional viruses include two of man, kuru and Creutzfeld-Jakob disease (CJD), and two of animals, scrapie and transmissible mink encephalopathy (TME). These slowly but inexorably progressive transmissible diseases share certain features which include:

1. transmission by injection of infected organs
2. an incubation period of months to years
3. a clinical course which is regular, protracted, and always fatal
4. similar pathology restricted to brain and consisting of gliosis and degeneration of neurons with vacuolation in dendritic and axonal processes leading to the reduction of the gray matter to a spongiform state
5. lack of inflammatory reaction
6. lack of cerebrospinal (CSF) changes associated with CNS infections
7. lack of detectable host-defense responses

HISTORY AND BACKGROUND

The modern era of slow infections of the nervous system began in 1954 when Sigurdsson, a veterinary pathologist, proposed certain criteria to define a group of chronic infectious diseases of sheep. These included: (1) a prolonged initial period of latency lasting for months or years, (2) a regular protracted course after clinical signs had appeared, ending in serious disease or death, (3) limitation of infection to a sin-

TABLE 25-1   EXAMPLES OF SLOWLY PROGRESSIVE CNS INFECTIONS OF MAN CAUSED BY CONVENTIONAL VIRUSES

Disease	Virus
Subacute sclerosing panencephalitis	Paramyxovirus—measles
Subacute encephalitis	Herpesvirus hominis, adenovirus
Progressive congenital rubella	Togavirus—rubella
Progressive multifocal leukoencephalopathy	Papovavirus—JC virus, SV40-PML
Rabies	Rhabdovirus—rabies

gle host species, and (4) localization of anatomic lesions in a single organ or tissue system.

Scrapie

Sigurdsson's original definition pertained to certain diseases of sheep which were recognized in Iceland, including scrapie, visna, and maedi. Visna and maedi are caused by conventional viruses. Scrapie, the prototype of the unconventional transmissible agents, occurs in nature primarily in sheep and occasionally in goats. The occurrence of scrapie was known throughout Europe 200 years ago, and today the disease is still widely distributed in Europe, America, and Asia. Sheep are generally affected between the ages of 2 and 5 years, and the pattern of occurrence suggests the existence of both genetic and transmissible components. Excitability and nervousness, the earliest signs of the disease, are followed by incoordination, ataxia, wasting, and inevitable death. Lateral transmission of scrapie occurs in nature, acquired by lambs from ewes even without suckling and by susceptible strains of sheep occupying pastures previously inhabited by scrapied animals.

Scrapie can be transmitted by inoculation, an observation made originally 80 years ago. No lateral transmission occurs following inoculation (experimental) scrapie. Experimental scrapie has been transmitted to goats, mice, rats, hamsters, gerbils, mink, and monkeys. Sheep, goats, and mice may be infected with scrapie by the oral route as well.

Transmissible Mink Encephalopathy (TME)

The other example of a naturally occurring subacute virus encephalopathy of animals is provided by TME, a condition pathologically similar to scrapie but with a somewhat less indolent course. Since this disease was first recognized on ranches where carcasses of scrapied sheep were fed to mink and since scrapie may be experimentally transmitted to mink, it is presumed that TME is scrapie manifested in a different species. It is interesting that, although sheep scrapie can be transmitted to mice as well as mink, neither TME nor experimental scrapie of mink can be transmitted to mice.

NATURE OF THE UNCONVENTIONAL TRANSMISSIBLE AGENTS

Although in vitro replication of the unconventional viruses has thus far been unsuccessful or of uncertain result, all have been transmitted by inoculation, and in the process investigators have derived most of the current information concerning these agents. The need for specialized experimental animals and the protracted incubation period of months and even years have made the characterization of these agents very difficult. The following are some of the characteristics derived from studies such as these and shared by most known unconventional viruses.

No recognizable virus particles have been seen on electron microscopic study of involved infected cells or in preparations concentrated by density gradient banding and shown by the results of inoculation to possess a high infectivity.

There is no evidence of antigenicity, and infected hosts do not make a recognizable immunologic response to these transmissible agents. No inflammatory response is demonstrable. Recently it has been shown that cell fusion may be induced by infectious scrapie and CJD derived from brain preparations. The results were found to parallel fusion induced by a conventional virus (Sendai).

Highly infectious preparations are extremely resistent to ultraviolet and ionizing radiation, as well as to heat, proteases, nucleases, and formaldehyde. Although high infectivity is often found associated with membrane fragments, it is resistant to sodium deoxycholate and fluorocarbons. These bizarre properties have spawned innumerable theories concerning the nature of these agents; such theories range from "infectious membranes" to "infectious proteins" to viroids, small circular singlestranded but highly base-paired RNA molecules that cause several diseases in plants.

CLINICAL CONDITIONS OF MAN

Kuru

Clinical and Epidemiological Features Kuru is a subacute, endemic, fatal neurologic (brain) disease restricted to certain Melanesian tribes, primarily the Fore, consisting of about 35,000 individuals who reside in 169 isolated mesolithic villages in the eastern highlands of Papua, New Guinea. When Kuru was described by Gajdusek and Zigas in 1957, this disease was the most common cause of death among the Fore people, and in some areas the death rate due to kuru was as high as 2 to 3 percent annually and was barely balanced by the birth rate.

The initial signs of kuru are a mild cerebellar ataxia and tremulousness (kuru is the Fore word for shivering or shaking). Later these patients develop a progressive mild dementia, clonus, hyperreflexia, and the gradual loss of muscle control. Affected individuals become severely ataxic and develop difficulty standing, eating, and even speaking. No consistent laboratory abnormalities are demonstrable and the CSF usually remains normal throughout the course. Death follows in about one year from the onset of symptoms.

In the 1950s kuru was common in children (equally in males and females) and women and was found rarely in adult males. Searches for responsible conventional viruses were negative, as were investigations of genetic factors and probes for toxic environmental substances. Ultimately, after kuru was found to be transmissible to experimental animals, it was established that the mode of spread of kuru was related to contamination associated with the ritual cannibalism of dead relatives—a devotional rite of mourning attended, prepared, and performed largely by women (accompanied by small children). Gajdusek has estimated that about 10 persons were infected at the time of each death ritual which involved a kuru patient. Kuru has become a vanishing disease as, in the ensuing 20 years, cannibalism has diminished and finally ceased among the Fore. At no time was kuru ever transmitted laterally to outsiders residing in Fore villages.

Pathology The primary lesions in the brain of kuru-affected individuals are vacuolation, neuronal loss, and astroglial proliferation. The affected areas are largely the midline structures of the brain and the cerebellum. The cerebrum and spinal cord are not usually extensively involved in kuru. There is minimal or no evidence of inflammation. Ruptured and curled membranes and membrane fragments are evident within neuronal vacuoles.

Hadlow first pointed out the similar neuropathology evident in scrapie and kuru. These observations led to the inoculation studies of kuru which established that it was a transmissible disease caused by an agent with unusual physical and biologic properties and associated with a protracted incubation period. The pathologic similarity of scrapie and kuru has also led to the use of scrapie as a model for the study of kuru and CJD.

Creutzfeld-Jakob Disease (CJD)

Clinical and Epidemiologic Features CJD is a rare form of presenile dementia first described by Creutzfeld in 1920 and independently by Jakob one year later. The disease is found worldwide and may occur sporadically or in a familial form (10 percent) which involves several generations in a pattern suggesting the existence in these cases of autosomal dominant inheritance. Conjugal cases are rare.

The mean age of CJD patients is 55 years, and an equal proportion of men and women are affected. The annual incidence in most parts of the world is about 1 to 3 per million population.

CJD may have its onset with the appearance of a progressive dementia accompanied and followed by myoclonic jerks, ataxia, muscle wasting, extrapyramidal disturbances, deterioration of all of these, helplessness, and inexorable death.

Laboratory studies in CJD resemble those which accompany kuru. The CSF is usually normal. The electroencephalogram (EEG) reveals high-voltage slow paroxysmal bursts of electrical activity.

Pathology The pathology of CJD (neuronal vacuolation and loss accompanied by astroglial hypertrophy) is similar to that found in kuru and scrapie. In CJD, however, these changes are most severe in the cerebral cortex and less prominent in cerebellar and brainstem tissues.

In 1959, at about the same time that Hadlow noted the similarity of scrapie and kuru, Klatzo et al recognized and described the similarities between CJD and kuru.

EXPERIMENTAL TRANSMISSION OF THE SUBACUTE SPONGIFORM (VIRUS) ENCEPHALOPATHIES

Scrapie, TME, kuru, and CJD have all been successfully transmitted to a variety of experimental hosts with the subsequent evolution, after a long (variable) incubation, of clinical syndromes and pathologic changes characteristic of the naturally occurring diseases. The specific transmissible agents can thereafter be reisolated from inoculated animals. Table 25-2 lists the various animal hosts successfully inoculated with these agents.

The transmission of kuru into susceptible primates (chimpanzees) required a lengthy incubation period (up to 82 months) prior to the onset of symptoms. Repeated passages were accompanied by a shortened incubation period and the acceleration of symptoms. The transmissible agent was found present in the brains of kuru patients in titers of greater than 10^8 infectious doses per gram of tissue. Little or no virus has been detectable peripherally.

The presence of a transmissible filterable agent associated with CJD was demonstrated by the intracerebral or peripheral inoculation of susceptible hosts with suspensions of brain, liver, kidney, lymph nodes, lung, and CSF. Brain tissues from CJD patients explanted, maintained, and even passaged in vitro are still capable of transmitting the disease, though the existence of in vitro replication has not certainly been established. A variety of primates, hamsters, guinea pigs, and even cats have been found to be susceptible to CJD transmission.

SOME IMPLICATIONS

The transmissible spongiform (virus) encephalopathies share features which suggest that these agents and conditions are closely related. There has been speculation concerning the evolution of these conditions. As noted above, though passage of scrapie to mink is successful (induction of pathology and disease), this transfer results in the alteration of the original host range (loss of murine susceptibility). The ingestion by susceptible persons of tissue from scrapied sheep may result in the sporadic appearance of CJD. Indeed, it has been suggested that the high incidence of CJD found in Libyan Jews (30-fold higher than elsewhere) may be related to their custom of eating sheep eyeballs. The sporadic occurrence of CJD may have resulted in the introduction of this transmissible agent into a susceptible population (the Fore) whose habit of ritual cannibalism established the endemic occurrence of kuru. Existing differences in the host range of these various infectious agents may reflect changes on passage such as those associated with scrapie in mink.

It is noteworthy that the control of scrapie in the United States is by (uncompensated) slaughter of the affected herds. It is likely that ranchers who recognize the appearance of scrapie will market their flocks, with the effect that infected meat will be concentrated in markets for human consumption.

These transmissible agents which seem to possess unusual and even unique characteristics may represent a newly recognized group of viruses. If so, it is likely that other related viruses and diseases will be ultimately recognized. It has been suggested that Alzheimer's disease may belong in this group and other candidates may turn up, even unrelated to conditions of the CNS.

Though the existence of these stable unusual agents is now recognized, the prolonged incubation period and difficulty of detecting virus may interfere with the recognition of associated illnesses, epidemiology, and even of certain (nosocomial) risk. CJD has been transmitted with corneal transplants and also by silver stereotactic electrodes employed during neurosurgical procedures and cleaned between cases with formalin and alcohol. It is disturbing to note that an unexpectedly large percentage (5–7 percent) of CJD patients have undergone previous brain surgery. The potential certainly exists for the transmission with tissue and organ transplants and transfusions of these and perhaps other related viruses and conditions. Some appropriate cautionary measures have been suggested.

Slow virus infections: conventional viruses

Two additional conditions have been classified with slow virus infections though related to in-

**TABLE 25-2   ANIMALS SUSCEPTIBLE TO THE SUBACUTE SPONGIFORM VIRAL
ENCEPHALOPATHIES (INOCULATION)**

Disease	Susceptible animals
Kuru	Chimpanzee, New and Old World monkeys, mink, ferret
Creutzfeld-Jakob disease	Chimpanzee, New and Old World monkeys, hamster, guinea pig, mouse, ferret, cat
Scrapie	New and Old World monkeys, sheep, goat, mink, mouse, rats, gerbil, vole, hamster
Transmissible mink encephalopathy	New and Old World monkeys, sheep, goat, mink, ferret, hamster, raccoon, skunk

fection with conventional viruses. These are:
subacute sclerosing panencephalitis (SSPE) and
progressive rubella encephalitis.

SUBACUTE SCLEROSING PANENCEPHALITIS

In the past few years, evidence has accumulated to implicate measles virus or a closely related agent in the pathogenesis of SSPE.

Clinical Features   This condition occurs almost
exclusively in children and adolescents, usually
between the ages of 5 and 10 years and predominantly in males (3:1). Rarely, young infants
are affected, and a few cases have been reported in the adult population. It usually presents clinically with subtle changes in personality, impaired school performance, and a
gradual intellectual deterioration over several
weeks or months (Table 25-3). This is characteristically followed by myoclonic seizures, focal
neurologic deficits, and a gradual or sometimes
precipitous diminution in conscious level. The
progress of the disease may appear to be ar-

rested for periods of years, usually when the
patient is in coma, but almost invariably the
patients die, often of intercurrent infection. A
few older patients with well-documented disease have recovered, but this is extremely rare.

Epidemiology   The true incidence of SSPE in
this country is unknown, although crude estimates based on case reporting have suggested a
figure of approximately 200, or about 1 per million population. The majority of patients give a
history of uncomplicated natural measles at a
slightly younger age than average, and all remain well during the interim of months or
years until the onset of their CNS disease. Occasionally, no prior history of measles is elicited, or liver-attenuated measles vaccine is the
only known exposure to measles antigen. No
immunologic defects in either cell-mediated or
humoral immunity have been demonstrated in
these individuals in their responses to either
routine infections or immunizations. Their in
vitro lymphocytic responses to stimulation with
measles antigen are normal.

   Some recent epidemiologic data have sug-

TABLE 25-3   SUBACUTE SCLEROSING PANENCEPHALITIS (SSPE)

Clinical	Laboratory	
Personality change; declining school per-formance; intellectual deterioration often mani-fested by impaired memory, altered judgment, and inappropriate behavior; occasionally chorio-retinitis; impaired motor activity, gait difficulty, speech difficulty; myoclonic jerks progressing to repetitive, often sound-sensitive myoclonic seizures; gradual deterioration in consciousness; coma; death	*EEG*	Paroxysmal, synchronous spike discharges with interim suppression of electrical activity
	CSF	Usually acellular, sometimes modest mono-nuclear pleocytosis; normal total protein; in-creased gamma globulins (IgG); detectable measles antibody
	Blood	Markedly elevated measles antibody
	Brain	Specific immunofluorescence to measles antigen; SSPE virus recovered in tissue culture

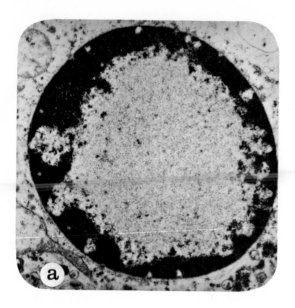

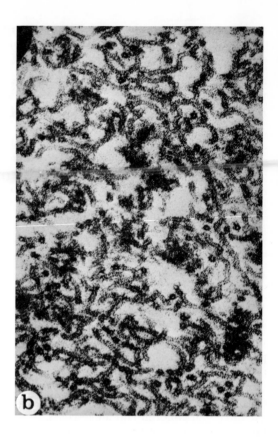

FIG. 25-1. **a.** Nucleus of a neuron from a patient with SSPE showing microtubular filaments. The nuclear chromatin is displaced to the periphery. ×16,000. **b.** Higher power view of microtubules, showing their irregular, convoluted appearance and circular contour on cross section. ×50,000.

gested that the disease may be more common in the Southeastern United States, that the primary measles infection seems to occur at a statistically younger age in such patients, and that rural rather than urban residents seem to be at greater risk.

Diagnosis The diagnosis can usually be suspected on the basis of history and the clinical findings of progressive changes in personality, myoclonus, and variable focal neurologic deficits, including impaired vision, sometimes as a result of chorioretinitis. A characteristic spike wave discharge is frequently seen on the EEG, and the CSF is usually normal, apart from a first-zone colloidal gold sol rise and sometimes a slight lymphocyte pleocytosis. The diagnosis is confirmed by the finding of measles antibodies in the CSF. Brain biopsy is no longer necessary for diagnosis, but should tissue become available, it may show perivascular round cell infiltration, neuronal degeneration, gliosis, and type A Cowdry intranuclear inclusion bodies when stained with hematoxylin and eosin (H & E) and examined with the light microscope. With the electron microscope, these inclusion

bodies resemble microtubular filaments, corresponding in size and configuration to the nucleocapsids of measles virus (Fig. 25-1). With appropriate fluorescent antibody staining, measles antigen can be demonstrated at these sites. Finally, when brain tissue is cultivated in the laboratory, serially passaged, and cocultured with either HeLa or Vero cells, complete infectious virus may be recovered. This SSPE agent is almost identical to rubeola virus but differs slightly in its behavior both in vitro and in vivo, in its affinity for the cell nucleus, and in certain of its molecular characteristics. It is not certain whether virus persistence in SSPE results from host factors, from the presence of genetic variants of the measles virus, or both. The SSPE agent appears to be a variant of measles virus, but how and why that defective virus emerges remains uncertain. Byington and Johnson have developed an animal (hamster) model of SSPE. Following inoculation of young (14–28 day) hamsters with the SSPE agent, a proportion of animals, recovered from the acute illness, develop a late-onset subacute encephalitis with myoclonus and lethargy reminiscent of SSPE.

Treatment   Numerous approaches to therapy for this condition have been attempted, but since its pathogenesis is so poorly understood, these attempts have quite predictably failed. At present, one can only provide general supportive care for the patient, attempt to control seizures, and try to allay parental anxiety by early diagnosis and proper explanation. The overall occurrence of measles infections in the United States has declined since the introduction of widespread immunization. Coincidentally there has been a definite reduction in the occurrence of SSPE.

PROGRESSIVE RUBELLA ENCEPHALITIS

A progressive rubella encephalitis has recently been described in four children as an apparent late sequel to congenital infection with this virus. The neuropathologic changes and clinical picture of spasticity, ataxia, intellectual deterioration, and seizures are remarkably similar to those seen in SSPE. It typically begins in the second decade of life in children with varying stigmata of congenital rubella, including mental retardation, cataracts, and deafness. The course is progressively downhill, with loss of motor and mental skills. Death occurred in two of the cases.

Elevated antibody titers to rubella virus were detected in the sera and CSF of patients who were tested, and rubella virus was removed from the brain tissue of one following cocultivation with CV-1 cells.

A clearer understanding of the incidence and pathogenesis of this condition must await further study of more patients with congenital rubella. It is tempting to conclude that rubella virus persists in neural tissue from the time of primary infection, rather than being acquired postnatally, but even this is uncertain.

FURTHER READING

Unconventional viruses: Subacute spongiform encephalopathies

BOOKS AND REVIEWS

Asher DM, Gibbs CJ Jr, Gajdusek DC: Pathogenesis of subacute spongiform encephalopathies. Ann Clin Lab Sci 6:84, 1976

Gajdusek DC, Gibbs CJ Jr: Kuru, Creutzfeldt-Jakob disease, and transmissible presenile dementias. In terMeulen V, Katz M (eds): Slow Virus Infections of the Central Nervous System. Springer, New York, 1977

Gajdusek DC: Unconventional viruses and the origin and disappearance of kuru. Science 197:943, 1977

Masters CL, Harris JO, Gajdusek DC, et al: Creutzfeldt-Jakob disease: Patterns of worldwide occurrence and the significance of familial and sporadic clustering. Ann Neurol 5:177, 1979

Hotchin J (ed): Slow Virus Disease, Progr Med Virol vol 18, 1974

SELECTED PAPERS

Bernoulli C, Siegfried J, Baumgartner G, et al: Danger of accidental person-to-person transmission of Creutzfeldt-Jakob disease by surgery. Lancet 1:478, 1977

Diener TO: Viroids: The smallest known agents of infectious disease. Ann Rev Microbiol 28:23, 1974

Gajdusek DC, Gibbs CJ Jr, Alpers M: Experimental transmission of a kuru-like syndrome to chimpanzees. Nature 209:794, 1966

Gajdusek DC, Gibbs CJ Jr, Asher DM et al: Precautions in medical care of, and in handling materials from patients with transmissible virus dementia (Creutzfeldt-Jakob disease). N Engl Med 297:1253, 1977

Gajdusek DC, Gibbs CJ Jr, Collins G, et al: Survival of Creutzfeldt-Jakob disease virus in formol-fixed brain tissue. N Engl J Med 294:533, 1976

Gajdusek DC, Gibbs CJ Jr, Rogers NG, et al: Persistence of viruses of kuru and Creutzfeldt-Jakob disease in tissue culture of brain cells. Nature 235:104, 1972

Gajdusek DC, Zigas V: Degenerative disease of the central nervous system in New Guinea: The epidemic occurrence of "kuru" in the native population. N Engl J Med 257:974, 1975

Gibbs CJ Jr, Gajdusek DC, Asher DM, et al: Creutzfeldt-Jakob disease (subacute spongiform encephalopathy): Transmission to the chimpanzee. Science 161:388, 1968

Hadlow WJ: Scrapie and kuru. Lancet 2:289, 1959

Kahana E, Alter M, Braham J: Creutzfeldt-Jakob disease: Focus among Libyan Jews in Israel. Science 183:90, 1974

Kidson C, Moreau M-C, Asher DM, et al: Cell fusion induced by scrapie and Creutzfeldt-Jakob virus-infected brain preparations. Proc Natl Acad Sci USA 75:2969, 1978

Klatzo I, Gajdusek DC, Zigas V: Pathology of kuru. Lab Invest 8:799, 1959

Lampert P, Hooks J, Gibbs CJ Jr, et al: Altered plasma membranes in experimental scrapie. Acta Neuropathol 19:81, 1971

Malone TG, Marsh RF, Hanson RP, et al: Evidence for the low molecular weight nature of scrapie agent. Nature 278:575, 1979

Manuelidis EE, Angelo JN, Gorgacz EJ, et al: Experimental Creutzfeldt-Jakob disease transmitted via the eye with infected cornea. N Engl J Med 296:1334, 1977

Marsh RF, Malone TG, Semancik JS, et al: Evidence for an essential DNA component in the scrapie agent. Nature 275:146, 1978

Porter DD, Porter HG, Cox NA: Failure to demonstrate a humoral immune response to scrapie infection in mice. J Immunol 111:1407, 1973

Roos R, Gajdusek DC, Gibbs CJ Jr: The clinical characteristics of transmissible Creutzfeldt-Jakob disease. Brain 96:1, 1973

Sigurdsson B: Observations on three slow infections of sheep. Br Vet J 110:7, 255, 307, 341, 1954

Slow Virus Infections: Conventional Viruses

Byington DP, Johnson KP: Experimental subacute sclerosing panencephalitis in the hamster: Correlation of age with chronic inclusion-cell encephalitis. J Infect Dis 126:18, 1972

Byington DP, Johnson KP: Subacute sclerosing panencephalitis virus immunosuppressed adult hamsters. Lab Invest 32:91, 1975

Fuccillo DA, Kurent JE, Sever JL: Slow virus diseases. Annu Rev Microbiol 28:231, 1974

Hamilton R, Barbosa L, Dubois M: Subacute sclerosing panencephalitis measles virus: Studies of biological markers. J Virol 12:632, 1973

Modlin JF, Jabbour JT, Witte JJ, et al: Epidemiologic studies of measles, measles vaccine, and subacute sclerosing panencephalitis. Pediatrics 59:505, 1977

Payne FE, Baublis JV, Itabashi HH: Isolation of measles virus from cell cultures of brain from a patient with subacute sclerosing panencephalitis. N Engl J Med 281:585, 1969

Weil ML, Itabashi HH, Cremer NE, et al: Chronic progressive panencephalitis due to rubella virus simulating subacute sclerosing panencephalitis. N Engl J Med 292:994, 1975

INDEX

QR   Principles of animal virology / edited
360   by Wolfgang K. Joklik. -- New York :
P965   Appleton-Century-Crofts, c1980.
1980   x, 373 p. : ill. ; 26 cm.

            Includes index.
            ISBN 0-8385-7920-5

1. Virology. 2. Virus diseases.   I. Joklik,
Wolfgang K.

MUNION ME   811002   810928 CStoC
R000431   LB /UPG   A*   81-B32
                             80-17809